Construction management

General Editor: Colin Bassett, BSc, FCIOB, FFB

Construction Management in Practice, *R. F. Fellows, R. Newcombe, D. A. Langford and S. Urry*

Related titles

Construction projects: their financial policy and control, *R. A. Burgess*

Building production and project management, *R. A. Burgess and G. White*

Project management and construction control, *G. Peters*

Construction management: planning and finance

David Cormican
formerly Lecturer in Civil Engineering Department, Queen's University, and Principal Lecturer, School of Building, Ulster Polytechnic

Construction Press
London and New York

Construction Press
an imprint of:
Longman Group Limited
Longman House, Burnt Mill, Harlow
Essex CM20 2JE, England
Associated companies throughout the world

Published in the United States of America
by Longman Inc., New York

First published 1985

British Library Cataloguing in Publication Data
Cormican, David
Construction management : planning and finance.
1. Construction industry – Management
I. Title
624'.068 HD9715.A2

ISBN 0-582-30516-0

Library of Congress Cataloging in Publication Data
Cormican, David, 1946-
Construction management

Bibliography: p.
Includes index.
1. Construction industry–Management. I. Title.
HD9715.A2C66 1985 624'.068 84–20054
ISBN 0–582–30516–0

Set in Plantin 10/12
Printed in Great Britain by The Pitman Press Ltd., Bath

Contents

Acknowledgements

Although complete acknowledgement of my intellectual debt to former colleagues, lecturers, authors and practitioners is not possible in the space available here, I express special thanks to the following for permission to make use of and adapt previously published works:

Mr Pierce Pigott, Head of Construction Division,
National Institute for Physical Planning
and Construction Research, Dublin.

The Queen's University of Belfast for permission to use and adapt past examination questions.

Mr Peter Roper, Editor, *Building Trades Journal.*

I wish to acknowledge the support of Professor A. Long, Head of Civil Engineering, The Queen's University, Belfast, who gave me much encouragement and time during the preparation of this text.

Lastly, my former postgraduate lecturers at the Department of Civil Engineering, University of Leeds, whose knowledge I have entered into.

David A. Cormican

Preface

Companies succeed or fail depending on the quality of their management. If managers make the correct decisions the organisation is succeeding. It is management and not the operatives who are ultimately responsible for the success or otherwise of a company.

In addition to the normal problems encountered by managers in other disciplines, construction managers by virtue of the nature of their industry have special problems to contend with, e.g.

- they are expected to set up a factory (site), produce and erect components (construct), make a profit and close in say a 1 to 3 year period;
- they will have to supervise technical staff in each new site who will probably not have worked together before, and therefore must mould a new team each time;
- they must obtain and control a labour force which fluctuates significantly during a contract;
- they must select and control proper plant and equipment, the requirements of which fluctuates significantly from month to month;
- they often cannot communicate with the design team or client before tendering;
- they often lack time after the award of a contract for detailed planning;
- they cannot mould their own market – usually what has to be built and when is decided by their clients;
- they are accountable for the performance of both domestic and nominated sub-contractors;
- the building industry holds the record for lowest profits of any manufacturing industry and is usually top of the bankruptcy league;
- their future work load is often at the mercy of Government stop–go policies; this creates uncertainty within the industry making finance for investment difficult to obtain and accurate long range planning nearly impossible;
- designers and clients are often vague about objectives and their designs are often subject to change at short notice.

In order for a company to survive in this highly competitive industry it must have managers of the highest calibre. In fact many managers of successful contracting companies justifiably consider themselves more capable and efficient than their counterparts in other spheres of industry.

The construction environment has changed considerably during the past two decades. In the past contract durations were longer, overall management tasks were simpler and the architect or engineer exercised considerable control over the contractor. Today the high cost of finance necessitates fast completion dates in order to obtain early return on invested capital, and massive public expenditure cuts always hit the contractor creating uncertainty and severe competition for work. In the past managers learned to cope through hard won experience and the application of rules of thumb and intuition were predominant. However, because of the fairly recent environmental changes, interest in the findings of other disciplines have resulted in a trend towards the study and application of the approaches and theories of a variety of the sciences, e.g. mathematics/operational research, accounting, economics and human relations.

There is undoubtedly a need for formal construction management education. This together with proper experience can best benefit the industry.

This book, which is divided into two main parts, covers quantitative material professionally recognised as belonging to the broad field of Building and Civil Engineering management. The style of the book is to present essential features of each topic in a simple, direct approach which is intended to avoid an over elaborate treatment. In this way it is hoped that the student will obtain a clear grasp of the essentials on which to build at later stages of study.

Teaching is highly personal and is heavily influenced by the backgrounds and interests of students in a variety of settings. In recognition of this, the book is designed as a pliable tool and not as a strait-jacket for the student. There are over 80 examples to be worked. Suggested solutions are included in order to encourage student self learning and confidence.

Part 1 considers the theory and application of basic operational research techniques to construction planning. Each technique is treated in a practical manner. The text concentrates on well-known techniques that have the potential of being applied in a meaningful way. Mathematical complexity has been kept to a minimum.

Part 2 describes methods used to guide financial decision making and control. Topics covered include break-even analysis, cash flow forecasting and investment appraisal. Worked examples and case studies are included to develop the rationale of solution methods.

Part 1

Construction planning studies

Chapter 1

Planning studies

1.1 General aspects

> Nowhere more than in construction is it true that, in business enterprise or work projects, forethought is necessary before useful decisions and consequent action can be taken – the basis of short-term effective action is long-term planning – without planning, a course of action becomes (if not completely aimless) a succession of random changes in direction (Brech, 1975, Ch. 12).

The main contractor's task is to convert the design drawings into reality. Sequence and methods to be used to execute the contract must be considered. Most construction projects are unique. Even the same building type to be constructed in a different locality will present new problems. The majority of projects will have many thousands of operations, many of which will be interdependent. A trial and error approach is no longer valid and planning is therefore vital. The amount of detailed planning or sophistication involved is likely to be a function of: (a) the size of firm; (b) complexity of project; (c) management expertise. A project manager in a small firm dealing with minor works may well be able to plan ahead mentally without recourse to drafting his ideas and plans on paper. However, for large complex projects this approach could be disastrous! Project management usually requires the services of planning engineers to advise on method, sequence, etc.

In practice the task of planning and controlling construction work is considerably more difficult than might be expected. Much happens to interfere with the smooth flow of work, e.g.:

- delays occasioned by weather;
- materials not being available;
- materials being rejected;
- breakdown of equipment;
- unwillingness to co-operate;
- inadequate detailing.

Other delays may be caused by external influences, i.e.:

- decisions of building control officers;
- lack of co-operation of statutory authorities.

A planning technique will not automatically solve these problems. The main task of site management is to use the programmes as a flexible guide in order to compare planned work against work actually achieved. In this way, corrective action can be taken at the earliest possible moment.

An important aspect of planning which is often overlooked is the fact that it disciplines those concerned. Planning is difficult. It is an intellectual operation requiring determined conscious effort. Ideally, problems will be solved before they crop up on site, with all the disadvantages of possible delay in sequence, costly remedies, etc.

Planning therefore is the process by which managers anticipate the probable effects of events that may change the activities and objectives of their business.

By planning they attempt to influence and control the nature and direction of the change and to determine what actions are necessary to bring about desired results.

Before a plan can be prepared the following should be done.

1. Decide what is the objective.
2. Find all the relevant facts and information and analyse them.
3. Consider the facts in the light of future trends and outside influences and use foresight.
4. Consider the work already in hand or plans being prepared and take account of this.

The characteristics of a good plan should include the following:

1. Be based on a clearly defined objective and must be practicable.
2. Be simple.
3. Be flexible.
4. Provide continuity of work.
5. Provide a balance of work where possible.
6. Provide for easy control by establishing standards.
7. Be orderly.
8. Exploit existing resources to the maximum before new resources are obtained.
9. Be definite in its requirements.
10. Be arranged in a series of steps to be taken in sequence, each step being a miniature objective in itself.
11. Be made in consultation with those concerned and have their approval.
12. Be made with the help of those who are to carry it out.
13. Incorporate all old and new plans to cover the total new objective.
14. Consist of one master plan incorporating all sub-plans, i.e. **Unity**.
15. Be arranged in stages – that which is definite, that which is probable and that which is tentative.

1.2 Advantages of planning

Advantages to the contractor

1. The fact that the job has been studied in detail in order to draw up a network or

bar chart means that the contractor knows more about the job, and this is usually reflected in improved and more systematic organisation of the project. Haphazard methods are thus avoided. Every site has its own particular problems and these need to be gone through in great detail to find the most efficient and economic answer.

2. A properly drawn up programme in conjunction with cost control can prevent loss of money and help to relieve the financial burden of the contractor. The information provided in the programme, together with the budget, will let the contractor know well in advance when provision has to be made for increased capital. In the same way, due to careful planning, on-site costs may be reduced as better all-round progress on the job means that the site agent, site accommodation, etc. are tied up for a shorter period of time.
3. Supply of labour required week by week for each operation can be gauged properly if a programme has been drawn up.
4. It is a simple matter to produce various schedules from the programme, e.g. materials, plant, subcontractor. This ensures correct notices given, delivery on time, in the right place and in the right quantity. It is easy to see from the programme if plant is being fully utilised.
5. A properly prepared programme can greatly assist co-ordination of subcontractors. It is important that the subcontractors 'dovetail' correctly into the sequence of a building contract. Draft programmes should always be submitted to subcontractors for their comments and approval. They in turn should be required to inform the main contractor of any changes, etc.
6. A programme lays down a preconceived plan not only for the whole job but also for the various stages and in the case of networks illustrates clearly their interdependence.
7. The programme provides a standard against which actual work may be measured.
8. Records of progressing may be kept of actual work carried out to provide information for handling future contracts.

Advantages for the client

The client will know exactly how long it will take to erect the building and what length of time his capital will be unproductive while tied up in construction work. At the same time, he has a guide upon which he can work for engaging staff or purchasing stock, equipment or furniture, etc. for use in the new building.

Advantages for the architect or engineer

The programme will normally be prepared by the contractor in close consultation with the architect. After the contractor has prepared a clear, concise picture of the construction in the form of a programme (e.g. network and/or bar chart) and targets have been laid down for the various operations, then a draft should be submitted to the architect or engineer for his consideration. With this information on hand the architect or engineer will know the anticipated rate of progress for all main operations. For a contractor to complete his work on time it is essential that programming the supply of working drawings by the design team receives careful attention. A large job may have thousands of drawings and these are not, of course, all issued *en bloc* at the start of the contract. If there is a delay in the issue of drawings, then the contractor may claim an extension of time as laid down by the Joint Contracts Tribunal (JCT) and Institution of Civil Engineers (ICE) conditions of contract. A properly drawn up programme will inform the architect/engineer when each operation is to be carried out so that he knows when each drawing will be required.

Advantages to other parties connected with the contract

Other people connected with the contract will include consultants, subcontractors, specialists, suppliers and the local authority. All these will derive benefit from a properly prepared programme. They will know, well in advance, in what stages and when the work is to be carried out and will thus be able to plan accordingly.

1.3 Programming – initial considerations

> A well prepared programme is essential to every construction project. Many activities have to be carefully defined and given a time scale, and it is necessary not only to marshal and list the information but also to display it visually in terms of the contract's objectives and the calendar. The working sequences and the relationships between individual activities must be clearly conveyed in this visual presentation. For this work the industry needs broadly defined standards of construction programmes (and programme use) to which everyone can work (Chartered Inst. of Building, 1980).

Before the actual formulation of the programme the following should be considered:

1. The planning technique to be used.
2. The number and type of programmes required.
3. The purpose.

Planning technique to be used

This will depend on various factors, e.g.:

(a) the expertise of the management team;
(b) the complexity of the job;
(c) degree of managerial experience on similar jobs;
(d) size of the firm;
(e) attitude of management to planning 'techniques';

(f) time period allowed between award of contract and commencement of site operations.

These techniques fall broadly into the following categories:

(i) Bar or Gantt charts and location–time charts.

(ii) Network analysis — Critical path analysis (CPA) / Precedence diagrams

(iii) Line of balance techniques.

Number and types of programmes required

It is necessary to know what sort of programme is wanted, by whom and for what specific purpose, e.g. take into account the stage at which the programme is being made: (a) pre-tender; (b) pre-contract; (c) stage programmes; (d) short-term plans; (e) weekly.

Obviously, the degree of detail will vary depending on whether the programme is to be used by senior management for head office control, for site management or for the architect or client.

Purpose

Is the purpose to:

(a) help in the preparation of an estimate;
(b) to show the sequence and ideally the interdependence of operations;
(c) for progressing purposes.

The answers to these points will enable the type of programme best suited to the particular problems and needs of difference levels of management to be chosen.

1.4 *Planning level*

Plans made at the upper levels of an organisation generally cover a long period of time and affect everyone in the organisation. These are known as major, strategic and/or policy plans. The attainment of one major plan requires the preparation and implementation of many minor plans. Minor plans cover a shorter time period and are more detailed than major plans. On the operational level, a master plan or pre-contract plan must be underpinned by short-term detailed plans such as monthly, weekly and stage programmes (Table 1.1). Through these, progress can be monitored and if necessary early corrective action taken to ensure work is progressing according to the master or pre-construction plan.

Table 1.1 Construction planning

PLANNING LEVEL

	POLICY	PRE-TENDER	PRE-CONSTRUCTION (MASTER PROGRAMME)
PURPOSE	Lay down strategy indicate objectives establish priorities within the company's resources	Reduce the risk of inaccuracy influencing preparation of tender	Analysis of production, i.e. amount of work to be undertaken, methods adopted and timing of its execution
OBSTACLES	Cyclical nature of the industry	Uncertainty Lack of detailed information. Lack of firm committment.	Expertise of management, time-span covered, availability of information, costs involved
ITEMS FOR CON-SIDERATION	Type of work to be carried out Financial record Organisation structure Development programme Sales and marketing information	Pre-tender report Methods state-ment, main schedules and Bill of Quantities Site organisation structure Subcontracting arrangements	Breakdown of major operations Commencement date for each operation Period for each operation Sequence of operations Phasing of operations Gang sizes per operation Plant requirements Material delivery dates Date when drawings and information must be available to execute each operation Names of sub-contractors involved Names of suppliers Holiday periods Provision to record progress Schedules
PARTIES INVOLVED	Senior company management	Estimator, purchasing plant dept., planning staff, contracts manager	Contracts manager planner, estimator plant dept., purchasing dept., architect/ engineer
MIS-CELLANEOUS			Consider: locality, access, existing services, ground conditions, storage, protection, adjacent buildings, bench marks, nearest tip, photographs

SHORT TERM	WEEKLY	STAGE
Enables management to take early corrective action in order to maintain progress Forward view of future work	Ensure progress is in accordance with detailed planning Opportunity to check resource requirements	Coordinate closely related operations in order to achieve continuity
Weather, availability of materials, rejection of materials, breakdown of equipment, inadequate detailing decisions of building control	As before	As before
Examine master programme Work to be included Critical dates Details of unfinished work Relevant correspondence List of revisions and variations and priorities Bill of Quantities	Examine short term programme, sequence of operations, operatives, gang sizes, plant materials Analysis of technical queries Detailed method statements Allocation of resources	As for master programme
Planner, site management	Planner, trades and specialist foremen	Planner, contracts manager, site manager relevant general foremen
Ensure that short term programme is in accordance with the master programme	Planning meetings and simple charts are important	Must tie into master programme

Chapter 2

Pre-construction procedure

2.1 Introduction

(This chapter is adapted from Byrne, 1967). The strategy to be adopted for the conduct of a construction project should be formulated well before commencement of site operations. A meeting will be arranged to announce the award of the contract and to acquaint all concerned with the facts. This will assist in co-ordination and co-operation from the start. The estimator will usually act as chairman at the meeting as he will be fully aware of the decisions taken at the pre-tender stage. From this meeting onwards the planning engineer will take over responsibility for pre-construction planning.

The personnel attending this first meeting will normally be the estimator, planner, projects manager, buyer, site manager for the project, contracts surveyor and possibly the work study engineer.

Many of the delays which occur on site can be traced back to someone's failure to do one or more of the following:

1. Double-check the notice required by suppliers for deliveries of materials.
2. Forecast the dates by which nominations are required from the architect or engineer.
3. Place orders for material and direct subcontractors' work.
4. Determine the off-site pre-fabrication times required by specialists on manufactured components of fittings.
5. Get agreement on the times subcontractors are required to commence and complete on site.
6. To notify local authorities.

In order that nothing is left to chance the planning team (under the direction of the contracts manager) will co-ordinate their efforts, dealing item by item with the following procedural checklist.

2.2 Procedural checklist

1. Carry out a site investigation, obtain subsoil information, prepare report and block plan, take photographs, etc.

2. Check available drawings required for the initial planning stage, i.e. 1–100 scale plans and alterations, block plan, 1–20 sectional details and schedules.
3. Obtain a full set of finished drawings as soon as possible.
4. Submit details to the Water Authority for temporary water supply.
5. Check Bill of Quantities (BOQ) to see that main items agree with drawings and calculate net labour costs.
6. Submit block plan to plant manager, and indicate dimensions of site accommodation. State commencement date.
7. Obtain from the estimating department a brief summary of the types of equipment and methods proposed that have been included in the estimate.
8. Prepare a list of queries for engineer's or architect's attention.
9. Abstract domestic trades (e.g. carpentry and joinery) from the BOQ and submit these to the joinery shop, etc.
10. Check through correspondence available in connection with the project to date.
11. Abstract all subcontract quotations and arrange to submit further inquiries in order to finalise the various subcontracts as soon as possible.
12. Prepare a schedule of all prime cost items and obtain the names of nominated subcontractors and suppliers.
13. Prepare a list of technical queries to engineer. This can be compiled throughout the planning procedure with the information obtained from the engineer to suit the progress of the overall programme. A copy will be retained in the head office and the other one for the site manager.
14. Arrange for the completion of necessary forms to comply with the statutory requirements.
15. Obtain all permits for hoardings, pavement crossings, etc. including preparing applications for telephone installations, electricity supply, sewer connections, water supply, etc.
16. Prepare a preliminary list of programme elements and produce the initial logic network and/or bar chart.
17. Commence preparation of the programme calculation sheets, i.e. abstract all the relevant items from the BOQ under separate headings for each programme element. Should be prepared in duplicate with one copy for the head office and the other copy for the site manager.
18. Obtain the net labour included in the BOQ for each programme element by multiplying the amount of each item in the bill by net labour values.
19. Convert the total money values for each programme element into man-hours.
20. Superimpose on the site layout drawings, the position of all drain trench excavations and foundation trench excavations to obtain an accurate picture of the state of the ground during excavation work.
21. Prepare detailed method statement. This will be discussed in greater detail during the programming and site layout preparation. Decisions will include:
    (a) type of equipment used;
    (b) method of carrying out the work;
    (c) size of gangs to be employed;
    (d) amount of net labour included in the estimate for the work involved.

22. Plot the decisions made by the planning team, i.e.: (a) contracts manager; (b) site manager; (c) estimator; (d) planner, on each separate programme calculation sheet. If a network is used convert to a bar chart.
23. At this stage it may be advantageous to invite the architect or engineer to attend the planning meeting.
24. Write to all subcontractors requesting them to attend the preliminary planning meeting. (Give approximately 2 weeks' notice.)
25. Write to all nominated and domestic suppliers to obtain delivery details, notice required, etc. and details of any drawings they may require. A standard format would be very useful.
26. Continue the procedure as detailed in item 22, co-ordinating the various subcontractors and specialists for the remainder of the programme elements.
27. Plot the dates for plant requirements, subcontractor's drawings, and indicate in the relevant positions allocated on the master programme chart.
28. Using the programme derive the following schedules: (a) labour; (b) plant; (c) materials; (d) subcontractors.
29. Decide on commencement date and plot the calendar on the master programme.
30. If necessary smooth resources to achieve optimum rhythm, utilisation, balanced gangs, etc.
31. Submit prepared schedules to relevant department within the firm (e.g. plant dept., purchasing dept.) for action.
32. Submit schedule of drawings required to the engineer or architect as soon as possible.
33. Ensure distribution of programmes to engineer/architect, site manager, general foreman, contracts surveyor, plant department and certain subcontractors.
34. Prepare in detail the first *stage* or *short-term* programme.
35. Take each item in the short-term programme, and ensure that all materials and plant have been requisitioned and ordered.
36. Complete the short-term programme in greater detail.
37. Send relevant programmes to site, including adequate supply of forms and planning calculation sheets.

Two important items stated in the above list are worth more detailed consideration. These are:

(a) site layout;
(b) method statements.

These are not prepared as isolated items since they are interdependent. A team effort is required for their preparation. The contracts manager should seek the distilled wisdom from all the planning team. A work study engineer would be a useful member of the team here.

2.3 Site layout

The decisions to be taken can be divided roughly into four groups, although they are interdependent to some extent.

1. Provision of adequate access roads for the transport of equipment on to and about the site. This is an extremely important consideration because it will normally be linked with the plan of construction and in some cases may actually control the progress of construction. Ideally there should be one-way traffic, i.e. entrance and exit, as this stimulates the flow of traffic and reduces bottle-necks.

2. Location of major plant. Many items of major plant require special provisions to be made for their installations and usually this is done at an early stage of construction. They should be positioned where they will cause least hindrance to adjacent construction, while providing the best service. Tower cranes are an example of this. If they are static they require special bases for the towers which need to be located where they will not obstruct foundations but will also have the capacity to handle required weights at the perimeter of the building.

Concreting plant should, ideally, be positioned at the centroid of concreting operations. This will reduce transportation to the minimum.

3. Storage areas. Areas must be set aside for the storage of materials. It is necessary to hold a 'buffer' stock on site in case of delays, etc. These should be stored as near as possible to the final position of incorporation in the structure. Again the aim is to reduce unnecessary travelling and double handling with least hindrance to other site activities.

4. Site accommodation. This must conform to the requirements of the construction (working places) regulations. The location is important and is dependent on individual site requirements, but there are several principles which should be observed whenever possible:

(a) A situation where there is natural light.
(b) The agent and general foreman would require a location where the contract can be viewed from the window.
(c) Offices need to be away from dust and noise of machinery.
(d) Timekeeper's office needs to be near site entrance.
(e) Materials controller's office needs to be in a position where the entry of lorries on to and off the site can be controlled.

Special problems related to site layout

Confined sites

Usually a problem in city centres and other built-up areas. The site may be confined by adjacent buildings, main streets or busy one-way traffic systems. The main problem is lack of space for storage, accommodation, plant movements, etc. Methods used to alleviate the situation are:

1. Site accommodation erected on gantries over the footpath areas.
2. Use of two- or three-storey site offices.

3. Use of ready-mix concrete.
4. Use of a climbing tower crane.
5. Buffer stock of materials kept to absolute minimum.
6. All waste materials immediately removed from site.
7. Material deliveries restricted to off-peak times.

Adjoining property can cause difficulties. If demolition work is involved, sewers may have to be diverted, underpinning and shoring may be required.

Tall buildings
Main problem is unproductive time due to slow vertical travel of labour, plant and materials.

Methods used to alleviate the situation are:

1. Establish site accommodation, e.g. canteens, toilets, etc. at, say, every sixth floor.
2. Install high-speed passenger lifts/hoists.
3. Ensure tower crane is fully utilised.
4. Provide temporary services outlets for 110 V supply.
5. Use of strategically placed chutes to remove rubbish.

In order to reduce the possibility of claims for damages by neighbours, photographs should be taken of their property before commencement of construction operations and agreement reached as to the condition of their property.

2.4 Method statements

It is wise to think widely in the initial stages of the optimum method to be adopted or plant to be used. The fact that one particular method has been successful on a previous project does not justify its exclusive use on all other similar projects. The method statement should include work of any consequence. It takes the form of a schedule. The method used to accomplish each major operation will be described together with plant, labour and duration. The method statement usually forms the basis for the programme preparation.

Chapter 3

Bar charts and location–time diagrams

3.1 Introduction

The bar chart is probably the best known of all planning techniques. It was developed by H. L. Gantt, an American pioneer of scientific management. It basically features a plan of a project split into logically related individual activities each represented graphically by scaled lines (Fig. 3.1).

The length of a bar is proportional to the scheduled time for the execution and completion of that activity. Progress then may be properly monitored by ensuring that the set of bars or activities is completed on schedule.

One shortcoming of the bar chart, though, is that sequential relationships between activities are not completely prescribed. The method also lacks an efficient tool for the control of schedule, resource and cost of a project. Its effectiveness, therefore, is restricted generally to the programming of simple and small projects only, especially when it is used independently.

Later refinement work on the technique has led to the development of the Milestone method, which is basically a linked bar chart (Fig. 3.2) and is also the first attempt to

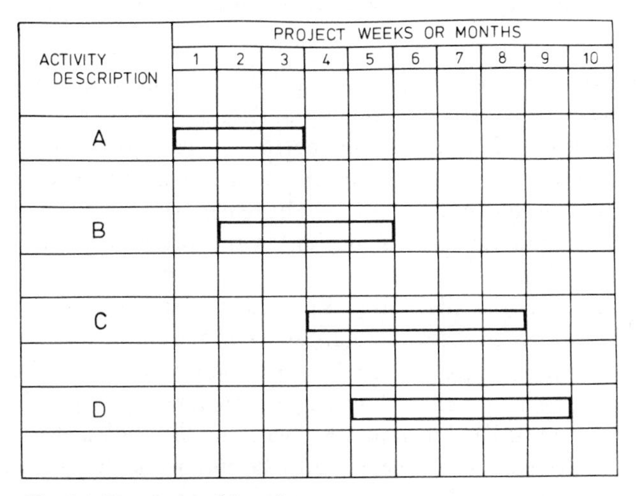

Fig. 3.1 Conventional bar chart

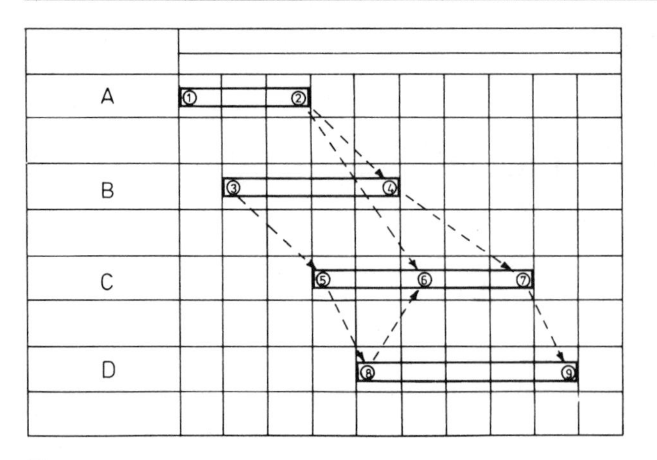

Fig. 3.2 Milestone technique: arrows indicate interdependence between activity milestones

indicate precise interrelationships between activities. This technique has enabled a reduction in the number of bars drawn on the chart. Graphically, this represented a significant advance in communication over the original chart.

3.2 Preparing a bar chart

When preparing a bar chart for the erection of, for example, a concrete-framed building, the work may be divided up into the following main stages:

1. Preliminaries.
2. Substructure (which consists of work below ground level, often taken as work below the damp-proof course).
3. Superstructure (which includes the main carcass above ground level).
4. Finishings (Fig. 3.3).

Each of these main stages or operations must then be divided up into sub-operations and to do this a good knowledge of trade sequence is most important. In some cases the exact order is sometimes debatable. The order of sub-operations may vary slightly. When the order of operations is being decided, however, the most economical sequence both in time and money should be chosen. It must always be remembered that it only needs two conflicting trades trying to work at the same time, or one trade to arrive late, for the complete contract to be disorganised.

For this reason it is most important that all the work is plotted by the main contractor in a logical and practicable sequence so that operations follow in the correct order.

In order to compile a programme schedule all the information contained in the contract documents, etc. should be used in an intelligent manner to provide a workable scheme. The amounts of work must be known together with the man-hours or the machine-hours required to do the work. From these figures the number of man-weeks will be calculated and it must then be decided how many men it is necessary to engage on each operation. Absenteeism must be allowed for. The best method is to compile a schedule of sub-operations.

After this has been done and before the programme is drawn up, the manager must decide upon the following points.

1. In which order and in what manner is the work to be carried out.
2. Which operations are fixed in relation to time and are in strict interlocking sequence.
3. What operations have a degree of latitude in time as to when they are carried out, and if so how much.

The programme should be carefully prepared and, as far as possible, should be carried out as laid down. It should only be necessary to make changes in the programme when problems arise which are beyond the control of the contractor, such as bad weather, or because time has been saved or because it is impossible to reach the prescribed targets. The scheme should therefore be flexible enough so that changes can be effected without too much inconvenience.

All major items should be recorded on a master programme chart and in some cases certain minor features of the construction will be grouped together. This will later be split down with the aid of 'short-term and weekly' programme and progress sheets.

It is normal to record each operation in the approximate order that it will be carried out on the site. A typical job might be split up into unit operations. The corresponding columns against each item would then be completed. In each case the 'man/weeks' would be divided by the number of men to be employed on the operation. This would provide suitable information to plot on a programme chart.

For building up a programme when operations are more complicated a programme calculation sheet may be used (Table 3.1).

3.3 Key operations

In every stage of construction work one operation will control the time required to complete that work. This operation is determined by setting out the minimum times required to complete each operation within the stage using the optimum gang size for each. The longest operation is known as the key operation and this will control the speed of construction within that stage. The manning of this operation will control the natural rhythm of all other operations and will establish a cycle time for a particular stage. Once the time period for the key operation has been calculated all other operations must be brought into phase, i.e. *balanced*. The total time period for other gangs should ideally be made the same as that of the key operation by adjusting manning levels (Tables 3.2 and 3.3 and Fig. 3.4).

On most framed buildings, once the substructure is completed, the controlling operations are: (a) the frame; (b) the floors; (c) the walls.

These operations, to a large extent, control the speed of erection of the building as a whole and the order of work is very definitely laid down. Until the frame has been erected it will be impossible to construct the floors, and until these two operations are

FACTORY & OFFICE BLOCK
MASTER PROGRAMME

STAGE		OPERATION	APPROX QUANTS	SUB.- CONTRCTR.	N if nominated	SUPPLIER	N if nominated
	1	Site installation					
Preliminories	2	Demolition		Regan Bros.			
	3	Setting out					
	4	Bulk excavation					
	5	Exc col. bases					
Sub - Structure	6	Hardcore					
	7	Con to col. bases					
	8	Ground beam					
	9	Gnd. floor slab					
	10	Conc. column					
	11	Conc. beam & floor slab					
	12	RC walls, lift well ctc					
Super - Structure	13	Ext. bk panel wall					
	14	Windows				Craig & Co	N
	15	Partition walls					
	16	Roof insulation & asphalt					
	17	Glazings		McCune & Co			
Site work	18	Drainage				Irelands	
	19	Conc. paving					
	20	Joiner 1st fixing					
	21	Mech. installation		Hadon & Co	N		
	22	Elec. installation		Hadon & Co	N		
	23	Plumbing					
	24	Heating					
	25	Plasterer					
	26	Floor screed					
	27	Wall tiling					
	28	Floor tiling					
Finishings	29	Patent flooring					
	30	Sanitary fittings		Smyths Ltd	N		
	31	Joiner 2nd fixing					
	32	Factory fitments					
	33	Office units				Starfit Ltd	N
	34	Ironmongery				Starfit Ltd	N
	35	Balustrades					
	36	Painting					
	37	Cleaning up & hand over					
	38						
	39						

Fig. 3.3 Bar chart for concrete-framed building

completed the panel walls or curtain walls, as the case may be, cannot be constructed. In some cases mechanical equipment – perhaps a crane – will control the rate of progress.

The same thing applies to a block-constructed building. In the case of a brick dwelling house, for example, once the foundations are in, the brick carcass controls the speed of erection, with other trades, such as the carpenters fixing the first-floor joists,

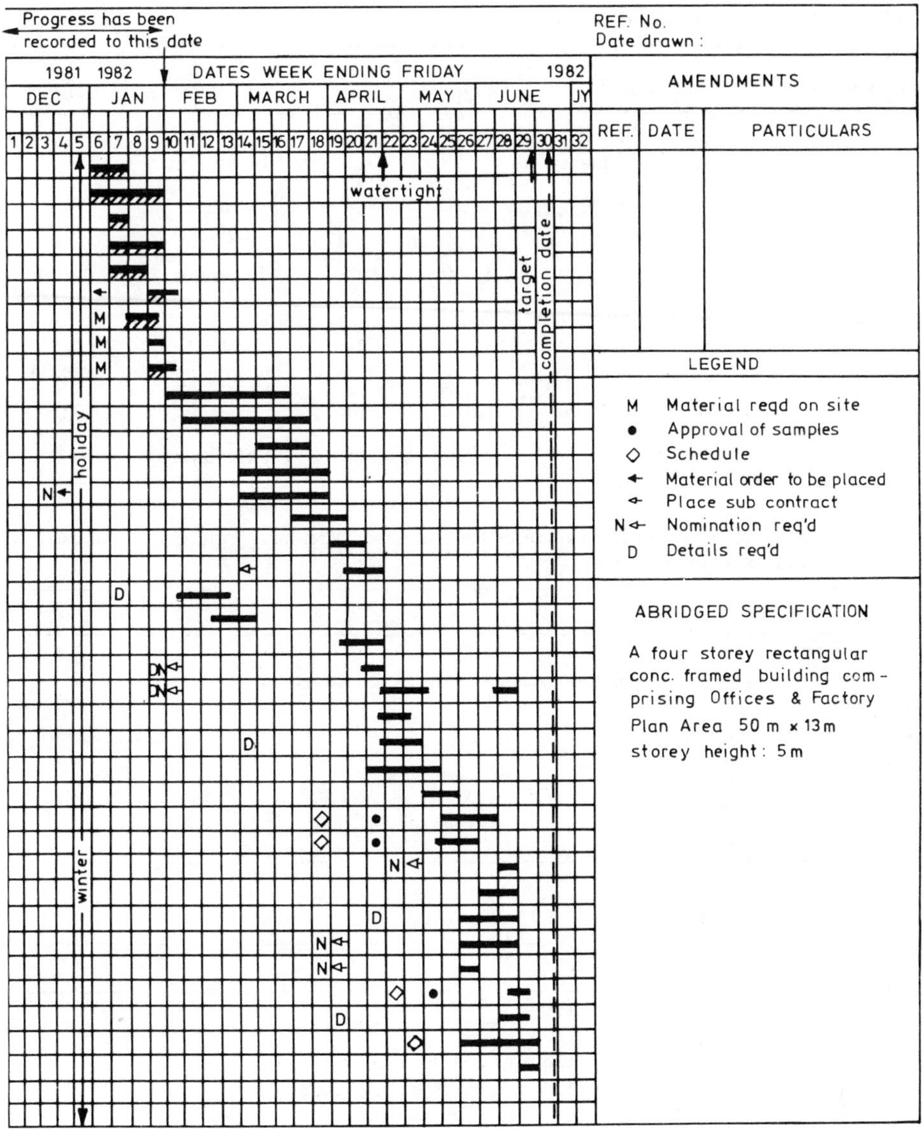

Progress has been recorded to this date

REF. No.
Date drawn :

1981 1982 DATES WEEK ENDING FRIDAY 1982

DEC JAN FEB MARCH APRIL MAY JUNE JY

1 2 3 4 5 6 7 8 9 10 11 12 13 14 15 16 17 18 19 20 21 22 23 24 25 26 27 28 29 30 31 32

watertight

target

completion date

winter holiday

AMENDMENTS

REF.	DATE	PARTICULARS

LEGEND

M	Material reqd on site
●	Approval of samples
◇	Schedule
◄	Material order to be placed
◁	Place sub contract
N◁	Nomination req'd
D	Details req'd

ABRIDGED SPECIFICATION

A four storey rectangular conc. framed building comprising Offices & Factory
Plan Area 50 m × 13 m
storey height : 5 m

etc. dovetailing in (Fig. 3.5). These key operations need extra-special consideration by the person who is drawing up the programme.

Once the controlling trades are on the job and the building is a few floors high then the subordinate operations come into the picture. Their order of working is not normally so easy to define. Every job should be treated on its merits. In some cases the subordinate operations such as plumbing, plastering, marble fixing, flooring contrac-

Table 3.1 Programme calculation sheet

Op. number	*Operation*	*Qty*	*Unit*	*Trades*	*Constant*
1	Site installation (huts, services plant)			Carpenter Electrician	
2	Setting out site			Site engineer	
3	Excavate foundations	1,500	m^3	Excavator	10 m^3/hr
4	Concrete to foundations	500	m^3	Ganger	4 m^3/hr (mixer)
5	Gravel bed to drains	2,010	m		0.05
6	Drainlaying	2,010	m	Drainlayer	0.25

Table 3.2 Balancing gangs. Detached houses 10 in number. Man-hour requirements for each main trade (per house)

	Man-hours	
Labourers	550	
Bricklayers	1,200	*Key operation, determines the cycle time
Carpenters	350	
Plasterers	480	
Roof tiler	85	
Plumber	150	
Electrician	80	
Painter	220	

Table 3.3 Balance the gangs

	Gang size	*Hours*	
Labourers	5	110	
Bricklayers	9, and 3 labs	130	
Carpenters	3	116	Aim for similar
Plasterers	4, and 2 labs	120	hours keeping gang
Roof tiler	1		sizes realistic
Plumber	1, and app.	100	
Electrician	1	80	
Painter	2	110	

tors, etc. can start work at intervals of a few days between each trade. In other cases they can perhaps start work at almost the same time, each commencing on different floors. In this respect a steel-framed building has an advantage in that the concrete floors need not necessarily be constructed in numerical order. It may be advantageous for the contractor to construct the ground floor first – if the building has a basement – and then to leave the first and second floor for the time being and to construct the

Total hours	*Total weeks*	*No. of tradesmen*	*No. of labourers*	*Others*	*Programme (weeks)*	*Remarks*
		3	3			Plus lorry and driver
		1			1	
			2		2	
150	4.3		2			Hydraulic excavator
125	3.6	2	2			10/7 Mixer
100	3					Drott
500	14	4			3.5	

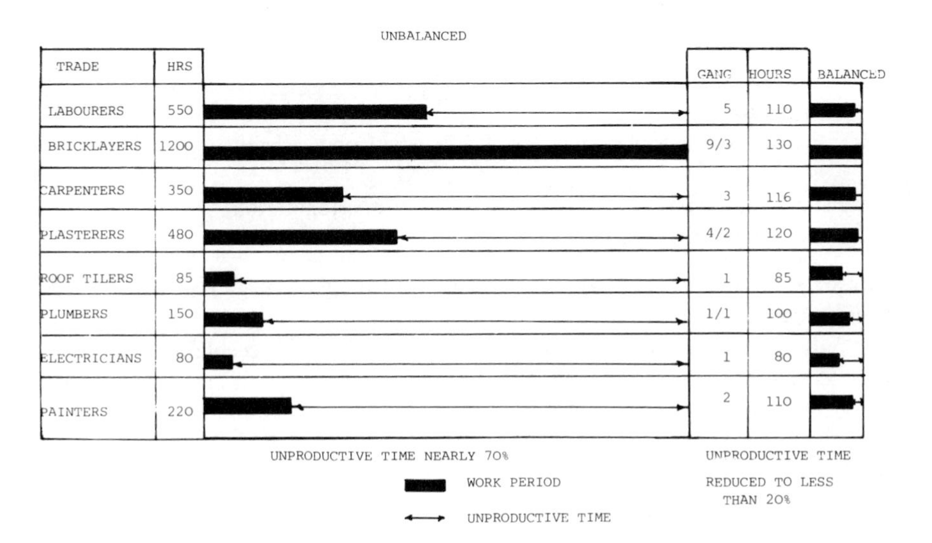

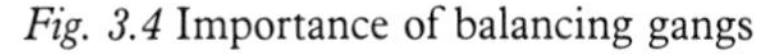
Fig. 3.4 Importance of balancing gangs

third floor. In this way, the bottom portion of the building will receive a certain amount of protection from the weather.

In some instances it may be worth while for a contractor to lay a temporary layer of asphalt over the floor, half-way up a building, so that the structure is weather-tight and then the finishing trades may proceed on the lower floors. In this way speed of erection can often be considerably increased.

A disadvantage often levelled at the bar chart technique is that interdependencies of activities are not shown. This criticism is largely true, but may not be a major disadvantage under the following conditions where:

(a) the project is not too complex;
(b) the programme has been systematically prepared as already outlined;

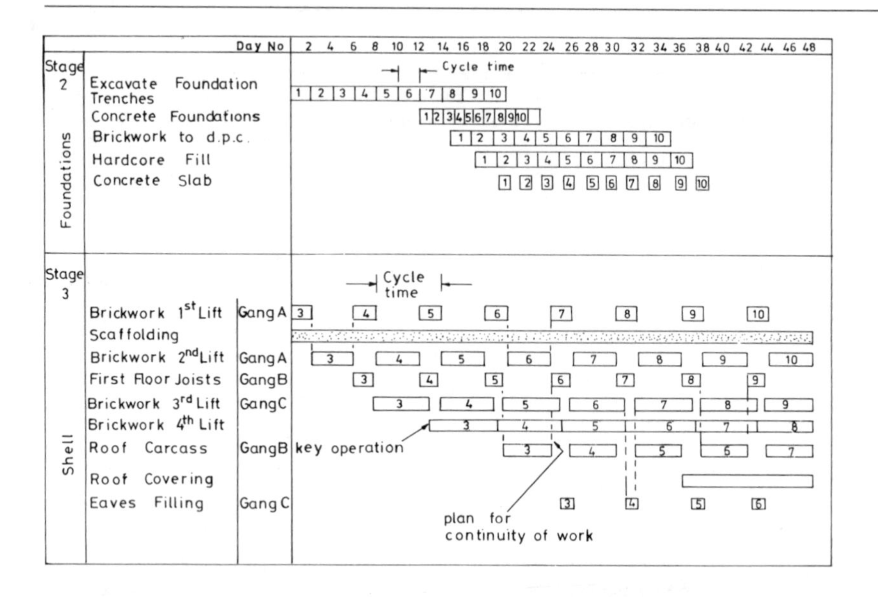

Fig. 3.5 Each stage or phase of construction may have different cycle times

(c) the site agent has been intimately involved in its preparation;
(d) the site agent is fully experienced.

This last point is important. An experienced agent may often know intuitively the interdependence of activities.

The building industry is notoriously conservative. The full acceptance of this type of simple planning may automatically facilitate the introduction (at a later date) of more accurate and powerful techniques.

3.4 Bar chart – the main advantages

The main advantages of the bar chart are:

1. Simplicity.
2. Ease of recording progress.
3. Ease of rescheduling.
4. Schedules can be directly produced from it.
5. Dates for ordering materials, notices required, etc. can easily be seen.
6. Short-term programme can easily be derived from the main chart.
7. Suitable technique for the smaller contractor.

3.5 Bar chart example – construction of a bridge

The project involves the construction of a bridge with central pier over a future road

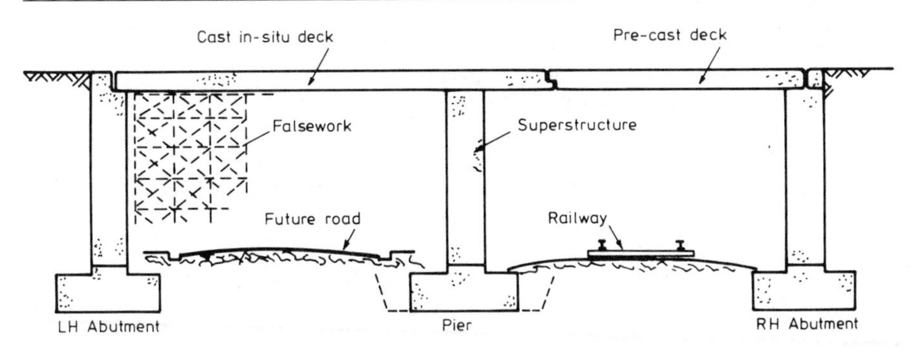

Fig. 3.6 Bridge construction

and existing railway (Fig. 3.6). The abutments, central pier and left-hand deck are cast *in situ*. In order to reduce construction time and falsework the right-hand deck is pre-cast.

Constraints

1. Work will commence on the left-hand abutment since this section of the bridge will require the longest duration.
2. Excavating plant can be released from left-hand abutment after 1 week.
3. Erection of falsework can commence immediately after the completion of the left-hand abutment superstructure.
4. Allow 2 weeks for curing of cast *in situ* deck before removing falsework.
5. There is a danger of overloading the cast *in situ* cantilevered section of the bridge deck, therefore placing of pre-cast deck to be delayed by 2 weeks.

Procedure

1. Prepare list of operations.
2. Establish expected durations.
3. Prepare a bar chart.
4. Derive required schedules.

1. Prepare list of operations

The list of operations and their detail is largely a matter of judgement. Activity listings for a short-term programme will be in much greater detail than listings for a master programme. Suggested listings for the bridge contract are:

Left-hand abutment	Foundations
	Superstructure
Centre pier	Foundations
	Superstructure
Right-hand abutment	Foundations
	Superstructure

Cast *in situ* deck	Falsework
	Permanent construction
Pre-cast suspended span	Erect beams
	Concrete infill

Other operations could be included such as *parapets, hand-railing, paving*, etc.

2. Establish expected durations

These are determined by applying factors against quantities. The planner must bear in mind that quantities stated in the BOQ are net, e.g. considering the foundations to the left-hand abutment. See Fig. 3.7.

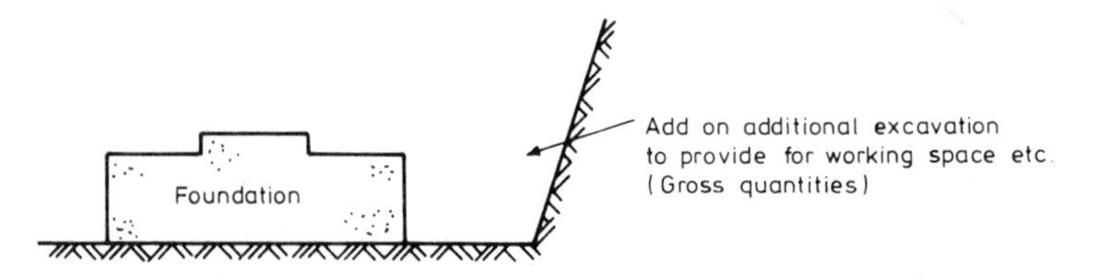

Fig. 3.7 Foundation

The operation of, for example, *foundation* is made up of a number of sub-operations such as: excavation; blinding concrete; formwork; reinforcement; concrete base; remove formwork.

Considering – excavation. The volume in the BOQ may be 800 m³. This is net volume. The planner checks the gross volume to be say, 1,000 m³. Available plant is next selected and its output factor applied to this volume to determine the duration, i.e.:

Plant output/day = 200 m^3

$$\text{Duration} = \frac{\text{Volume}}{\text{Output}}$$

$$= \frac{1{,}000}{200} = 5 \text{ days}$$

Other sub-operations, for example *reinforcement*, are treated similarly. Total weight in foundations to left-hand abutment = 200 kg. The rate per 50 kg for this job (including cutting to length, bending and fixing) is say 5 hours. Therefore the duration is

$$\frac{2{,}000}{50} \times 5 = 200 \text{ man-hours or 5 man-weeks}$$

i.e. 5 weeks for one man or five men for 1 week, etc.

3. Prepare a bar chart

When each operation duration has been determined it is listed in construction sequence on the chart adjacent to the operation description. Particular attention is paid to any constraints or contract conditions that may influence construction progress or sequence. The bar chart is drawn as in Fig. 3.8.

An advantage of the bar chart is that it can facilitate simple progress recording. Progress can be marked in coloured ink or dotted lines. If contract time is half-way through week 3 then the history of the job can clearly be seen, i.e.:

- Foundations to left-hand abutment started on time but overran by half a week.
- Superstructure was half a week late in starting but is 50 per cent completed.
- Foundations to centre pier started late but finished on time.
- Superstructure to centre pier and foundations to right-hand abutment should have started at beginning of week 3.

3.6 Location – time charts

The normal bar chart or Gantt chart method is suitable for most structural and building applications. However, for the programming of construction works such as motorways, railways and drainage schemes it requires some modification. Rate of progress, cut and fill operations, location of likely restrictions to production can most readily be recorded on a *location–time chart.*

Most common restrictions are:

- access or egress to works;
- haul routes to borrow pits;
- temporary and permanent crossing for haul roads, main roads, railways, watercourses, etc.;
- variations in site conditions.

3.7 Example: Roadway construction

A length of roadway is to be constructed as shown in Fig. 3.9. Chainages, quantities of cut and fill and culvert positions are indicated on the diagram.

Prepare a *location–time chart* based on the following planning data:

1. Earthworks plant output is 2,000 m^3/week.
2. Installation of culverts in advance of embankment construction will take 2 weeks.
3. Allow 2 weeks' delivery time for culvert materials.
4. Road drain installation can proceed at a rate of 400 m^2/week.
5. Road base can be laid at a rate of 3,000 m^2/week.
6. Road surfacing on bays can be laid at a rate of 4,800 m^2/week.
7. All labour and plant resources involved in drainage work, road base and surfacing must be utilised continously.

The solution is shown in Fig. 3.10.

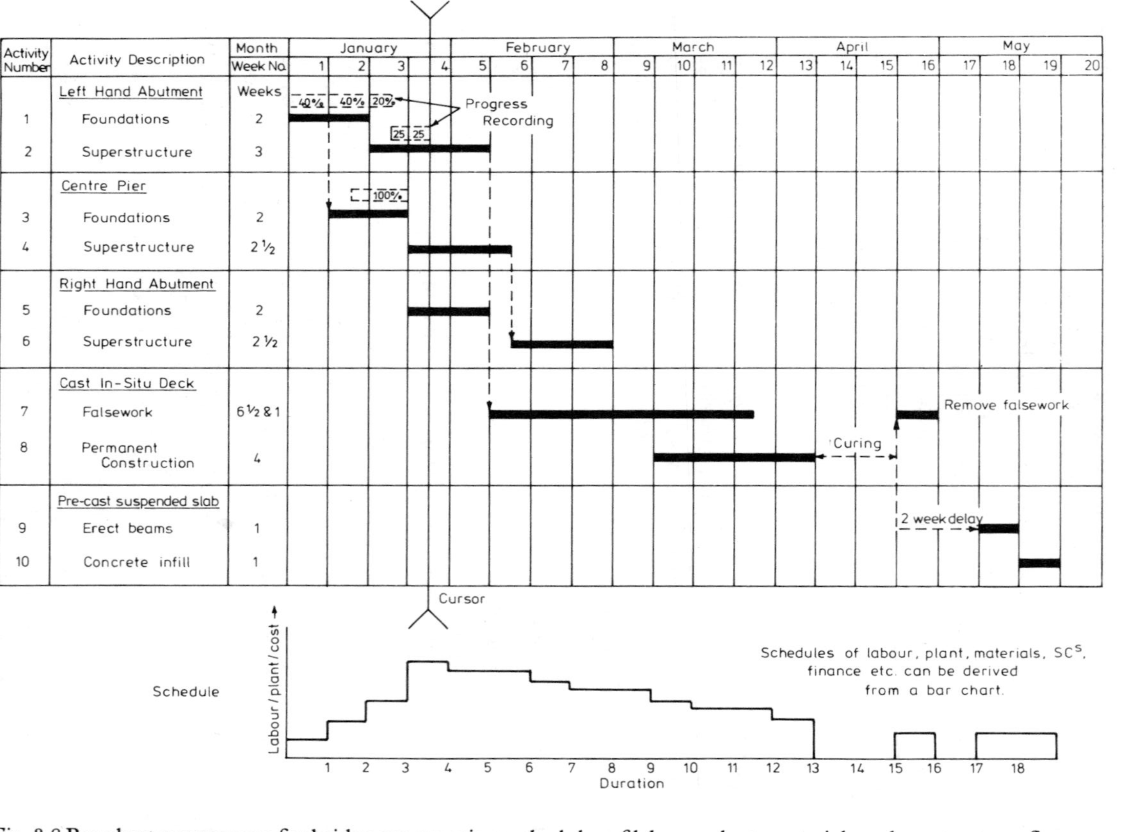

Fig. 3.8 Bar-chart programme for bridge construction: schedules of labour, plant, materials, sub-contractors, finance, etc. can be derived from a bar chart

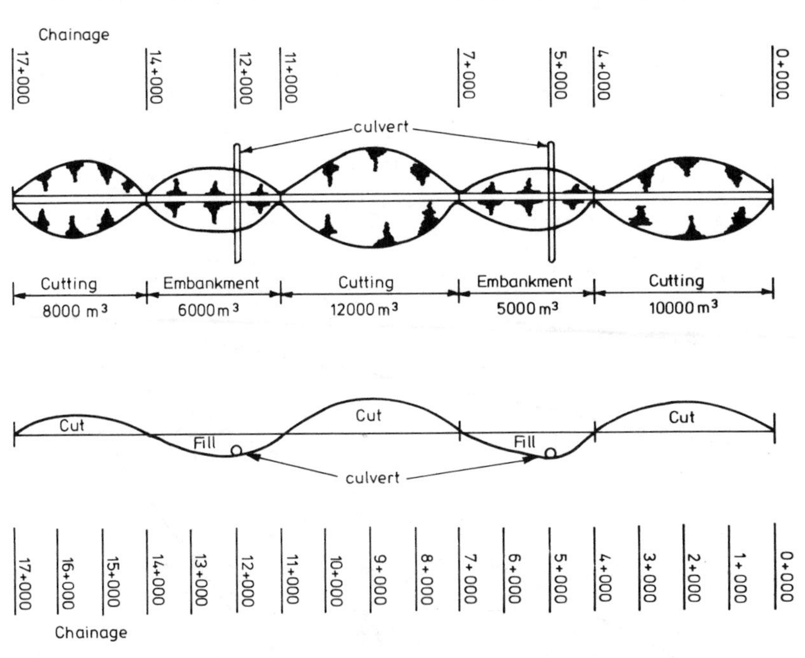

Fig. 3.9 Roadway construction

3.8 Worked examples: Bar charts and location – time diagrams

Q.3.1.

Figure 3.11 illustrates an irrigation scheme which a Middle East country is proposing to develop by constructing an impounding embankment across a wadi through which water runs during the short rainy season. Two irrigation canals, one partially lined, the other fully lined, will conduct impounded water from the reservoir so formed.

Two contractors have been appointed for the works on the understanding that once his plant has been mobilised on site each contractor will have continuous working for his plant.

1. *Earthworks contractor* – will excavate and trim canals and construct the embankment.
2. *Paving contractor* – will line the canals and waterproof the upstream face of the embankment with asphalt.

The earthworks contractor has a hydraulic back-hoe excavator and dumpers with outputs:

Trenching for canal (lined) in soft material 400 m^3/day
Trenching for canal (unlined) in rock material 110 m^3/day.

This equipment can also be used to provide additional filling for the embankment

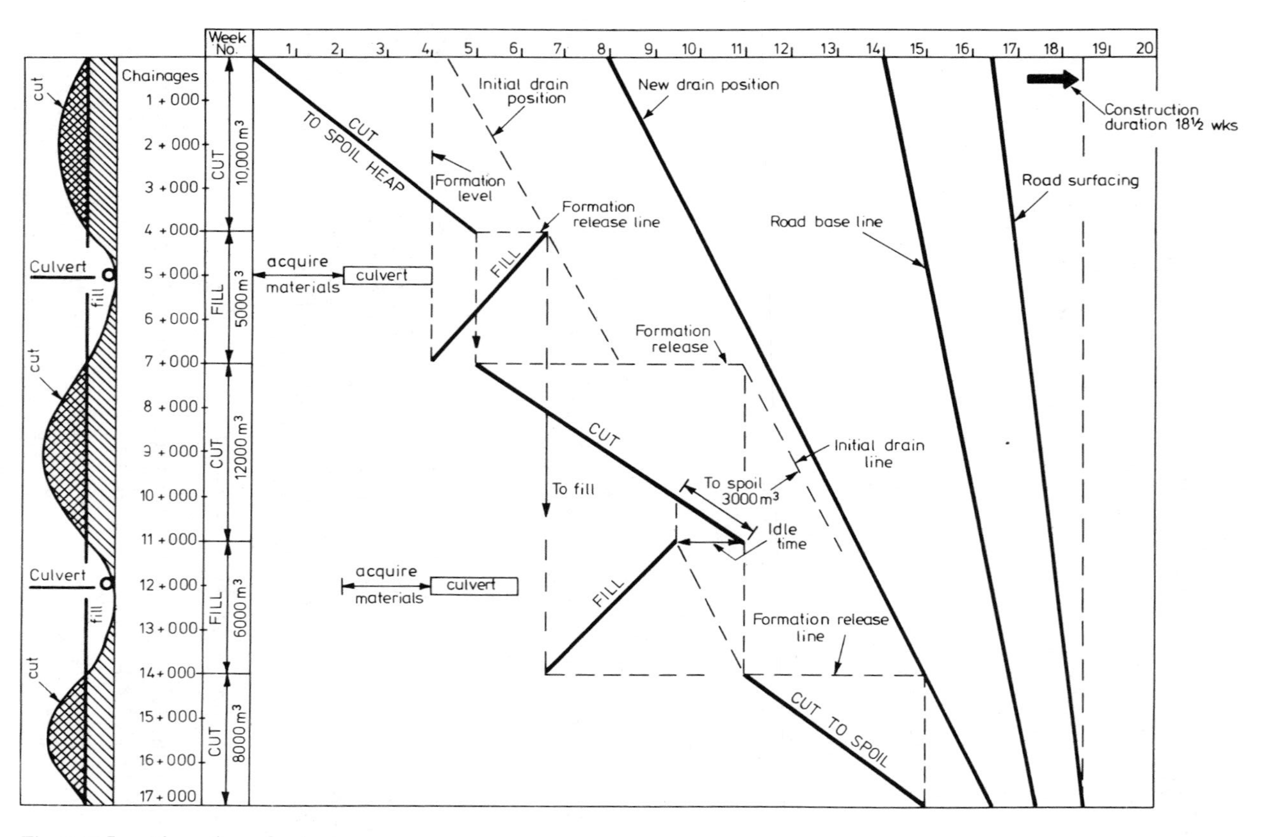

Fig. 3.10 Location–time chart

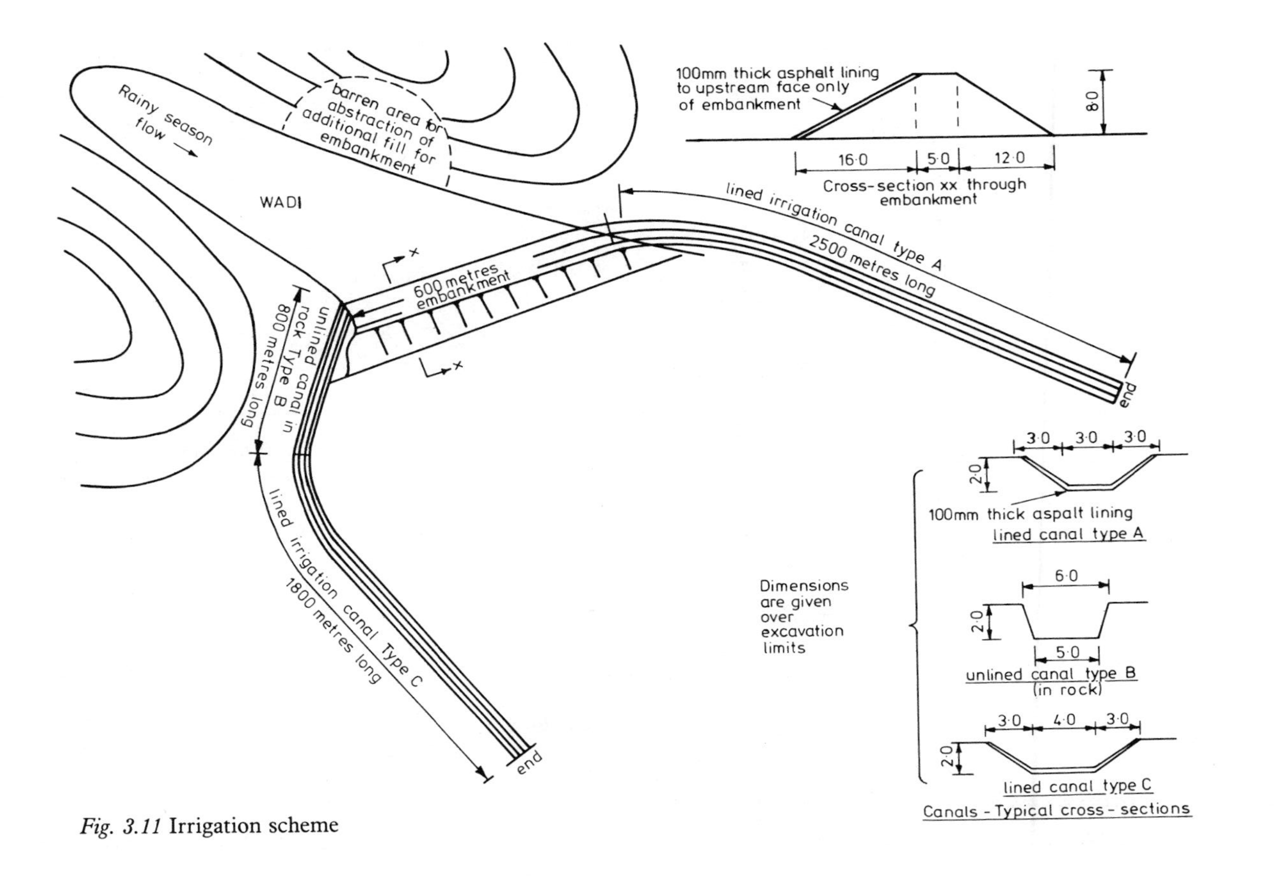

Fig. 3.11 Irrigation scheme

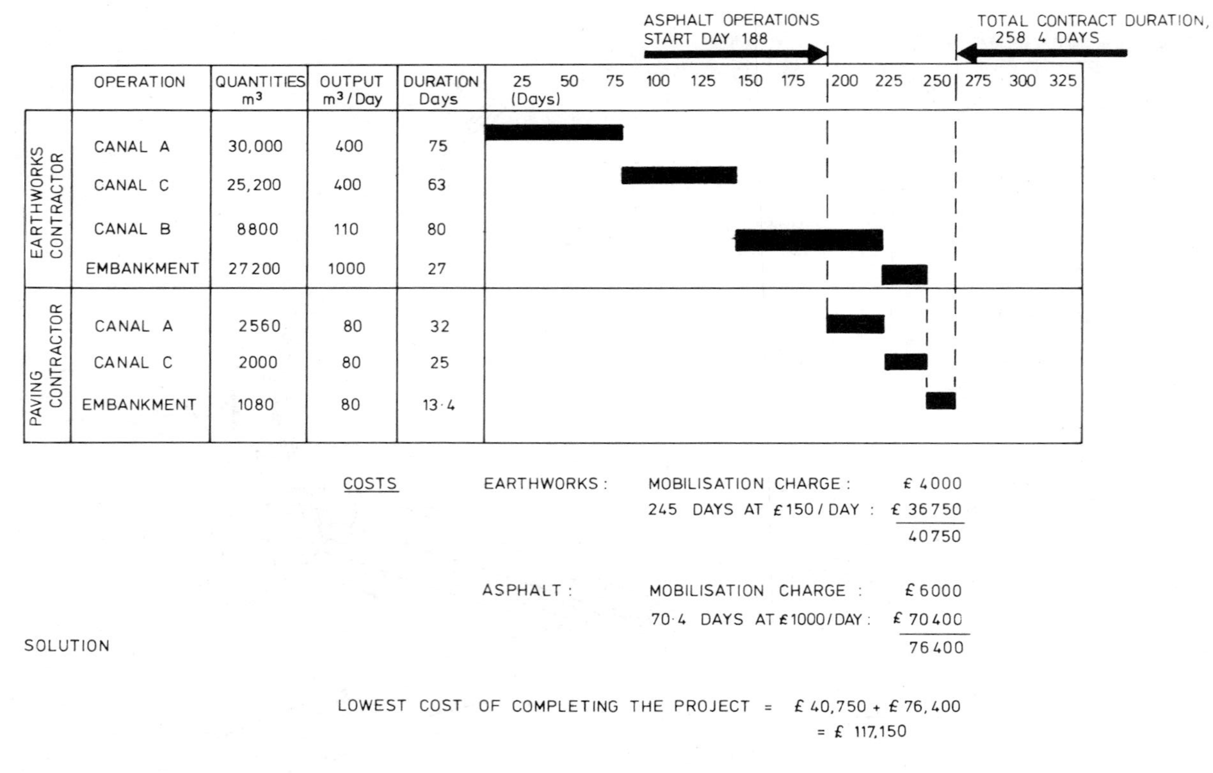

ASPHALT OPERATIONS START DAY 188
TOTAL CONTRACT DURATION, 258 4 DAYS
OPERATION
QUANTITIES m3
OUTPUT m3/Day
DURATION Days
25 50 75 100 125 150 175 200 225 250 275 300 325
(Days)
EARTHWORKS CONTRACTOR
CANAL A 30,000 400 75
CANAL C 25,200 400 63
CANAL B 8800 110 80
EMBANKMENT 27200 1000 27
PAVING CONTRACTOR
CANAL A 2560 80 32
CANAL C 2000 80 25
EMBANKMENT 1080 80 13·4
COSTS
EARTHWORKS: MOBILISATION CHARGE: £4000
245 DAYS AT £150/DAY: £36750
40750
ASPHALT: MOBILISATION CHARGE: £6000
70·4 DAYS AT £1000/DAY: £70400
76400
SOLUTION
LOWEST COST OF COMPLETING THE PROJECT = £40,750 + £76,400
= £117,150

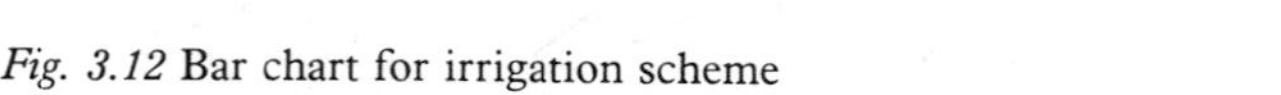
Fig. 3.12 Bar chart for irrigation scheme

from the borrow area at a rate of 1,000 m³/day. It costs £4,000 to mobilise and £150/day thereafter.

The paving contractor's asphalt equipment costs £6,000 to mobilise and its working cost, including materials used, is £1,000/day. It has an output per day of 80 m³ asphalt. Using bar chart programming determine the lowest cost of completing the project. When will the asphalt lining start relative to the starting of earthworks?

The following assumptions may be made:

1. All material excavated from canals must go to the embankment.
2. Bulking of excavated materials can be ignored.
3. Costs of compaction equipment are included in the excavation team costs.
4. Asphalt lining to the embankment upstream slope cannot commence until the embankment has been completely filled.
5. Volume of asphalt per linear metre of canal = excavation girth × 100 mm.
6. All dimensions in Fig. 3.11 are in metres except where otherwise stated.
7. The canal construction sequence will be A–C–B.

The solution is shown in Fig. 3.12.

Q.3.2.

A motor vehicle manufacturer is promoting the construction of a new vehicle test track 6 m wide by 5,200 m overall length which consists of two straights each 1,700 m long connected at both ends by circular curves, without transitions, each 900 m long. Figure 3.13 illustrates the track layout, earthworks required and an ancillary pipe culvert which crosses two of the embankment seats.

It is proposed to employ three contractors on the works.

Contractor A to lay the pipe culvert which has an overall length of 1,300 m. The contractor estimates his laying output to be 20 m/day.

Contractor B to carry out the earthworks which require the excavation of a total of 45,500 m³ from cuttings and placing in embankments. Earth-moving outputs have been assessed as follows:

- Haul not exceeding 350–1,000 m³/day.
- Haul exceeding 350 m but not exceeding 600–900 m³/day.
- Haul exceeding 600 m but not exceeding 800–850 m³/day.
- Haul is in every case measured from the mid-point of the cutting to be excavated to the mid-point of the embankment in which the excavated material is placed.

Contractor C to lay the reinforced concrete (RC) pavement on the excavated or filled formation. On straight sections he can construct 600 m²/day but on curved sections output is reduced to 300 m²/day because of form-laying difficulties.

(a) *Using the location–time chart method determine the time of completion for the project as a whole.*

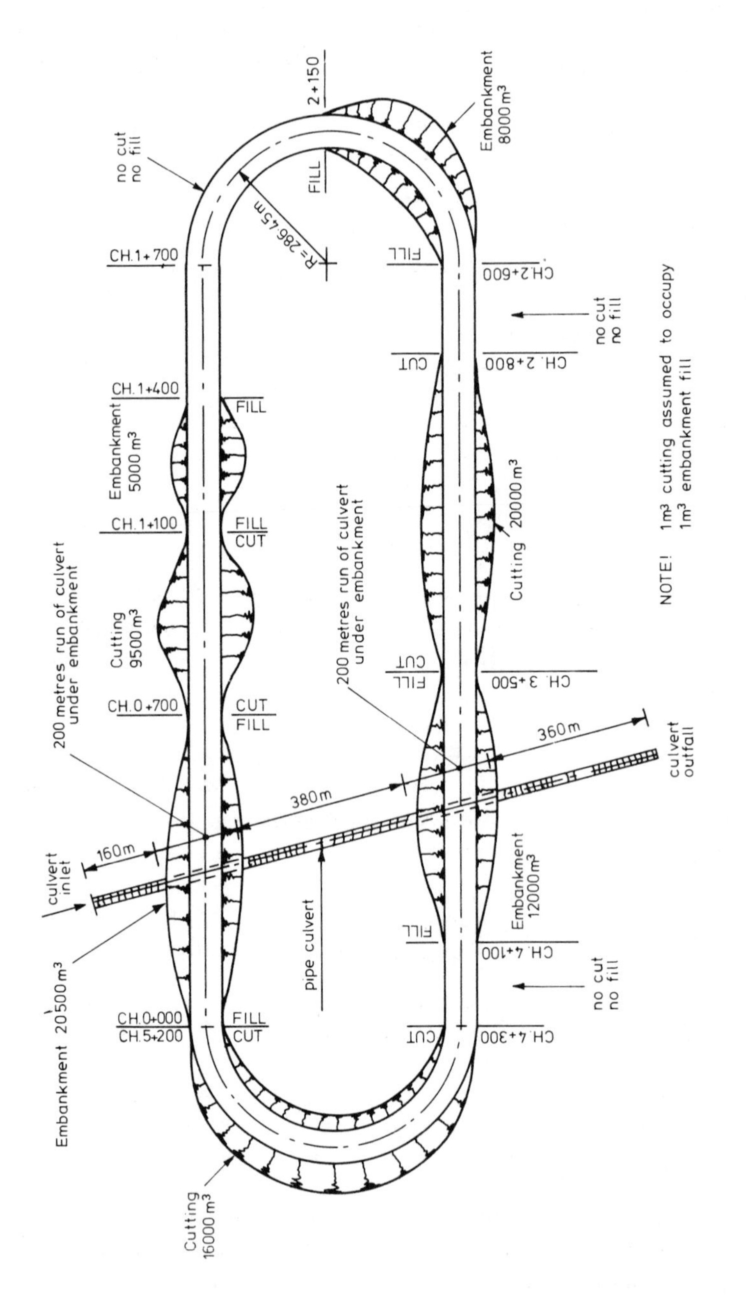

Fig. 3.13 Vehicle test track

The following assumptions should be made:

1. Contractor A must start at the outfall and proceed continuously to the culvert inlet.
2. Possession of any embankment section can only be taken by the earthworks contractor when the culvert section immediately under the embankment section is complete.
3. Possession of any section of cutting or embankment cannot be taken by the paving contractor until all the earthworks in the section are complete.
4. Possession of a no-cut no-fill section cannot be taken by the paving contractor until hauling through that section by Contractor B has ceased.

Each contractor must have continuity of work without interruption once his plant has been mobilised on site.

Mobilisation is to occur at the earliest possible date.

Earth-moving operations commence in cutting:

CH.2 + 800 – CH.3 + 500 and then proceed to cutting
CH.4 + 300 – CH.5 + 200 before completing in cutting
CH.0 + 700 – CH.1 + 100

Paving commences at CH.1 + 400 and proceeds in a clockwise direction around the track.

(b) *What would be the effect on contract duration if the paving contractor was unable to commence operations until the completion of all cut and fill operations?*

The solution is given in Fig. 3.14.

Q.3.3.

Figure 3.15 shows a longitudinal section on the line of a proposed railway development. The earthworks plant for excavation of cuttings and filling of embankments has the following outputs:

2,400 m^3/day on an average haul of 200 m
2,000 m^3/day on an average haul of 300 m
1,600 m^3/day on an average haul of 400 m

Embankment filling is to start on section AB but only after completion of the pipe culvert which will take 8 days from project commencement date. Viaduct CD has a time for completion of 40 days from project commencement date.

(a) *Using a location–time chart method determine the minimum number of days in which the ballast-laying operation can be carried out.*

The following assumptions should be made:

1. The viaduct is not required to enable earth-moving plant to cross the river.
2. One day's work is lost when the earthworks plant breaks down during transfer from cutting DE to cutting FG.
3. Ballasting starts at A on release of earthworks formation on embankment AB and proceeds continuously at constant output.

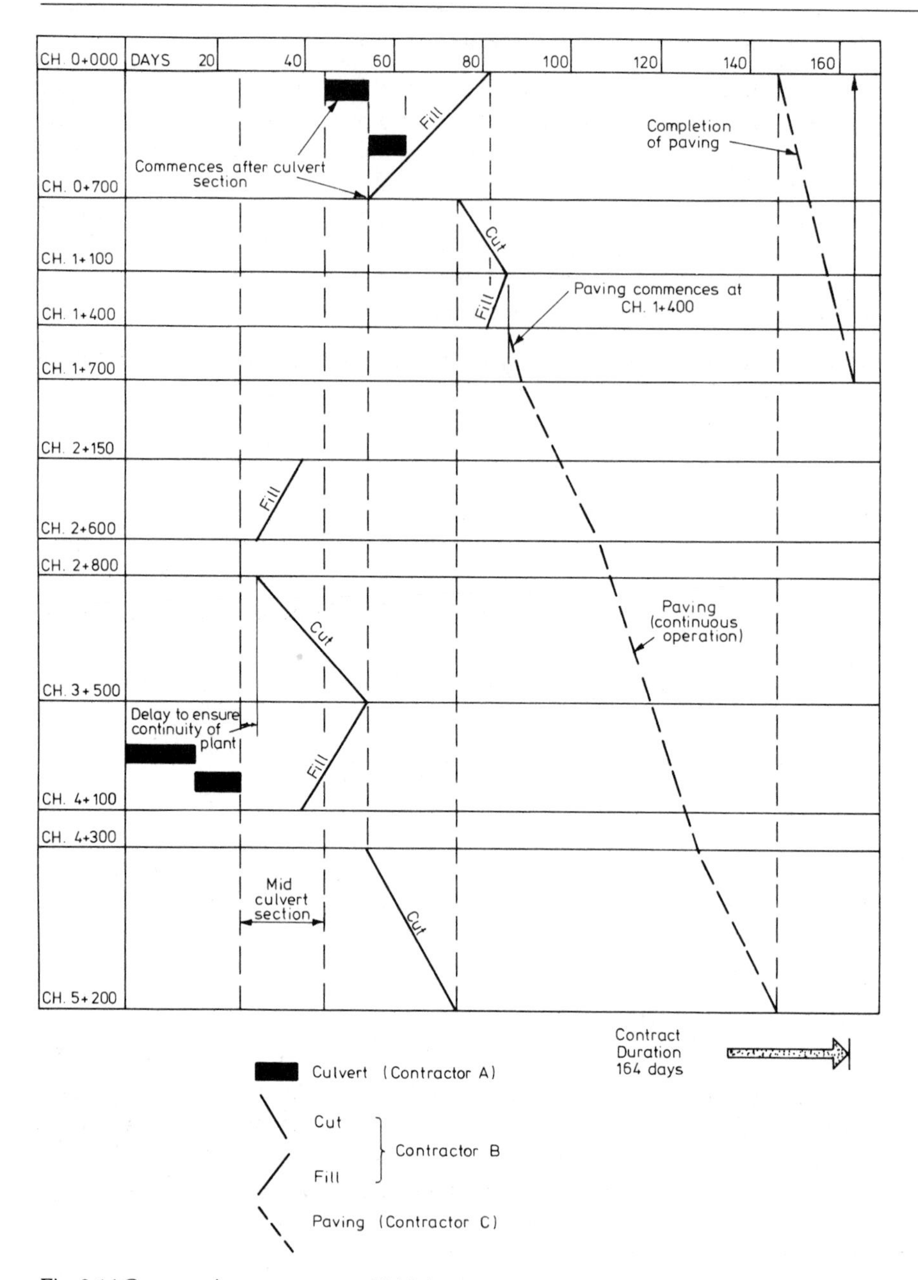

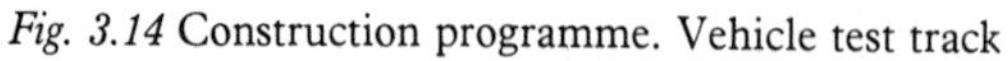

Fig. 3.14 Construction programme. Vehicle test track

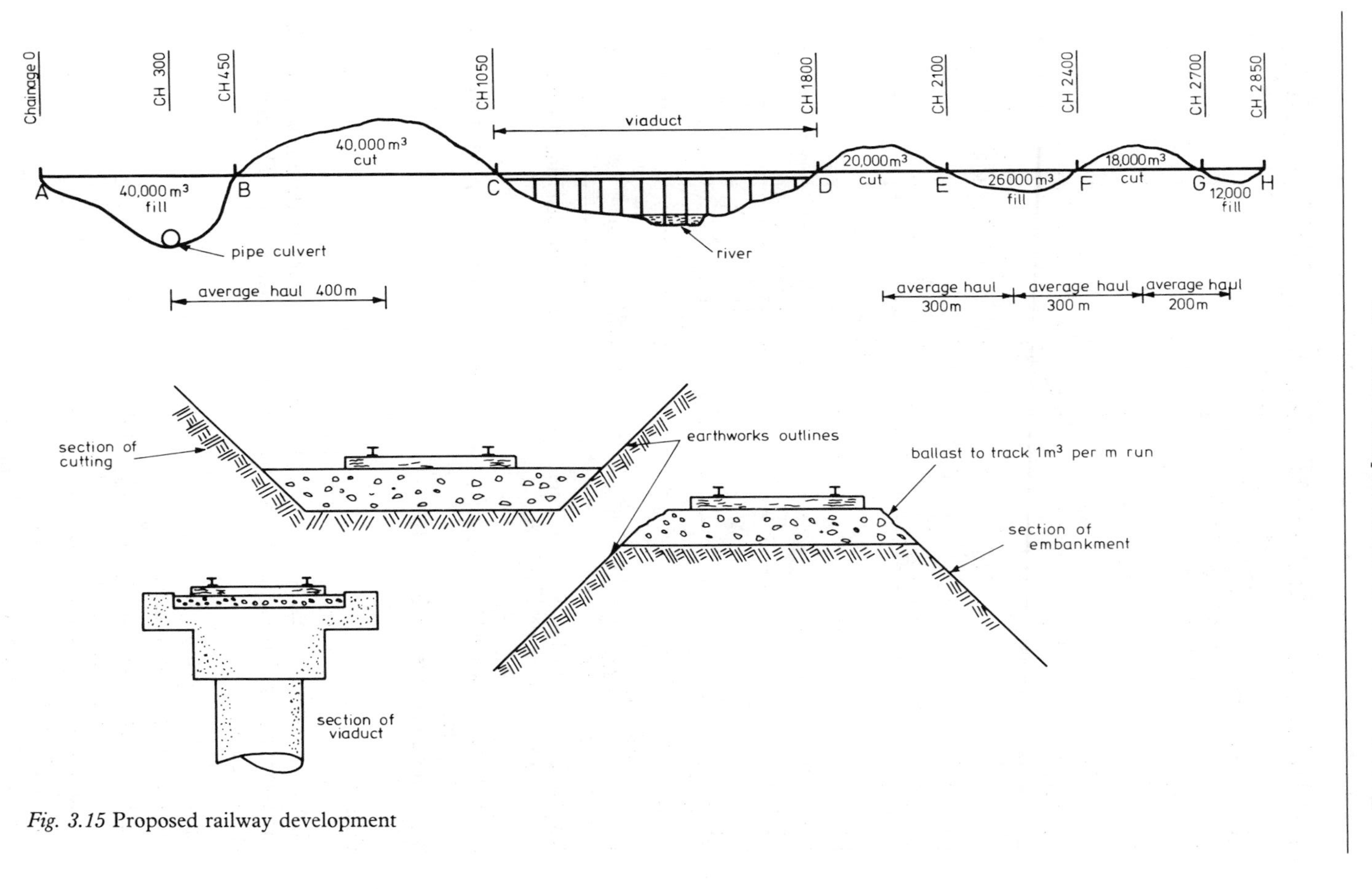

Fig. 3.15 Proposed railway development

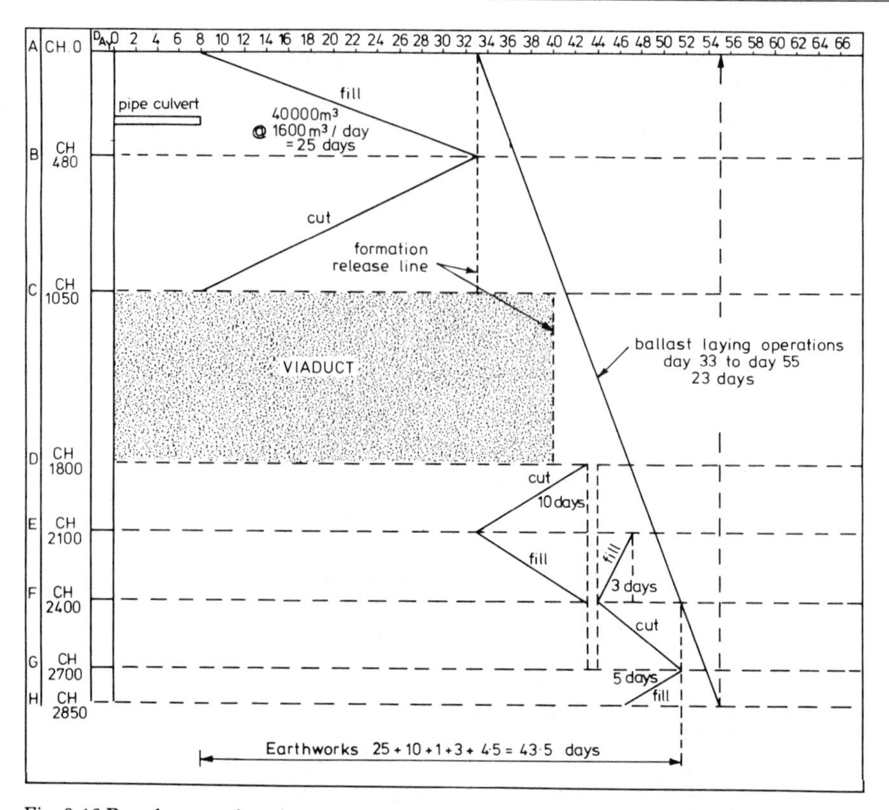

Fig. 3.16 Bar chart used to determine minimum number of days in which ballast-laying operations can be carried out

(b) *On what day should track-laying operations commence if they can proceed at a rate of 150 m/day?*

The solution is given in Fig. 3.16.

Q.3.4.

Figure 3.17 illustrates excavation works required for a water distribution system. The contractor has available a number of identical hydraulic excavators. The daily output of an excavator is:

(a) on bulk excavation for reservoirs – 100 m^3;
(b) trenching for pipelines:
  on 300 mm diameter mains – 50 m run;
  on 450 mm diameter mains – 40 m run;
  on 600 mm diameter mains – 30 m run.
(c) trenching for overflow drains:
  on 225 mm diameter drain – 40 m run;
  on 300 mm diameter drain – 30 m run;
  on 450 mm diameter drain – 25 m run.

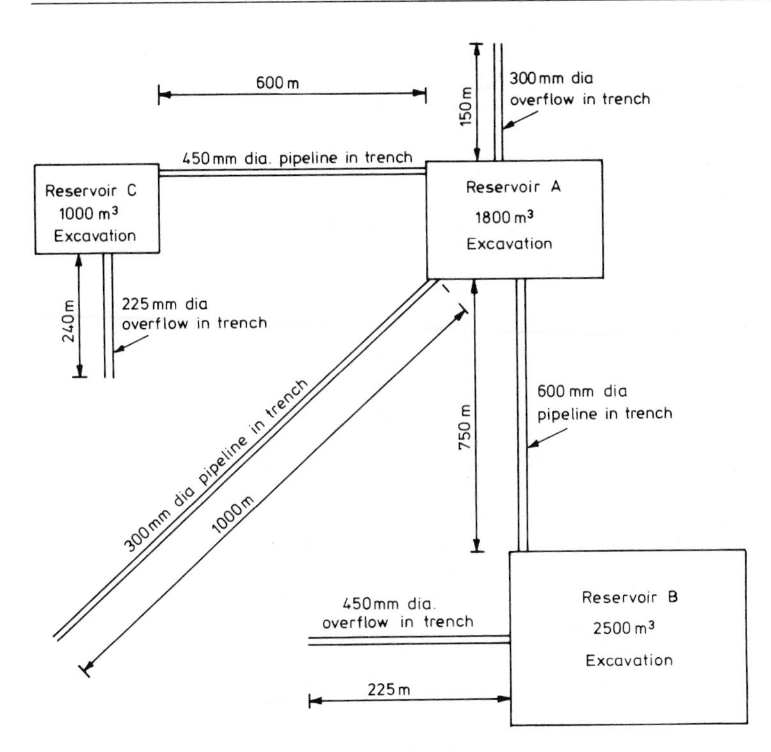

Fig. 3.17 Excavation works for a water distribution system

The following constraints apply to procedure:

1. Bulk excavation at any reservoir site must await completion of the overflow trench at that site.
2. Trenching for a pipeline can proceed only when the bulk excavation at one end of the pipeline has been completed.
3. If a section of the works is defined as a reservoir site, a pipeline or an overflow drain, assume an excavator can operate on one section only at a time.
4. It takes one day to deliver an excavator to the works area and only one delivery vehicle is available.

Determine, using any method of programming you wish, the shortest time to complete the system. Assume the maximum number of hydraulic excavators in use at any time is limited to two. Assume work must start at site A and time is measured from despatch of the first excavator to site.

The solution is given in Fig. 3.18.

	OPERATION	PIPE (Diam) (mm)	QTY (m)	OUTPUT (m/day)	DUR. (DAYS)
SITE A	OVERFLOW	300	150	30	5
	RESERVOIR		1800	100	18
	PIPELINE	300	1000	50	20
SITE C	OVERFLOW	225	240	40	6
	RESERVOIR		1000	100	10
	PIPELINE	450	600	30	20
SITE B	OVERFLOW	450	225	25	9
	RESERVOIR		2500	100	25
	PIPELINE	600	750	30	26

Fig. 3.18 Bar chart for excavation work: water distribution system

Chapter 4

Networks I: Critical path analysis (CPA)

4.1 Evolution of CPA

The historical approach to represent dependencies by arrows grew quite naturally out of the milestone technique (Fig. 4.1). Here arrows are shown linking the milestones (or events) in the different bars to indicate interdependency between activities. The logical development that followed was to replace the bars themselves with further arrows (Fig. 4.2). A further step was to discard the 'time scale' of the bars in order to separate programming from scheduling. What resulted from these was the early concept of network analysis – the diagrammatic representation of the sequences of interdependent activities (called a 'plan') taking the form of an 'arrow diagram' where arrows of any length represented 'activities' and circular nodes at the heads or tails of these arrows represented milestones or 'events'. This method of arrow diagramming was later given the name of critical path method (CPM) or critical path analysis (CPA).

Historical development of the critical path method

The mid-1950s saw an explosion of interest in the CPM problem. By 1955 and 1957, the Imperial Chemical Industries and the Central Electricity Generating Board of

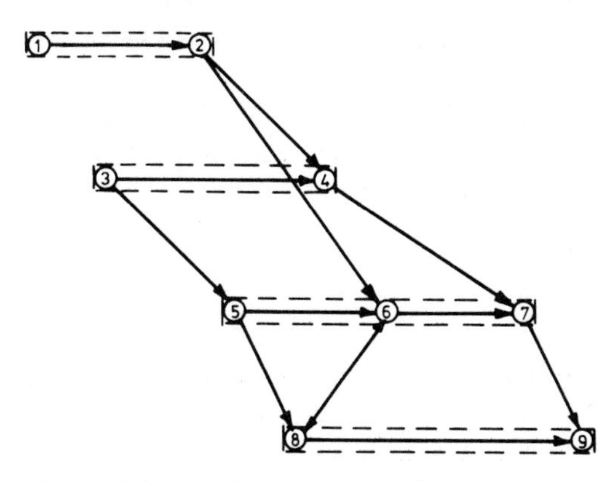

Fig. 4.1 Evolution from milestone diagram

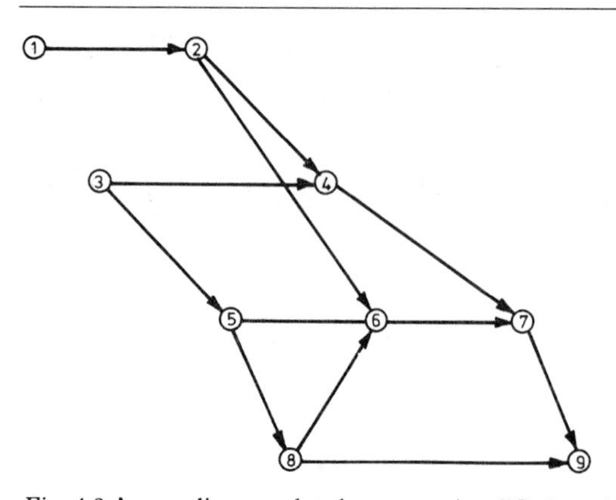

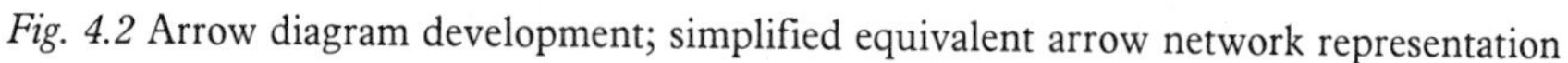

Fig. 4.2 Arrow diagram development; simplified equivalent arrow network representation

Britain were already using a basic technique which was capable of identifying 'the longest irreducible sequence of events' – more correctly known as 'the critical path' in CPA context. Subsequent success in the implementation of the concept by these firms has claimed time saving of over 40 per cent. Unfortunately this early management innovation faded into obscurity mainly due to lack of effort to publicise it.

At about the same time similar research work was under way in the USA and France. In the USA, Mr Walker of E. I. du Pont de Nemours Co. working in conjunction with J. E. Kelley of Remington Rand, Univac Computer Division, developed the CPM method which was essentially computer-based.

During this period CPM development work was being conducted in the US military organisation. In particular, the Special Project Office of the US Army, under Admiral Raborn, in conjunction with management consultants Booz-Allen and Hamilton and the Lockheed Missiles System Division, devised a similar method for the control of the Polaris project. The technique was later named project evaluation and review technique or PERT. This method succeeded in reducing the development time of the Polaris missile by over 2 years, a time saving of some 45 per cent.

Independent studies in scheduling and graph theory were also carried out by Monsieur B. Roy in France at about the same period as CPM and PERT. His work has led to the development of the method of potentials, more popularly known later as precedence diagram network.

Since 1958, further work has been undertaken mainly in the USA by C. E. Clark on PERT, J. W. Manchly on CPM and J. W. Fondall on non-computer-based networks, the main aim being to consolidate and improve the techniques.

4.2 Process of network planning

A CPA network is a diagrammatic representation of a plan for a particular project that shows the sequence and relationship of activities required to achieve the end objectives. Each arrow represents one activity and the relationship between activities is represented by the relation of one arrow to the others; each circle or node represents an event. The length of arrow bears no significance; it merely represents the passage of time (for the execution of a particular activity) in the direction of the arrow. The start of all activities leaving a node, however, depends on the completion of all activities entering that node.

The process of network planning can be summarised into the following steps:

1. *Project breakdown* – split the project work into unique activities.
2. *Activity listing* – prepare a list of all these activities.
3. *Activity constraints* – determine which activities must precede, succeed or may be done concurrently with any other activity specified so that sequential relationship between activities can be established.
4. *Network diagramming* – this produces a plan in the form of a network representing the operational sequences of all the activities in the project.

Certain laws must be adhered to in order to ensure correct network construction.

(a) The network must have definite points of beginning and finish.
(b) The network must be a logical representation of all the activities. Where necessary, 'dummy activities' are used for unique numbering and logical sequencing. Similarly, 'ladder construction' can also be used for overlapping activity sequencing.
(c) There must be no 'looping' in the network.
(d) The network must be continuous, i.e. without unconnected activities.

Examples

In Fig. 4.3 'excavate foundations' and 'concrete to foundations' are activities represented by arrows. The circles labelled 1, 2 and 3 represent events. The diagram means that activity 'concrete to foundations' follows activity 'excavate foundations'. It follows that 'concrete to foundations' may not start until 'excavate foundations' has been completed and event 2 has occurred.

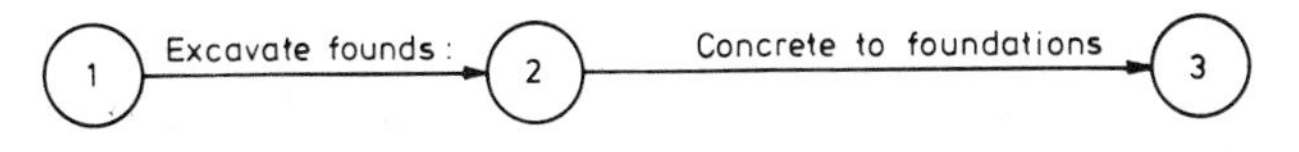

Fig. 4.3 Activities represented by arrows

Activity 'excavate foundations' could be described as 1–2, activity 'concrete to foundations' could be described as 2–3. Figure 4.4 means that activities 2–3, 2–4 and 2–5 follow activity 1–2. None of these activities can start until activity 1–2 has been completed. Figure 4.5 means that activities 17–18 and 17–19 follow activities 8–17, 12–17 and 16–17. Neither 17–18 nor 17–19 can start until activities 8–17, 12–17 and 16–17 have been completed. Event 17 cannot occur until activities 8–17, 12–17 and 16–17 have been completed.

Dummy activities

A dummy activity is represented by a dotted line. It can be considered as an activity

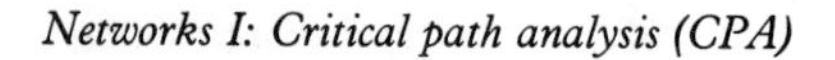

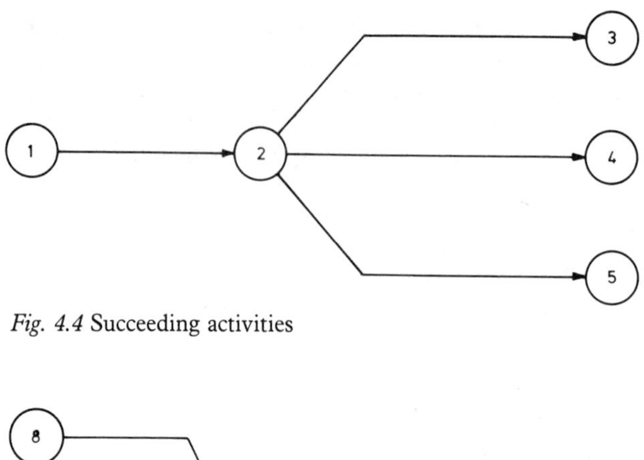

Fig. 4.4 Succeeding activities

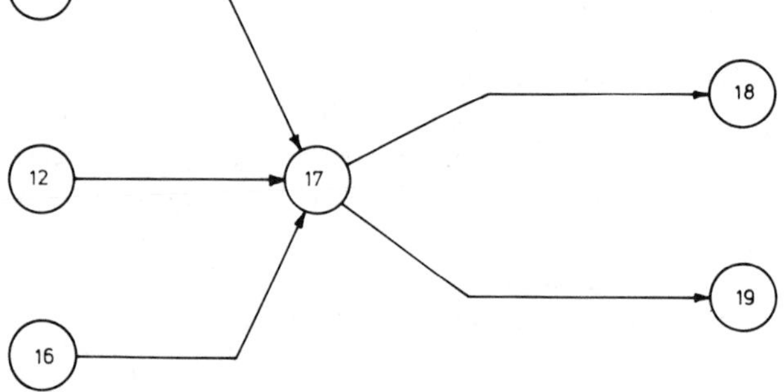

Fig. 4.5 Concurrent activities

which consumes neither time nor resources. Dummy activities are used for two reasons:

1. Unique numbering. Dummies enable activities, which can be carried out concurrently between the same two events, to be uniquely described by their starting and finishing even numbers.

In Fig. 4.6 activities A and B could both be described as 7–8. A dummy is therefore introduced.

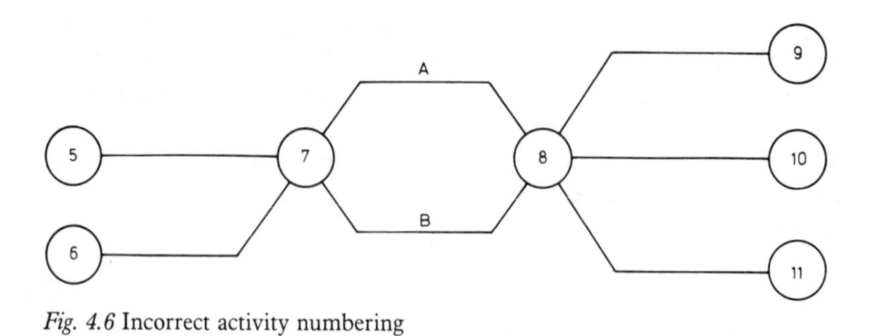

Fig. 4.6 Incorrect activity numbering

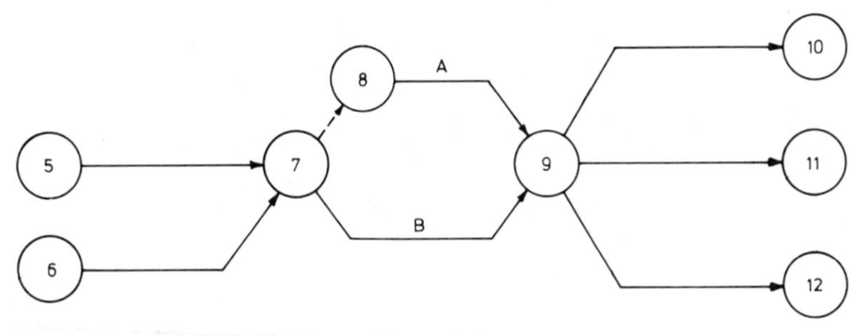

Fig. 4.7 Unique numbering (use of dummies)

In Fig. 4.7 the logic of the network is the same as in Fig. 4.6 but unique numbering of activities is achieved by using a dummy arrow. Activity A is now described as 8–9 and activity B as 7–9.

2. *Express logical relationships*. Figure 4.8 means that:

(a) Activity 5–7 cannot start until both activities 3–5 and 2–4 have been completed. Activity 4–6 cannot start until 2–4 is completed but it is not dependent upon 3–5.

(b) A similar situation occurs with activity 10–12. It cannot commence until the completion of both 5–7 and 8–10.

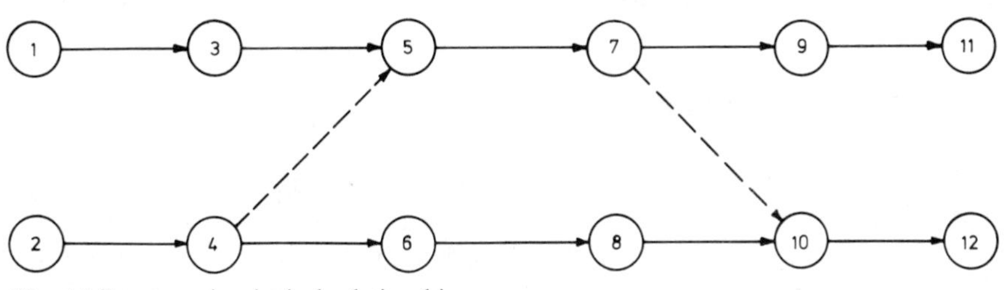

Fig. 4.8 Representing logical relationships

Looping

A loop in a network is a logical impossibility (Fig. 4.9). The network should be checked to ensure that the event number at the head of the arrow is larger than the event number at the tail. Activity 11–10 highlights this.

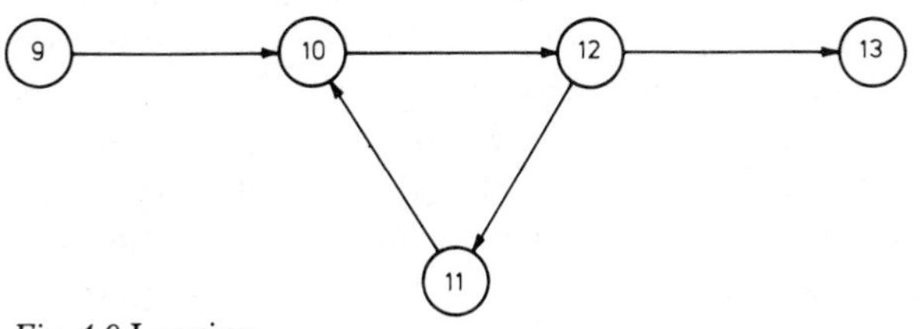

Fig. 4.9 Looping

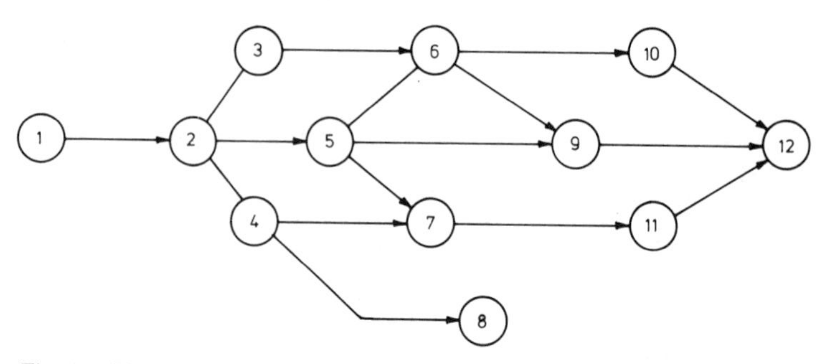

Fig. 4.10 Unconnected activity

Unconnected activities
The diagram should be checked to ensure that all activities are connected and that none are left 'dangling'. Figure 4.10 shows activity 4–8 dangling.

The logic of any network diagram specifies the construction process in a rigid manner. There is more than one way of constructing a network for a particular contract. It must also be appreciated that human behaviour, unlike a machine, is unpredictable. This means that actual site performance may be very different from what the planner envisages.

All planning processes must facilitate effective communication between the planner and site staff. One of the major difficulties in applying this technique is the communication/knowledge barrier that exists between planner and foreman-operatives.

4.3 Network computation

After completion of the network diagram (plan), what follows naturally is the estimation of activity project durations. Accuracy of an activity duration estimate generally depends on who and how experienced is the person performing the task. It is also a function of many other factors such as quality of labour, site location, site characteristics, climatic conditions, methods of construction, height of construction, safety regulations and human relationships.

Once duration estimates for every activity in the network have been obtained, it is then possible to proceed with scheduling calculations.

Numbering of events is done in ascending order from left to right. The circles (events) may be divided to show the early and late start times for each event, and the activity numbering. Analysis is carried out using simple addition for the forward pass, taking the largest number each time. This gives 'early start' times. The longest route through the network is the shortest time the contract will be completed. 'Late starts' are computed by means of a backward pass from right to left through the network. This is carried out by subtracting each activity duration time from its subsequent duration. The smallest number is chosen when more than one activity converges on an event. The critical path is not only the longest route in time, but also each event on that route has early and late start times identical. Other activities will have float or slack time.

The identification of these floats and other critical elements in the plan enables trouble areas to be positively revealed and thus allow management to focus their main attention on, say, some 10–20 per cent approximately of the project, which is the most constraining on the schedule. With careful utilisation of these 'critical activities' effective monitoring and control of the project in terms of time, resources and cost may be achieved.

If the project shown in Fig. 4.11 started at event time 0 the earliest time for reaching event 1 is 3 days.

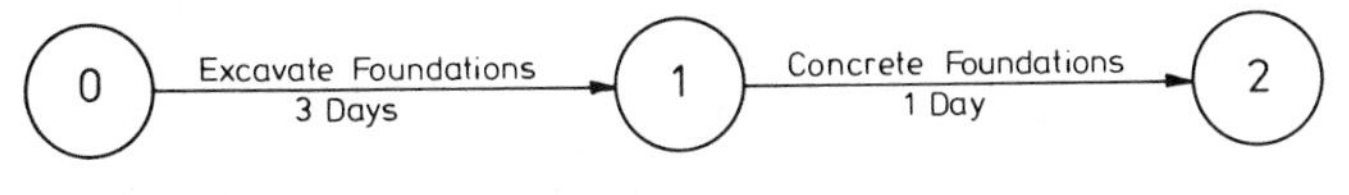

Fig. 4.11 Activity duration

Early event times T_E

The *early time* T_E for event 1 is 3 days. How early could event 2 be reached? 3 + 1, i.e. at the end of the fourth project day. To keep track of these results keep them in the event circle, as shown in Figs. 4.12 and 4.13. Where two or more activities converge on the one event as in Fig. 4.14, the earliest time for reaching event 13 is along the *longest path*.

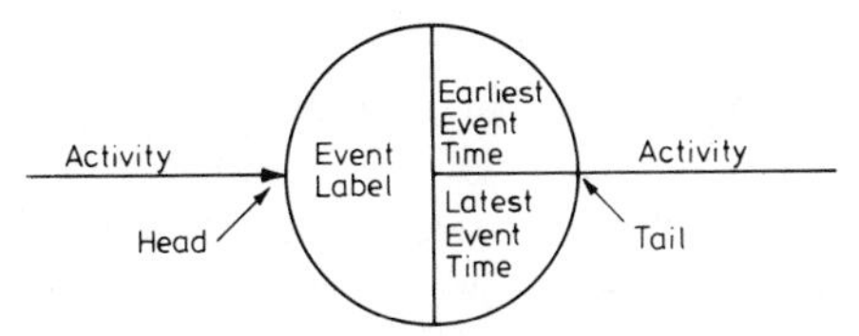

Fig. 4.12 Standard layout for recording data

Fig. 4.13 Recording data

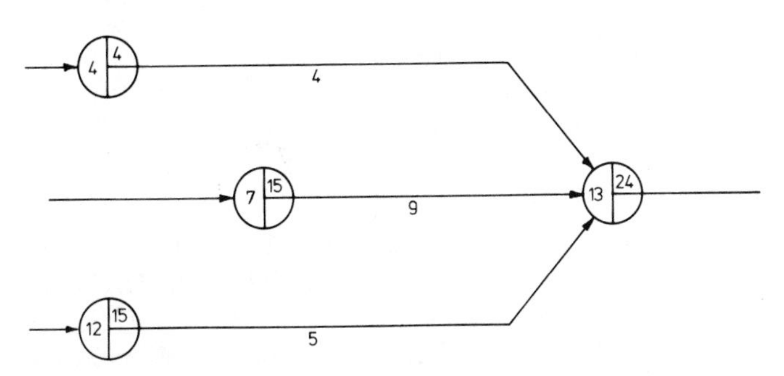

Fig. 4.14 Concurrent activity

For example, determine T_E of event 13 in Fig. 4.14. Three routes converge, i.e.

4–13 T_E of event 4 = 4 + duration (4) = 8
7–13 T_E of event 7 = 15 + duration (9) = 24
12–13 T_E of event 12 = 15 + duration (5) = 20
Therefore the longest path is through route 7–13 = 24
T_E of event is 13 = 24 days.

Discard the 8-day and 20-day solution; T_E is always the larger value when there is a choice between two or more values (Fig. 4.15).

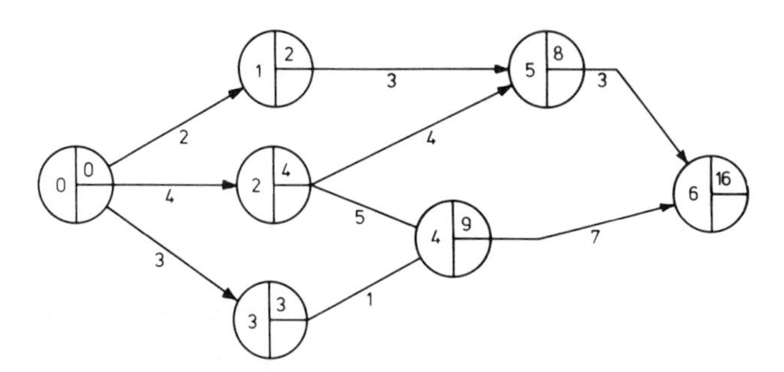

Fig. 4.15 Forward pass computation

Late event times T_L

Here, T_L for an event is defined as the latest time at which an event may be reached without delaying the computed project duration. To determine T_L work backwards through the network. By definition, the late event time at event 6 (Fig. 4.16) is 16, since the late event time for the terminal event equals the early event time for that event.

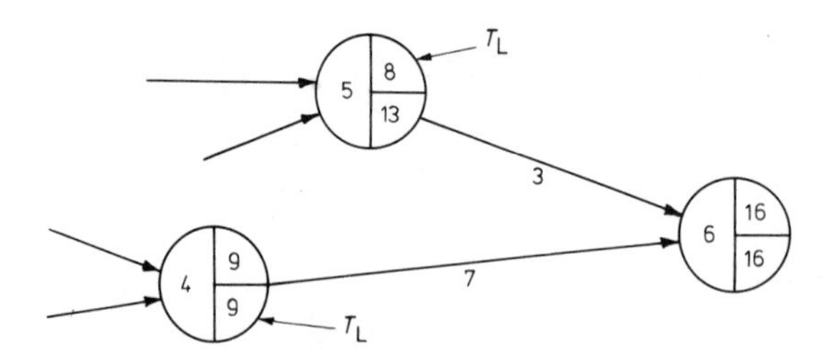

Fig. 4.16 Late time for terminal event

If event 6 is to be reached by time 16, event 5 must start no later than 16 less the duration of activity 5–6 (16 – 3). Thus the late event time T_L for event 5 is 13. The late event time for event 4 is 16 – 7 = 9. Where two or more arrow tails converge on the one event as in Fig. 4.17, the late event time T_L for event 2 is along the *shortest route*. For example:

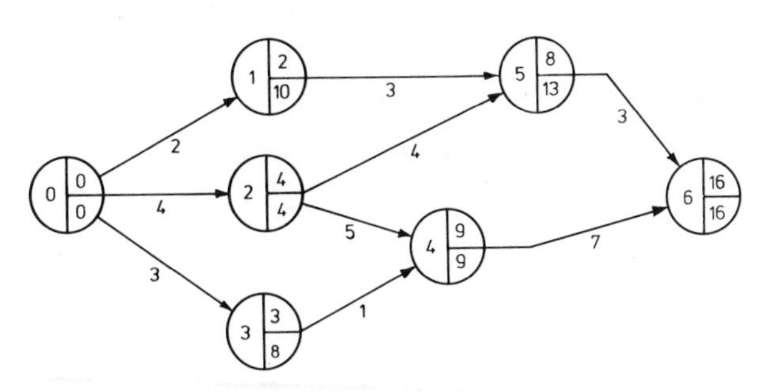

Fig. 4.17 Background pass computation

Two routes converge, i.e.
2–5, T_L of event 5 = 13 − duration (4) = 9
2–4, T_L of event 4 = 9 − duration (5) = 4

Therefore the shortest path is through route 2–4 = 4
T_L of event 2 = 4 days.

Activity time information

The source of activity time information is in the event time calculations. Each activity must be bounded by two events. The earliest time an activity can start is when the T_E for its starting (or *i*) event has been reached (Fig. 4.18).

Early start (*ES*) = T_E (event *i*)

If the earliest start is known what is the earliest time this event can be finished?

Early finish (*EF*) = Early start + duration (*ES* + *D*)

Having determined the early times for an activity the next stage is to determine the late times. The late finish is of course T_L for the finishing (*j*) event, i.e.:

$LF = T_L$ (event *j*)

The late start is therefore $LS = LF - D$.

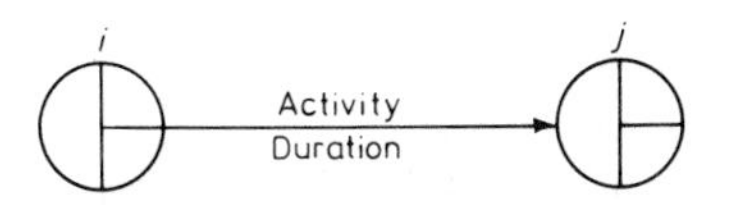

Fig. 4.18 *i–j* numbers

Before any calculations are made, certain information can be summarised in tabular form (Table 4.1). After the event times have been calculated, the additional information can be included as shown (early start and late finish) in Table 4.2. By adding the activity duration to the early start *ES* and subtracting it from late finish *LF* gives the

Table 4.1

Activity	Duration	Description

Table 4.2

Activity	Duration	Description	*ES*	*LF*

Table 4.3

Activity	Duration	Description	*ES*	$(ES + D)$ *EF*	$(LF - D)$ *LS*	*LF*

early finish time *EF* and late start time *LS* respectively (Table 4.3). The network in Fig. 4.19 has been analysed and tabulated in this manner (Table 4.4).

Critical path

The critical path determines the contract duration. It is the longest path into the last event. There is no free time associated with the critical path.

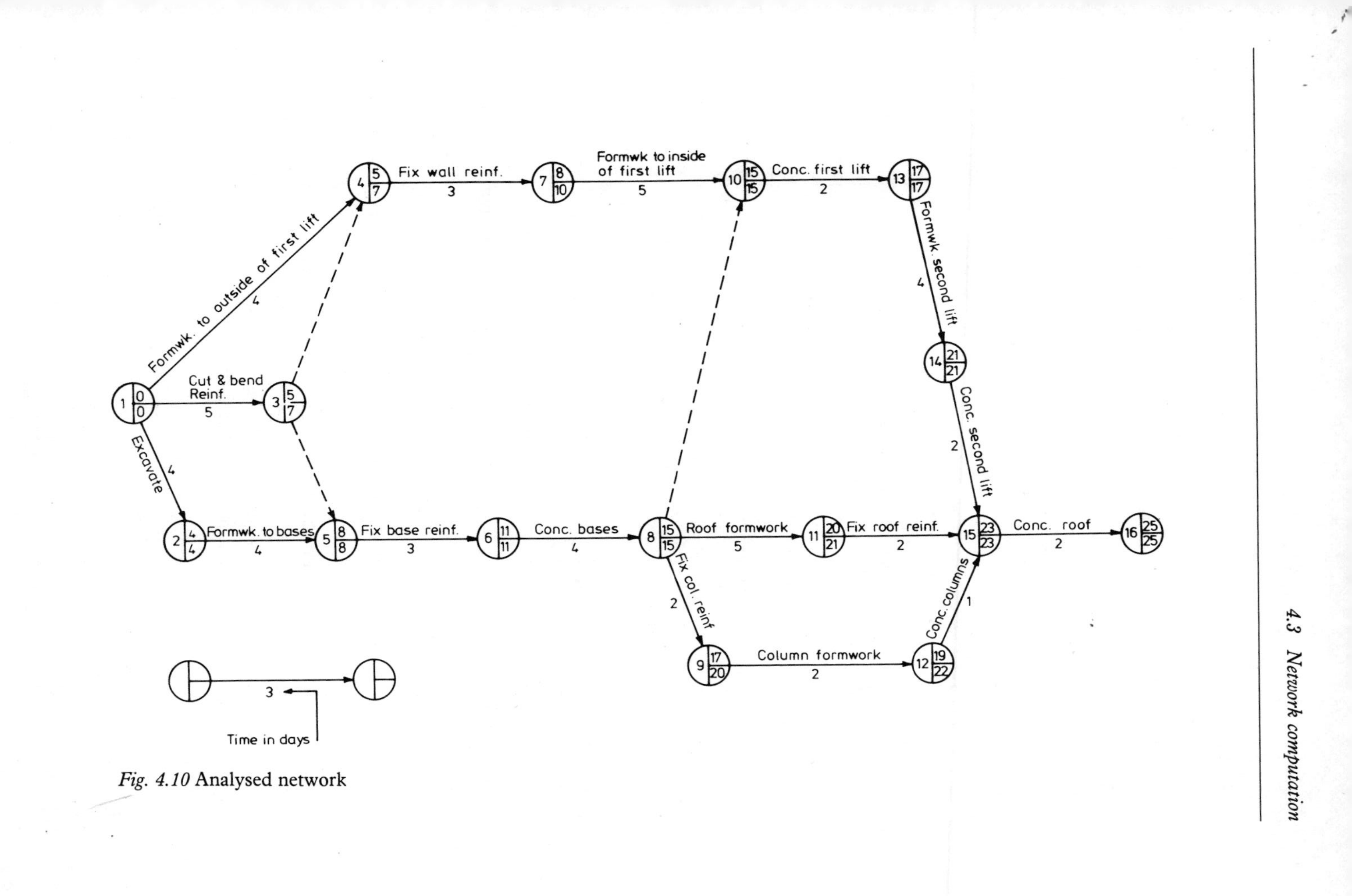

Fig. 4.10 Analysed network

Table 4.4

Activity	*Duration*	*Description*	*Early start (ES)*	*Early finish (EF)*	*Late start (LS)*	*Late finish (LF)*
1–2	4	Excavate	0	4	0	4
1–3	5	Cut and bend reinforcement	0	5	2	7
1–4	4	Formwork to outside of first left	0	4	3	7
2–5	4	Formwork to bases	4	8	4	8
3–4	—	Dummy	5	5	7	7
3–5	—	Dummy	5	5	8	8
4–7	3	Fix wall reinforcement	5	8	7	10
5–6	3	Fix base reinforcement	8	11	8	11
6–8	4	Concrete bases	11	15	11	15
7–10	5	Formwork to inside of first left	8	13	10	15
8–9	2	Fix column reinforcement	15	17	18	20
8–10	—	Dummy	15	15	15	15
8–11	5	Roof formwork	15	20	16	21
9–12	2	Column formwork	17	19	20	22
10–13	2	Concrete first lift	15	17	15	17
11–15	2	Fix roof reinforcement	20	22	21	23
12–15	1	Concrete columns	19	20	22	23
13–14	4	Formwork second lift	17	21	17	21
14–15	2	Concrete second lift	21	23	21	23
15–16	2	Concrete roof	23	25	23	25

4.4 Critical activities and float

There are three conditions which each critical activity must meet:

1. The earliest event time (T_E) and latest event time (T_L) at the activity start must be equal.
2. The earliest event time (T_E) and latest event time (T_L) at the activity finish must be equal.
3. The difference between early start (*ES*) and late finish (*LF*) must equal the duration.

There can be any number of critical paths. Each path must be continuous and there must be at least one.

The critical path is not always obvious, as shown in Fig. 4.20. Activities 8–11 and 11–15 meet all three conditions for critical activities.

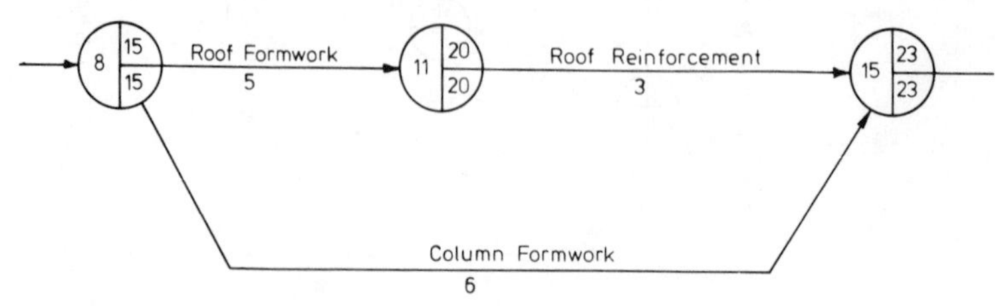

Fig. 4.20 Which path is critical?

Activity 8–15 is not, however, critical even though it spans two critical activities. It meets the first two conditions, i.e. T_E and T_L of activity start and finishes must be equal, but 23–15 is greater than the activity duration of 6. Activity 8–15 is not therefore critical even though it spans critical activities. Since it is not critical it must have some free time available.

The available working time is 8, i.e. (23 – 15) and with an activity duration of 6 has a latitude in scheduling of 2 days. This 2-day latitude is known as float.

There are four types of float used in critical path methods:

Total float (TF)	$= LET_j - (EET_i + D_{ij})$
Free float (FF)	$= EET_j - (EET_i + D_{ij})$
Independent float ($INDF$)	$= EET_j - (LET_i + D_{ij})$
Interfering float (IF)	$= TF - FF$

where LET_j is the latest event time j, EET_i the earliest event time i and D_{ij} the duration of i–j.

Total float. This is the amount of time by which the completion of an activity can exceed the earliest finish time without affecting the total duration of the contract. It is found by subtracting the earliest start time from the latest start time, or subtracting the earliest finish time from the latest finish time of an activity.

$$TF = LET_j - (EET_i + D_{ij})$$

Free float. Free float is a measure of the maximum time activity (i–j) may be delayed without affecting the start of successor activities. It differs from total float in that it measures the time available without delaying succeeding activities. An activity's free float can never be larger than its total float.

It is found by

$$FF = EET_j - (EET_i + D_{ij})$$

Independent float. This is a useful measure of scheduling freedom. Independent float of activity (i–j) is the maximum time this activity can be delayed without delaying successor activities, if all prior activities are finished as late as possible.

It is found by

$$INDF = EET_j - (LET_i + D_{ij})$$

Interfering float. This is the difference between total float and free float for any activity. If this type of float exists in an activity it indicates that the activity completion in this range does not change the duration of the project but decreases the floats of subsequent activities. If the interfering float is fully used succeeding activities in the chain will become critical and if exceeded the project duration will be increased.

$$IF = TF - FF$$

Negative float. This term is used when a project fails to maintain the planned progress because an activity cannot be completed within its allocated duration. The activity is then worse than critical and the project can only be completed in time if the succeeding critical activities are reduced in duration.

Example of float calculations

Consider activity 8–11 (stud partitions) in Fig. 4.21.

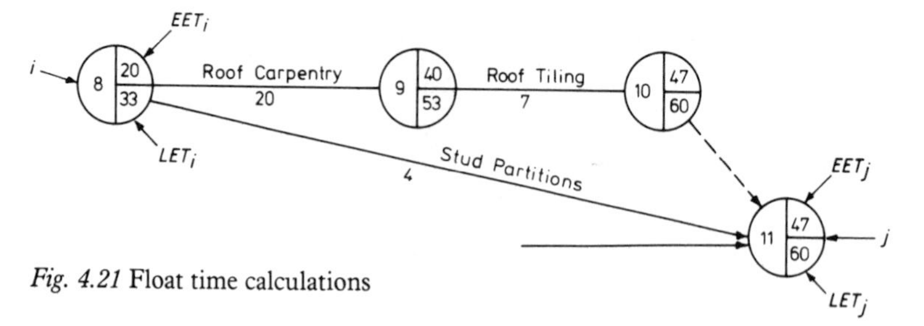

Fig. 4.21 Float time calculations

Total float = latest succeeding event time − (earliest preceding event time and duration)

$= LET_j - (EET_i + D)$

$= 60 - (20 + 4) = 36$ days

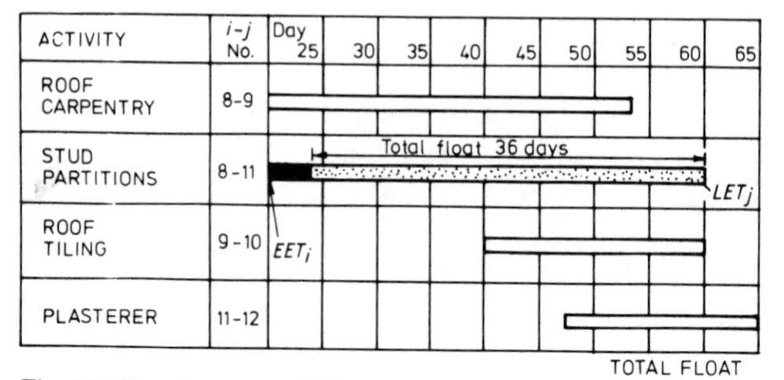

Fig. 4.22 Bar chart – total float

Free float = earliest succeeding event time − (earliest preceding event time and duration)

$= EET_j - (EET_i + D)$

$= 47 - (20 + 4) = 23$ days

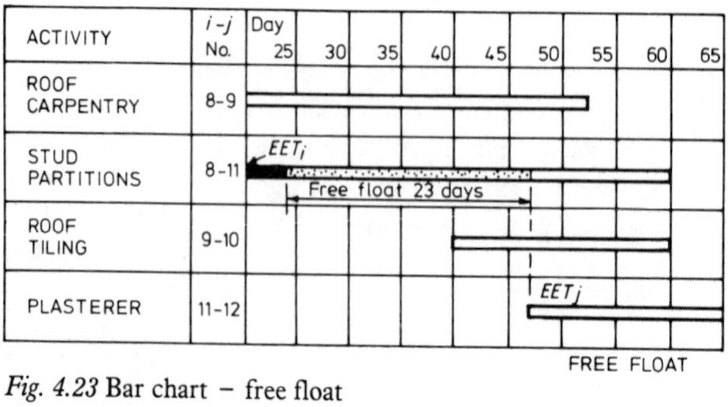

ACTIVITY
i-j No.
Day 25 30 35 40 45 50 55 60 65
ROOF CARPENTRY
8-9
STUD PARTITIONS
8-11
EET_i
Free float 23 days
ROOF TILING
9-10
PLASTERER
11-12
EET_j
FREE FLOAT

Fig. 4.23 Bar chart – free float

Independent float = earliest succeeding event time − (latest previous event times and duration)
$= EET_j - (LET_i + D)$
$= 47 - (33 + 4) = 10$ days

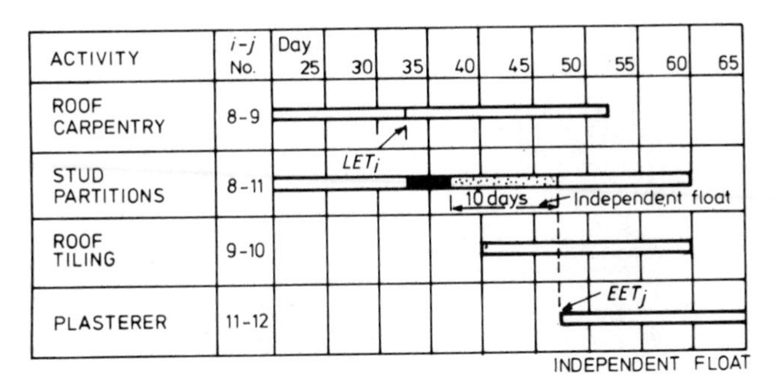

Fig. 4.24 Bar chart – independent float

Interfering float = total float − free float
$= 36 - 23 = 13$ days

ACTIVITY	i-j No.	Day 25	30	35	40	45	50	55	60	65
ROOF CARPENTRY	8-9									
STUD PARTITIONS	8-11									
ROOF TILING	9-10									
PLASTERER	11-12									

Total float
Interfering float
13 days
Free float
INTERFERING FLOAT

Fig. 4.25 Bar chart – interfering float

In practice the types of float most commonly required are total float and free float. The table of data from the network in Fig. 4.19 has been further analysed to provide this information (Table 4.5).

Table 4.5

Activity	Duration	Description	Early Start (ES)	Early Finish (EF)	Late Start (LS)	Late Finish (LF)	Total Float (TF)	Free Float (FF)	Critical Path
1–2	4	Excavate	0	4	0	4	0	0	*
1–3	5	Cut and bend reinforcement	0	5	2	7	2	0	
1–4	4	Formwork to outside of first left	0	4	3	7	3	1	
2–5	4	Formwork to bases	4	8	4	8	0	0	*
3–4	—	Dummy	5	5	7	7	2	0	
3–5	—	Dummy	5	5	8	8	3	3	
4–7	3	Fix wall reinforcement	5	8	7	10	2	0	
5–6	3	Fix base reinforcement	8	11	8	11	0	0	*
6–8	4	Concrete bases	11	15	11	15	0	0	*
7–10	5	Formwork to inside of first left	8	13	10	15	2	2	
8–9	2	Fix column reinforcement	15	17	18	20	3	0	
8–10	—	Dummy	15	15	15	15	0	0	*
8–11	5	Roof formwork	15	20	16	21	1	0	
9–12	2	Column formwork	17	19	20	22	3	0	
10–13	2	Concrete first lift	15	17	15	17	0	0	*
11–15	2	Fix roof reinforcement	20	22	21	23	1	1	
12–15	1	Concrete columns	19	20	22	23	3	3	
13–14	4	Formwork second lift	17	21	17	21	0	0	*
14–15	2	Concrete second lift	21	23	21	23	0	0	*
15–16	2	Concrete roof	23	25	23	25	0	0	*

Chapter 5

Networks II: Precedence diagrams

5.1 Comparison of precedence diagrams with CPA

The precedence diagram performs the same function as the arrow diagram; there are, however, important differences and similarities between them (Fig. 5.1):

1. Arrow network planning is essentially activity-orientated while precedence diagrams tend towards more generalised process representation, thereby possibly losing some of the unique features of the detailed approach inherent in the CPA technique.
2. At present there appears to be more computer packages available which are based on arrow network rather than on precedence diagrams.
3. Dummy and ladder construction are characteristically unproductive features of CPA, while in the precedence diagram the dummy is eliminated and overlapping activities are presented in a simpler and clearer sequence.
4. Due to its simplified approach in network representation, it is claimed that precedence networks are more easily understood by site personnel, more flexible in use and easier to amend than CPA. Precedence methods also open up possibilities of standardising activity codings, leading to development of integrated cost control systems.
5. CPA is easier to prepare than precedence diagrams but looks more complex, due mainly to the frequent uses of dummy and ladder construction.
6. The concept of activity lead or lag times can be included in a precedence network but this is not normally done in CPA planning.

5.2 The technique

A precedence diagram consists of rectangles and links (Fig. 5.2). Within the rectangles are activity descriptions. The links display the logic.

The calculations are very similar to that of CPA. Definitions of early start, latest start, early finish, latest finish and floats are identical. Typical activity codes are shown in Figs. 5.3 (a) and (b).

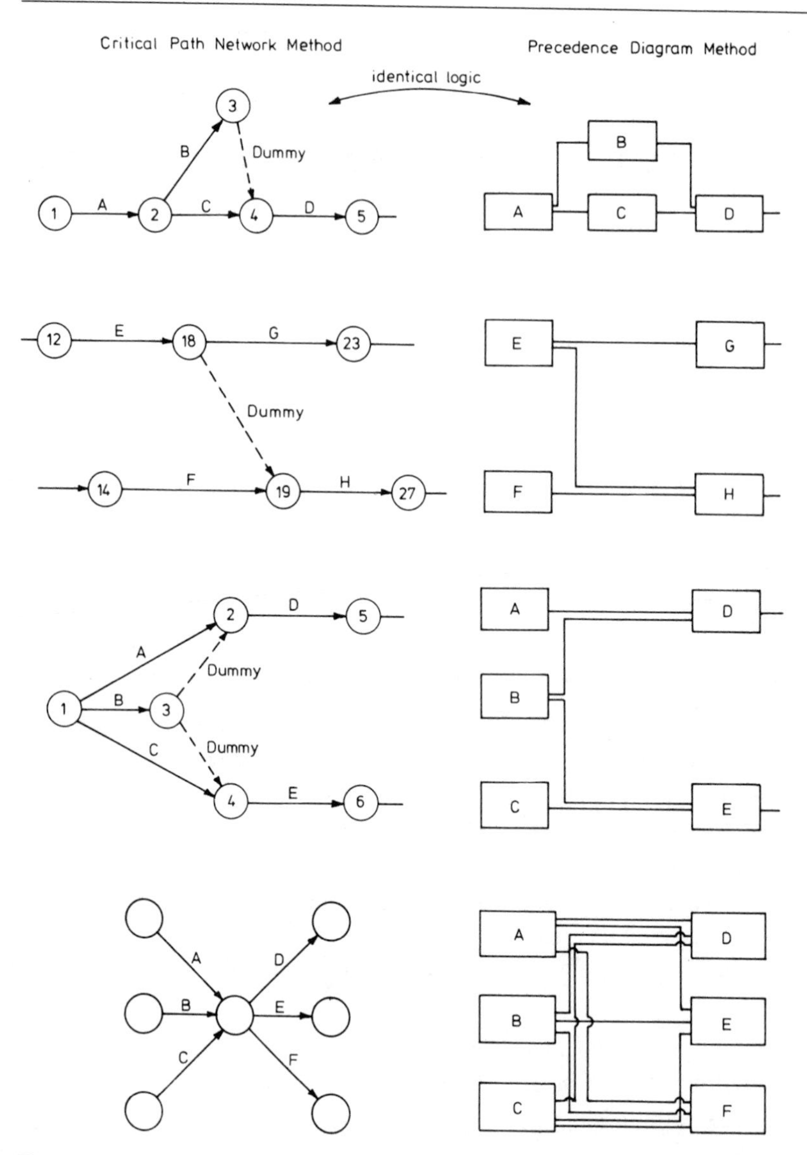

Fig. 5.1 Network comparisons

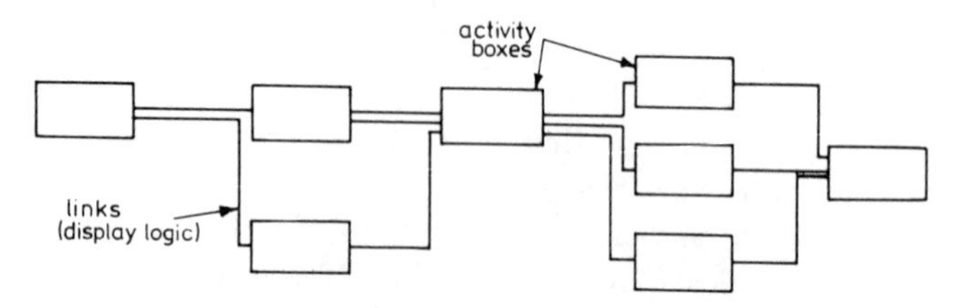

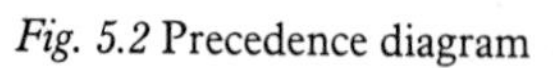

Fig. 5.2 Precedence diagram

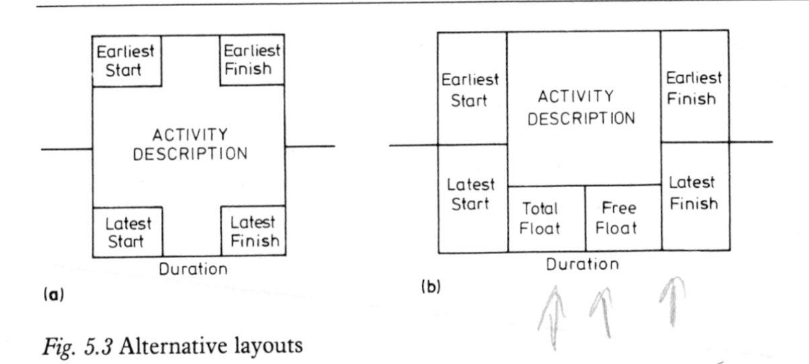

Fig. 5.3 Alternative layouts

Once the diagram has been drawn the calculations are carried out as follows:

1. Determine the duration of each activity and write these in the code box.
2. *Forward pass.* With the starting time at 0 calculate the earliest starting and finishing times.
   N.B. The earliest starting time for an operation is equal to the earliest finishing time of its preceding operation; where there is more than one such operation then the latest of the earliest finishing times must be used. The earliest finishing time is found by adding the operation duration to the earliest starting time.
3. *Backward pass.* Calculate the latest starting and finishing times. Start by making the latest finishing time for the last operation equal to the earliest finishing time for the operation.
   The latest finishing time is equal to the latest starting time of its succeeding operation, where it is connected to more than one operation the earliest (lowest) of the latest starting times must be used.
   The latest starting time is found by subtracting the operation duration from the latest finishing time. An analysed network is shown in Fig. 5.4.

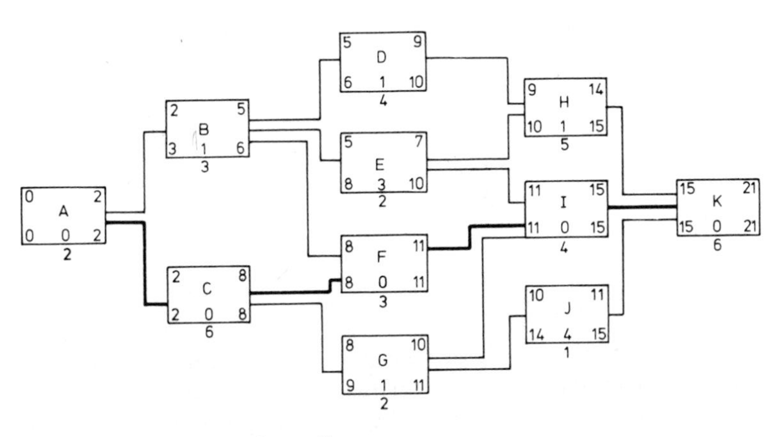

Fig. 5.4 Computed precedence diagram

5.3 Delays, leads and lags

In construction work it is quite common for a succeeding activity to be delayed for some time after completion of the preceding activity, e.g. curing time for concrete. Here the delay may be indicated on the link line (Fig. 5.5). Earliest start of B = latest of: earliest finish of C, or earliest finish of A + delay d, i.e. 33 + 9 = 42 or 38 + 6 = 44. Therefore earliest start of B = 44.

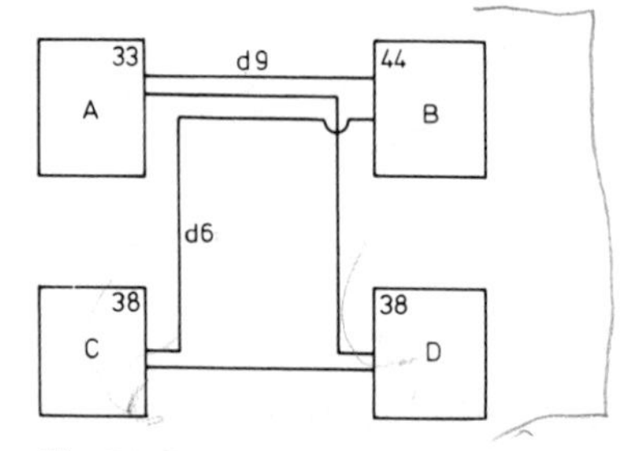

Fig. 5.5 Computing delays

Leads occur when an activity can start after only a percentage of its preceding activity(ies) has been completed. The lead factor is that amount of time that must elapse between the start of one operation and the start of its succeeding operations (Figs. 5.6 and 5.7).

When making use of lead factors care must be taken to ensure that they do not become overtaking activities. In Fig. 5.8 it would appear that the foundation concrete has been poured before the foundation trench has been completely excavated. This of course is illogical and sufficient time must be allowed to prevent this situation occurring.

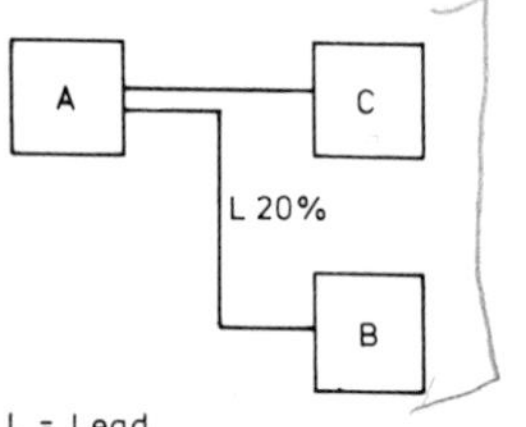

Fig. 5.6 Lead factors. Activity B may start when activity A is 20 per cent complete, although activity C must wait for completion of activity A.

Lag times (*F*)

If a lag factor or partial finish is specified, then the proportion of the duration remaining is determined and added to the earliest finish of the predecessor having the overlap with this activity. The earliest finish of this activity will be the later of this value, or the earliest start plus duration of the activity (see Figs. 5.9, 5.10 and 5.11).

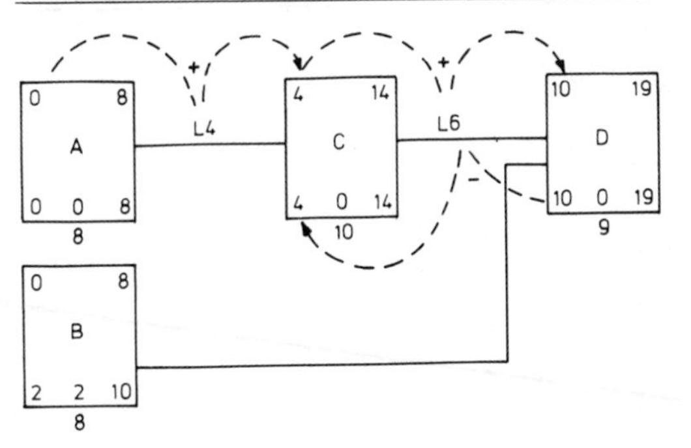

A
B
C
D
L4
L6

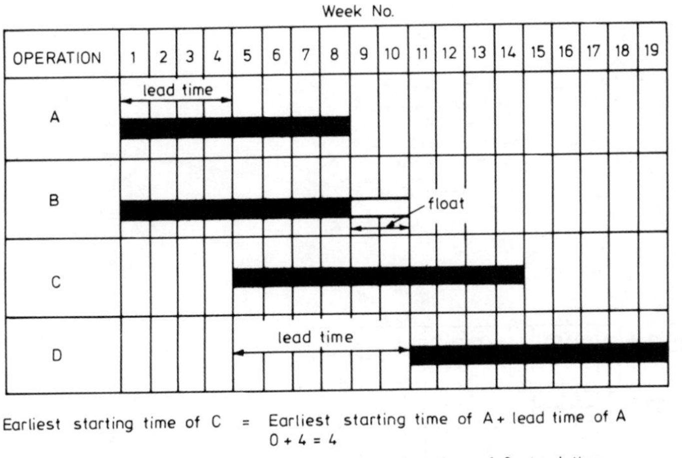

Week No.
OPERATION
lead time
float
lead time
Earliest starting time of C = Earliest starting time of A + lead time of A
0 + 4 = 4
Earliest starting time of D = (i) Earliest starting time of C + lead time of activity C (6)
(ii) or Earliest finishing time of B
(i) = 4 + 6 = 10
(ii) = 8
Therefore use 10

Fig. 5.7 Lead times

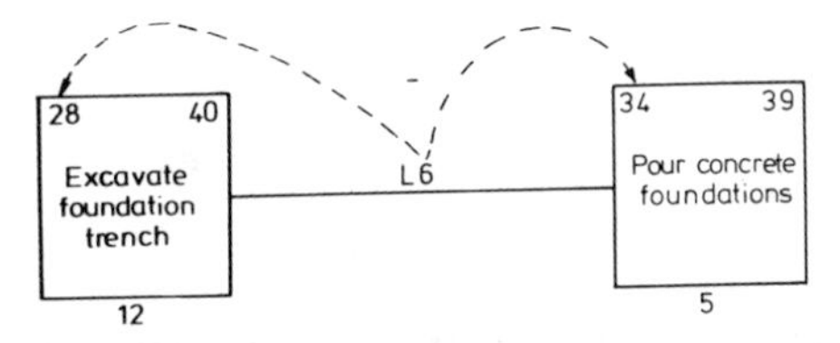

Excavate foundation trench
L6
Pour concrete foundations

Fig. 5.8 An overtaking situation

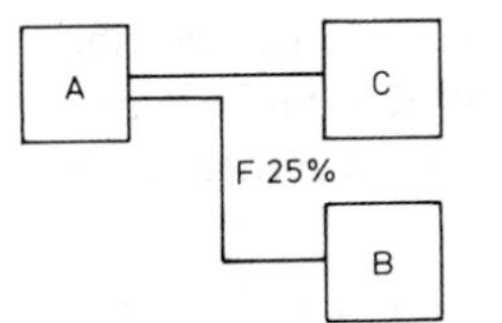

A
C
F 25%
B

Fig. 5.9 Lag factor. Activity B may start before activity A is complete, but will still have 25 per cent to complete when activity A is complete. The last 25 per cent of activity must remain until all of A is complete.

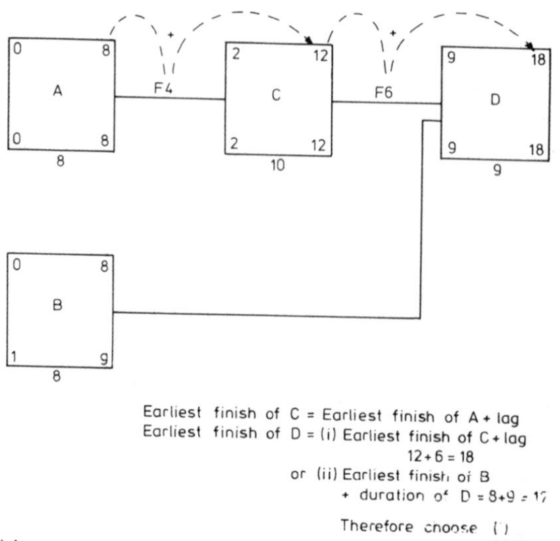
A
B
C
D
F4
F6
Earliest finish of C = Earliest finish of A + lag
Earliest finish of D = (i) Earliest finish of C + lag
12 + 6 = 18
or (ii) Earliest finish of B
+ duration of D = 8+9 = 17
Therefore choose (i)
(a)

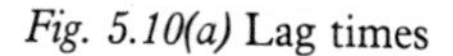
Fig. 5.10(a) Lag times

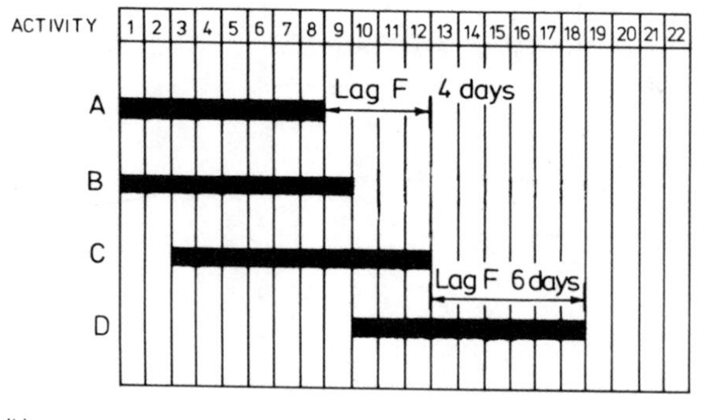
ACTIVITY
A
B
C
D
Lag F 4 days
Lag F 6 days
(b)

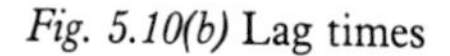
Fig. 5.10(b) Lag times

An alternative to the use of lead and lag factors is the ladder arrangement. This may be used to give a clearer picture of the construction sequence and often simplifies the planning effort. In Fig. 5.12, three housing blocks are being planned. Relationships and dependencies must be established. A bar chart has been derived from the network and continuity of work has been achieved by commencing foundations at their earliest times and floors at the latest times.

Combination – leads and lags

In Fig. 5.13, activity B may start when activity A is 3 days advanced, but will still have 4 days work left when activity A is completed.

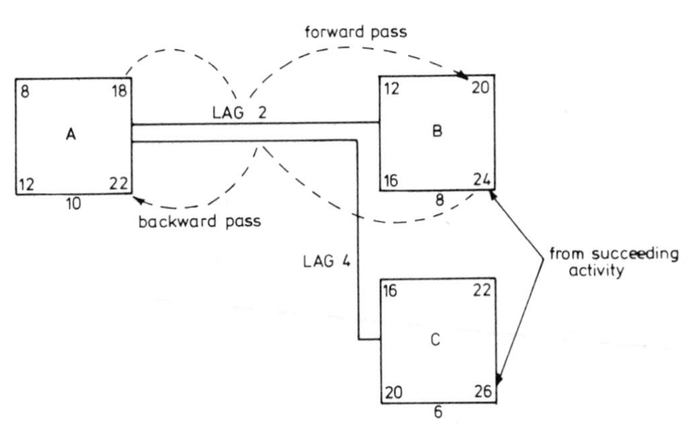

Fig. 5.11(a) Lag times – forward and backward pass

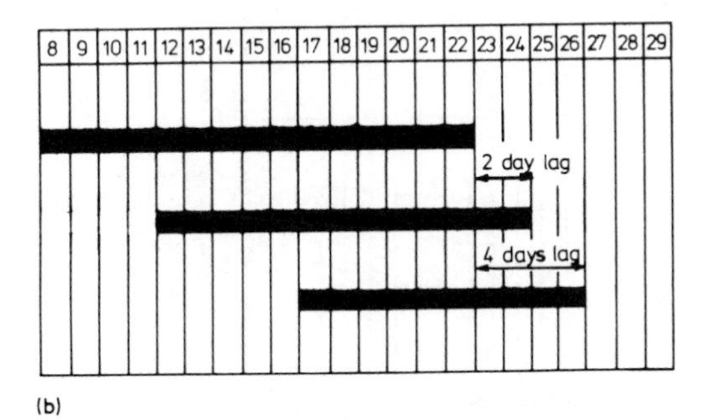

Fig. 5.11(b) Bar-chart representation

5.4 Float calculations

In order for a contractor to gain the full potential of any form of network planning he must understand the concept of critical activities and the various types of float. This has already been considered in the previous section dealing with CPA. It will be useful, however, to examine these aspects again within the precedence diagram context. A critical activity is one in which the earliest and latest dates are equal. The sequence of these activities from start to finish is known as the *critical path.* The shortest duration of the project is the sum of these critical activities on the critical path. The activities off the critical path have periods of time between their earliest and latest dates exceeding the activity duration. This additional time is known as float. The following are types of float which occur most frequently on a construction contract:

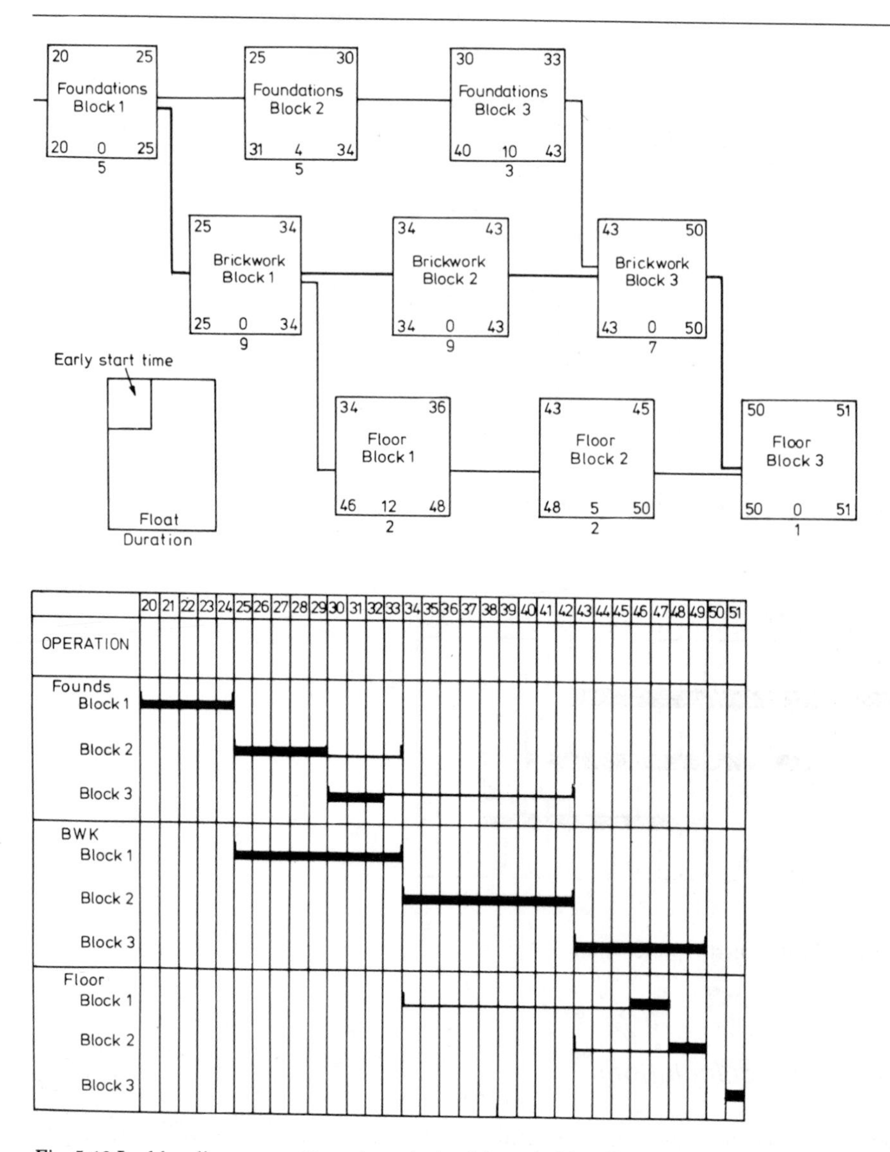

Fig. 5.12 Ladder diagram and bar chart derived from ladder diagram

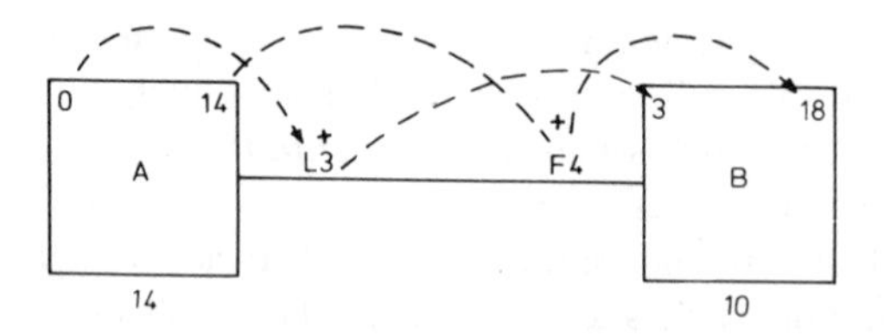

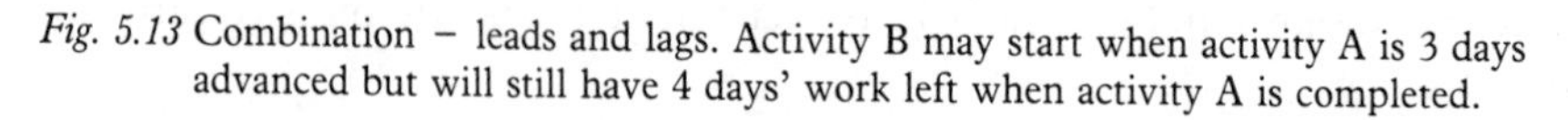
Fig. 5.13 Combination – leads and lags. Activity B may start when activity A is 3 days advanced but will still have 4 days' work left when activity A is completed.

Total float

This is the time by which an activity may be delayed without affecting the planned contract completion date (Figs. 5.14(a) and (b)). It is found by subtracting the earliest starting time of an operation from its latest starting time, e.g. considering activity B in Fig. 5.14(a).

$TF = 53 - 24 = 29 \quad \text{or} \quad 58 - 29 = 29$

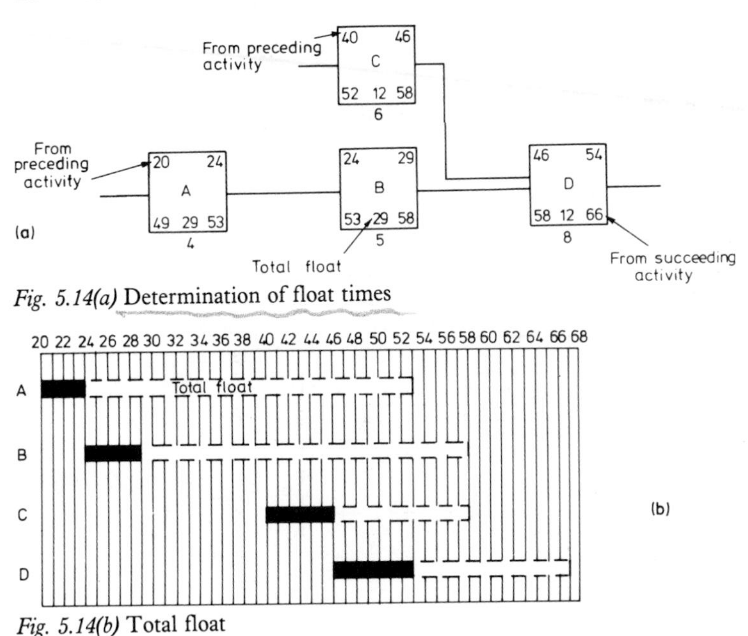

Fig. 5.14(a) Determination of float times

Fig. 5.14(b) Total float

Free float

This is the time an activity may be delayed without affecting any other activity. It is found by subtracting the earliest finishing time of an operation under consideration from the earliest starting time of its succeeding activity. In Fig. 5.14(a) the free float available to activity B is found by subtracting the earliest finishing time of B (29) from the earliest starting time of D (46) = 17. The bar chart is shown in Fig. 5.14(c).

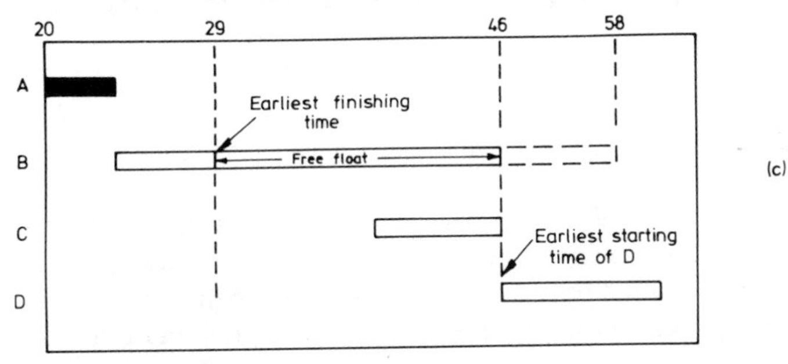

Fig. 5.14(c) Free float

Interfering float

This is the difference between total and free floats for an activity, or the difference between the latest finishing time of a preceding activity and the earliest starting time of a succeeding activity under consideration.

For example, in Figs. 5.14(a) and (d) considering activity B:

$$IF = TF - FF$$
$$= 29 - 17 = 12$$

or

$$IF = \text{latest finishing time of B} - \text{earliest finishing time of D}$$
$$= 58 - 46 = 12$$

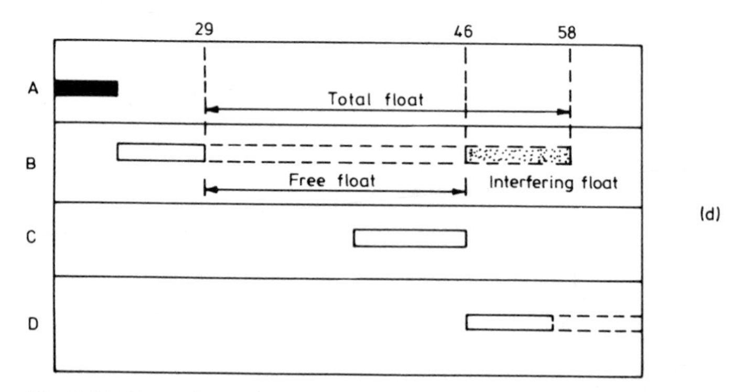

Fig. 5.14(d) Interfering float

Independent float

This is the amount of float which an activity will always have irrespective of how early or late its preceding or succeeding activities are. It is calculated by subtracting the latest finishing time of the preceding activity from the earliest starting time of the succeeding activity and subtracting from the answer the duration of the operation under consideration. For example, consider activity B in Fig. 5.15(a). The latest finishing time of activity A (10) is subtracted from the earliest starting time of operation E (25) = 15. Then the duration of operation B is subtracted from the answer of 15 (15 − 8 = 7).

5.5 *Crash costing*

Activity durations will normally have been determined on a particular method of work, labour gang size, etc., which gives the minimum cost. The contract duration obtained from the network may not be realistic when compared to the contract duration stated in the contract documents. The overall network duration must be 'compressed' to reduce the contract time.

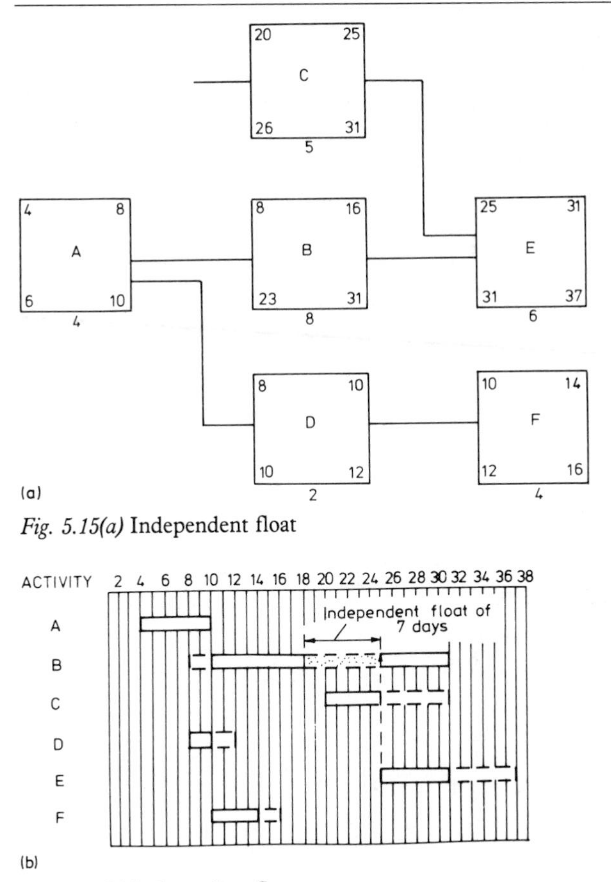

Fig. 5.15(a) Independent float

Fig. 5.15(b) Independent float

Since the critical path is the longest route through the network it is these activities to which attention is initially directed. It must be remembered that when shortening the contract duration the position of the critical path may change.

The planner and/or manager will select the actual activities to be shortened depending on cost and practicability. As an example, if two activities on the critical path can be shortened by 1 week and the costs of shortening are £500 and £750/week each, then the former obviously takes priority for crash costing.

Consider the networks in Figs. 5.16(a) and (b) with normal and crash durations as given in Table 5.1. As activities of the critical path are shortened, the float associated with activities which are not critical may be reduced, depending on the location of such activities relevant to the shortened activities. As further shortening takes place these non-critical items become critical.

Contract duration of the crashed network (Fig. 5.16(b)) is 14 weeks and there are three critical paths. Every activity in the network is now critical. Shortening either activity A or E would reduce the contract duration but in the middle area all three activities would have to be shortened simultaneously to effect further reduction.

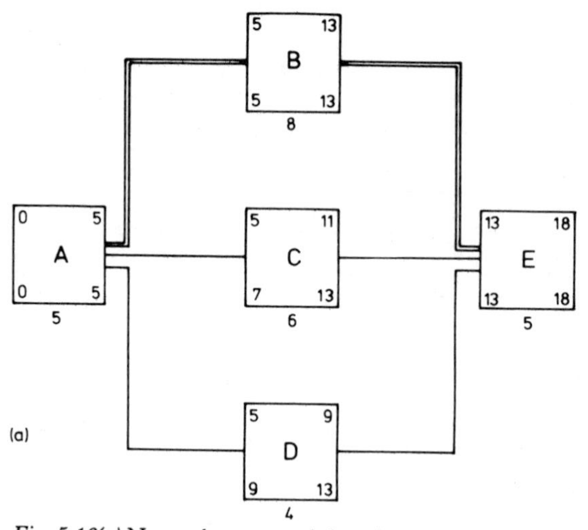

Fig. 5.16(a) Network at normal durations

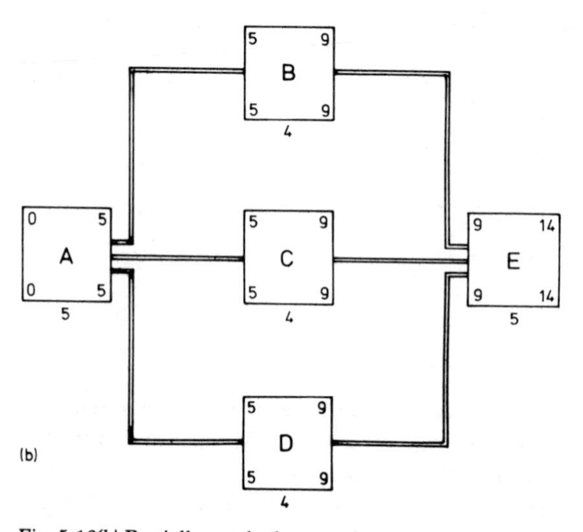

Fig. 5.16(b) Partially crashed network

Table 5.1

Activity	*Normal duration*	*Crash duration*
A	5	3
B	8	2
C	6	4
D	4	4
E	5	3

As a contract duration is reduced in this way there is usually an increase in productive resources, i.e. labour, plant, etc. to execute the work within the compressed time. These are known as direct costs. At the same time, since the contract will be completed sooner there will be a saving in overhead costs, i.e. rent, rates, management, salaries, insurances, etc. These are indirect costs and will reduce with a shortened contract time.

A contract should not be speeded up blindly. There is an optimum time–cost relationship. For each activity on the critical path the crash cost, crash time and normal cost should be ascertained (Table 5.2).

Table 5.2

Activity	*Crash time*	*Normal time*	*Crash cost* (£)	*Normal cost* (£)
A	3	5	2,400	2,000
B	2	8	4,500	2,500
C	4	6	4,500	3,000
D	4	4	2,000	2,000
E	3	5	3,500	1,000
			16,900	10,500

Calculate the cost slope, i.e. the cost per day of speeding up each activity.

$$\text{Formula: } \frac{\text{crash cost} - \text{normal cost}}{\text{normal time} - \text{crash time}}$$

Activity		*Cost slope* (£)	*Priority*
A	$\frac{2{,}400-2{,}000}{5-3} = \frac{400}{2}$	200	First
B	$\frac{4{,}500-2{,}500}{8-2} = \frac{2{,}000}{6}$	333	Second
C	$\frac{4{,}500-3{,}000}{6-4} = \frac{1{,}500}{2}$	750	Third
E	$\frac{3{,}500-1{,}000}{5-3} = \frac{2{,}500}{2}$	1,250	Fourth

Activity D cannot be shortened

Procedure
To reduce the contract duration to, say, 12 weeks:

Shorten activity A by 2 weeks (least cost slope).
Shorten activity B by 2 weeks.
Shorten activities B and C by 2 weeks.

Contract durations and costs are given in Table 5.3. Crashed network is shown in Fig. 5.17.

Table 5.3

	Total cost (£)	*Total time saved (weeks)*	*Cost/week saved (£)*	*Contract duration (weeks)*
Shorten activity A	400	2	200	16
Shorten activity B	666	2	333	14
Shorten activity B and C	2,166	2	1,083	12
	3,232			

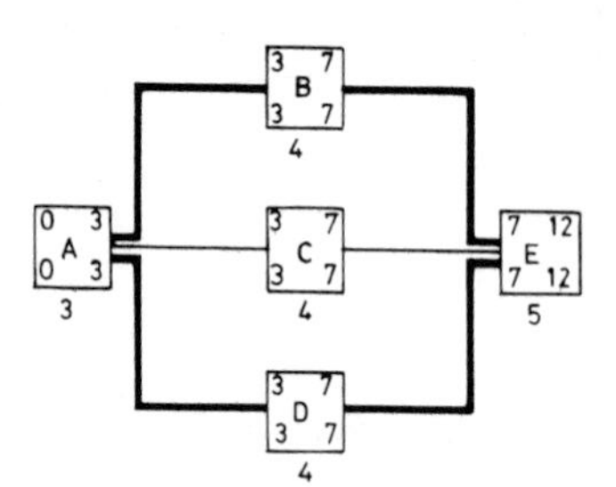

Fig. 5.17 Crashed network

Direct cost

The increase in cost with weeks saved is shown in Fig. 5.18 and Table 5.4. The costs of £10,500, £10,900, £11,566, etc. are determined by adding the cost of each shortening of the contract to the normal contract costs.

Table 5.4

Contract duration (weeks)	*12*	*14*	*16*	*18*	*20*
Normal contract sum (£)	10,500	10,500	10,500	10,500	10,500
Increase in direct costs (£)	3,232	1,066	400	—	—
Total of normal and direct costs (£)	13,732	11,566	10,900	10,500	10,500
Adjust indirect cost saving or loss (£)	−3,000	−2,000	−1,000	—	1,000
Contract cost according to duration (£)	10,732	9,566	9,900	10,500	11,500

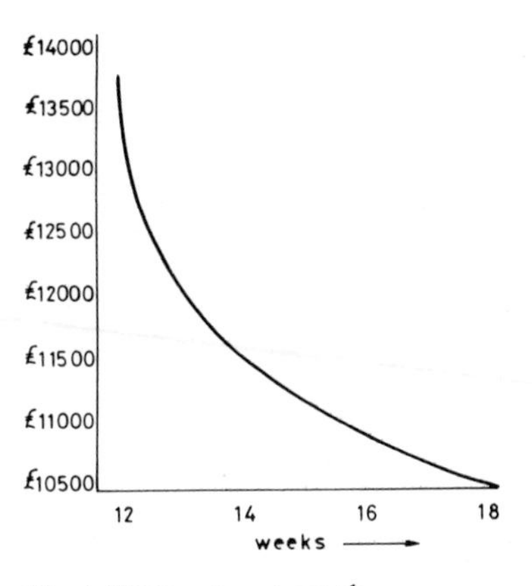

Fig. 5.18 Direct cost graph

Indirect cost/savings

If it is assumed that the indirect costs are £500/week, then over the *normal* contract duration this would be £9,000 (Fig. 5.19 and Table 5.4).

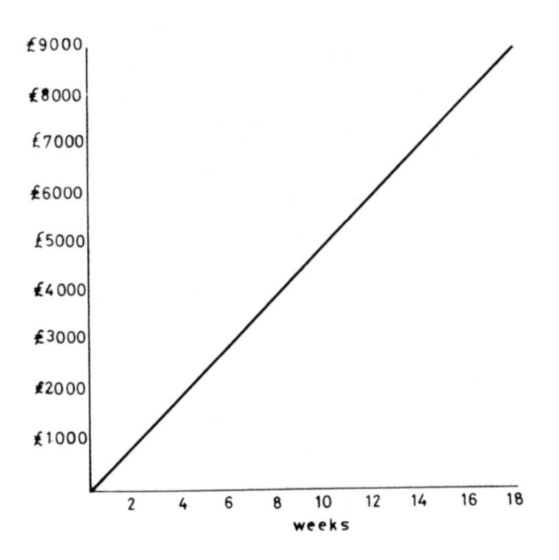

Fig. 5.19 Indirect cost graph

Procedure

The direct cost graph and indirect cost graph are combined in order to determine the optimised contract duration. This optimum time/cost relationship is 14 weeks' duration resulting in a cost of £9,566. Both a shorter and longer contract duration increases costs. This is clearly seen from Fig. 5.20.

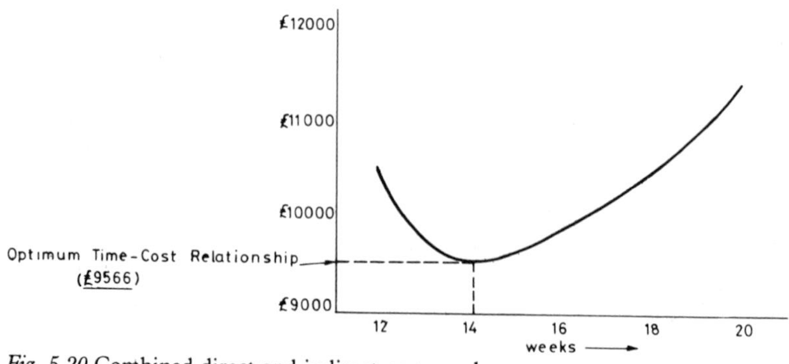

Fig. 5.20 Combined direct and indirect cost graphs

5.6 Resource scheduling

There is always a conflict of demand among resources required for construction work. Resource allocation is a means of determining within the constraints of the float available an optimum combination of activity schedules which will best achieve contract completion. By rescheduling activities within certain limits it may be possible to reduce the peak demand. The aim is to achieve a reasonably constant demand for a particular resource during construction.

Figure 5.21 shows a simple bar chart with each activity needing the same resource. The labour histogram indicates the peak resources as well as the pattern of demand. By rescheduling activities within their float times it is possible to smooth out the resource demands. As a guide, initial priority is given to those activities with least float during scheduling. The rescheduled chart and resource histogram is shown in Fig. 5.22. Resource scheduling enables management to plan for regular gang sizes, etc. to be used on site with minimum fluctuation in demand.

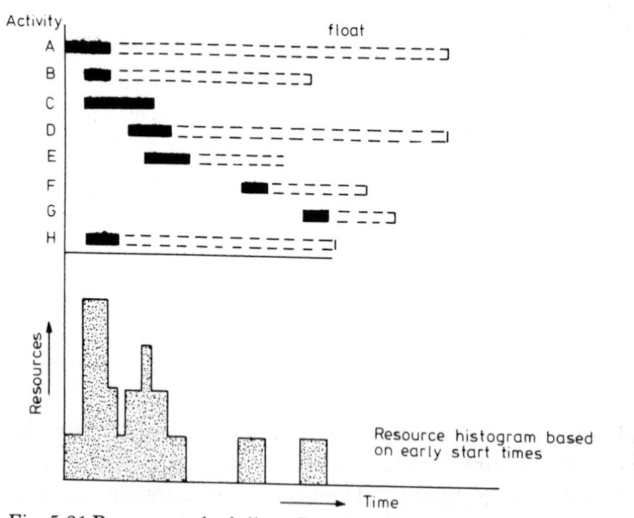

Fig. 5.21 Resource scheduling. Bar chart and resource histogram before resource scheduling

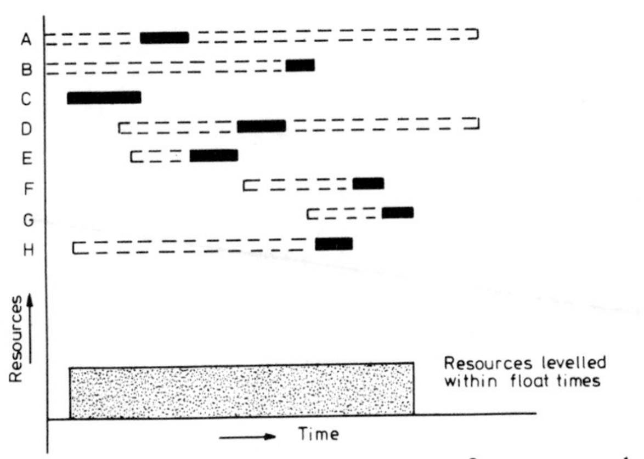

Fig. 5.22 Bar chart and resource histogram after resource scheduling

An analysed precedence diagram is shown in Fig. 5.23. A resource requirement graph is drawn from the network based on earliest start times (Fig. 5.24). It can be seen from the diagram that by manoeuvring activities within their earliest start and latest finish times a levelling of resources can be achieved (Fig. 5.25).

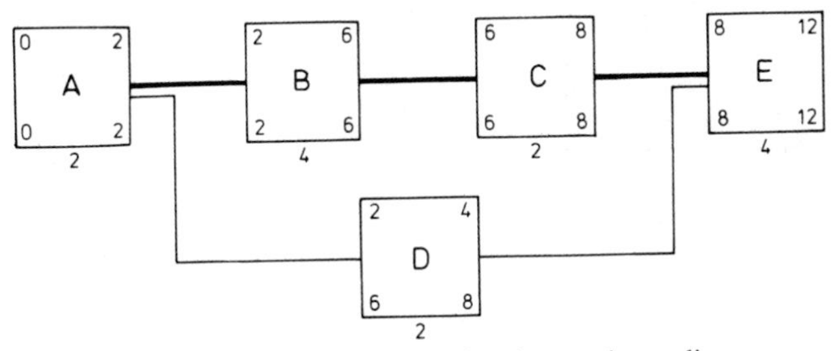

Fig. 5.23 Resource scheduling – analysed precedence diagram

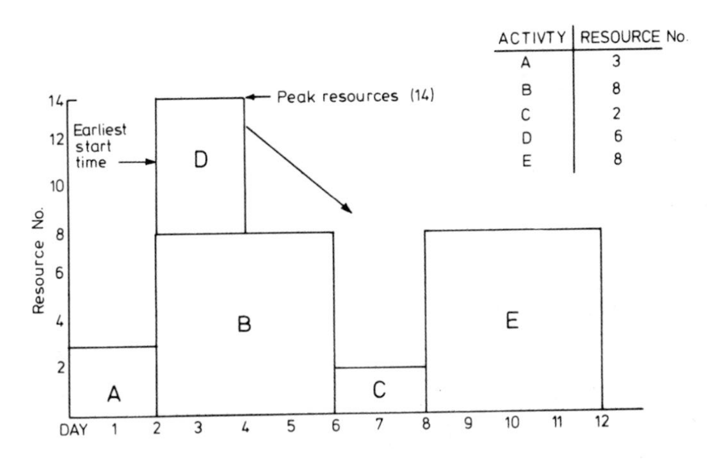

Fig. 5.24 Resource requirement graph

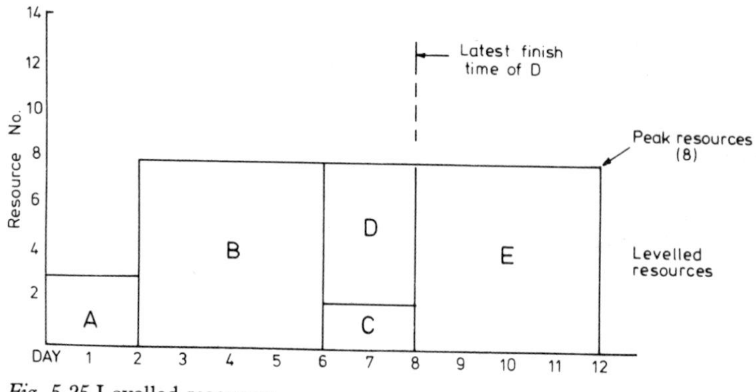

Fig. 5.25 Levelled resources

A further example, Fig. 5.26, will be used to illustrate the technique when there is a limitation on resources. (In this case the maximum number of resources available at any time is five.) A bar chart is derived from the analysed precedence diagram (Fig. 5.26). It can be seen from the resource schedule (Fig. 5.27) at the bottom of the bar chart that resources exceed the maximum available.

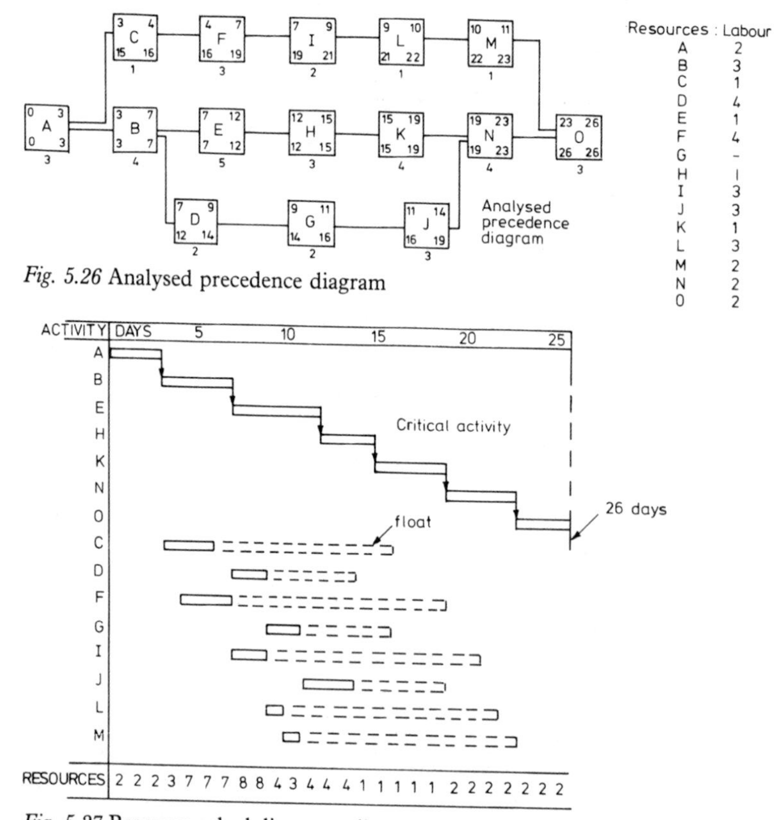

Fig. 5.26 Analysed precedence diagram

Fig. 5.27 Resource scheduling at earliest starts

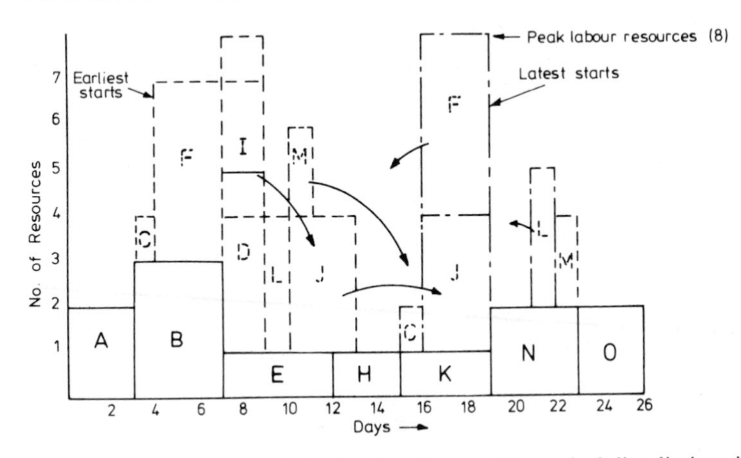

Fig. 5.28 Labour resource chart showing the possible rescheduling limits within activity floats

A labour resource graph is drawn from the network based on the extremes of early start and latest finish for non-critical activities. Scheduling within these extremes is logical and feasible. The rescheduled resource graph and bar chart is shown in Figs. 5.29 and 5.30, with a maximum resource requirement of five.

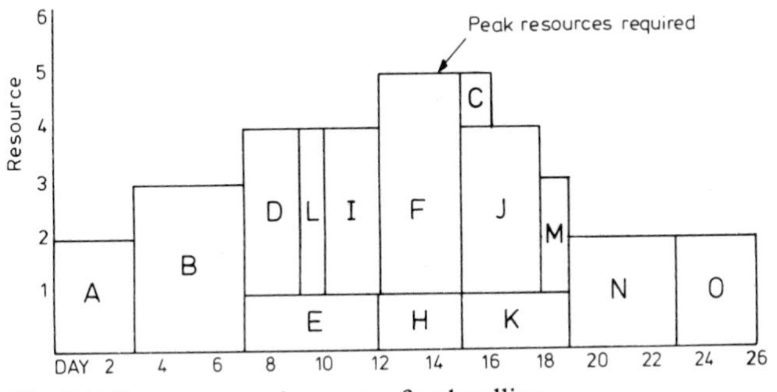

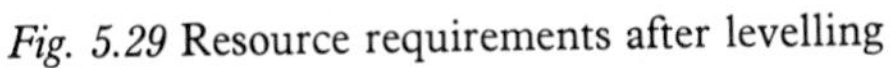
Fig. 5.29 Resource requirements after levelling

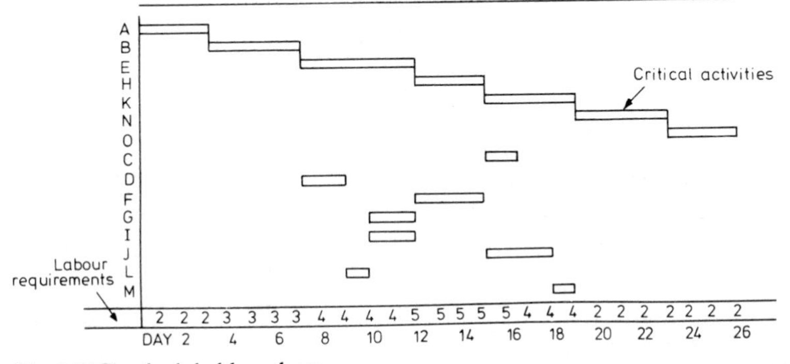

Fig. 5.30 Rescheduled bar chart

Initial allocations of resources may be insufficient to ensure planned progress. To determine the optimum number of resources it is usual to start with the smallest level of availability of resource and gradually increase this until the contract completion date is reached. Figure 5.31 shows a network for a small project together with resource requirements (Table 5.5).

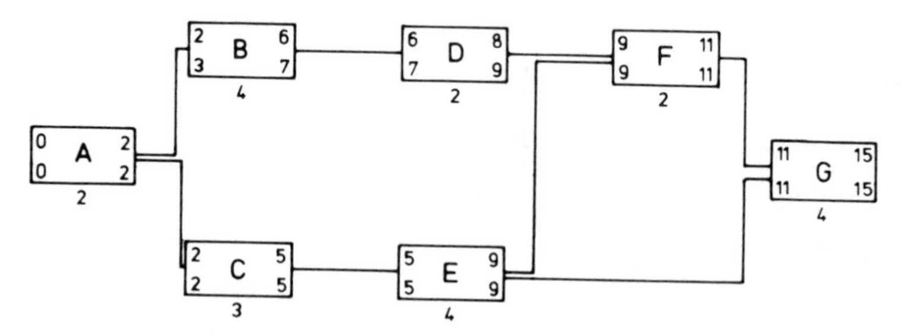

Fig. 5.31 Network for a small project

If there were no limitations on resources the contract could be completed in 15 days. However, assuming a resource limitation of three, the revised contract duration would be 20 days (Table 5.6). If the resource limitation is changed to five, the project could be completed in 16 days (Table 5.7).

Table 5.5

Activity	*Resource*
A	2
B	3
C	2
D	2
E	3
F	2
G	3

Table 5.6 With only 3 resource units available

Day	*1*	*2*	*3*	*4*	*5*	*6*	*7*	*8*	*9*	*10*	*11*	*12*	*13*	*14*	*15*	*16*	*17*	*18*	*19*	*20*
Activity																				
A	2	2																		
B			3	3	3	3														
C							2	2	2											
D										2	2									
E												3	3	3	3					
F																2	2			
G																		3	3	3
																		20 days		
	2	2	3	3	3	3	2	2	2	2	2	3	3	3	3	2	2	3	3	3

Table 5.7

Day	*1*	*2*	*3*	*4*	*5*	*6*	*7*	*8*	*9*	*10*	*11*	*12*	*13*	*14*	*15*	*16*
Activity																
A	2	2														
B			3	3	3	3										
C			2	2	2											
D						2	2									
E							3	3	3	3						
F											2	2				
G													3	3	3	3
														16 days		
	2	2	5	5	5	5	5	3	3	3	2	2	3	3	3	3

5.7 Uncertainty

Quite often in construction planning the person responsible for estimating how much time each activity in the network will take has little experience with exactly that particular construction form. When the planner estimates that a certain activity takes 7 weeks, he is really saying that he thinks it will take 7 weeks. From experience he knows that such things as adverse weather, poor materials delivery, machinery breakdown, etc. may affect the activity duration. This uncertainty that exists in construction work is a fact of life and there is no way of removing it. There are, however, methods by which the probability of an activity being completed as planned can be assessed.

The originators of PERT selected a probability distribution appropriate for most projects on which they would use the PERT network technique. They choose the beta distribution (Fig. 5.32) because:

1. There is only a small probability of completing the project in the most optimistic time.
2. The probability of completing the project in the most pessimistic time is small.
3. There is only one most likely time. It is closer to the most optimistic time, implying that the planner feels that the most likely time is near the most optimistic time than nearer the most pessimistic time.
   (a) *Most optimistic time* – the particular time estimate that has a very small probability of being reached. Probability of 1 in 100.

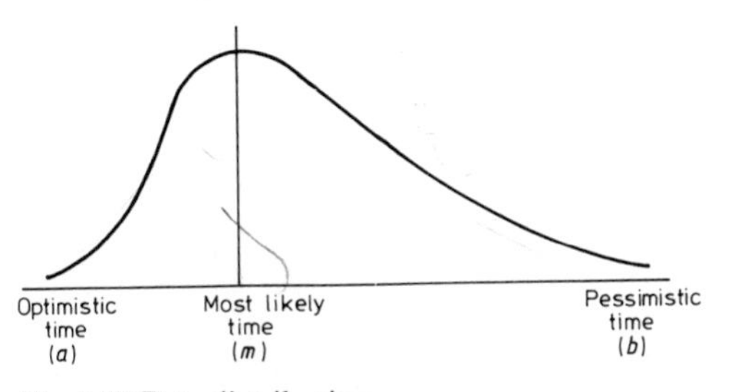

Fig. 5.32 Beta distribution

(b) *Most pessimistic time* – again this time estimate has a very small probability of being achieved. Probability of 1 in 100.
(c) *Most likely time* – this represents the time the activity would most often require if the work were done again and again under the same conditions.

These three time estimates are combined into a single workable time value using a weighted average. Here, m will be weighted heavily since there is a greater chance that the project will be completed near the most likely time than near the pessimistic (b) or optimistic (a) time, i.e. the expected time for the activity:

$$t_e = \frac{a+4m+b}{6}$$

If three time estimates are taken for an activity (Fig. 5.33), these can be plotted on the beta distribution: $a = 2 \quad m = 6 \quad b = 16$. Using the weighted average method:

$$t_e = \frac{2+(4 \times 6)+16}{6} = \frac{42}{6} = 7$$

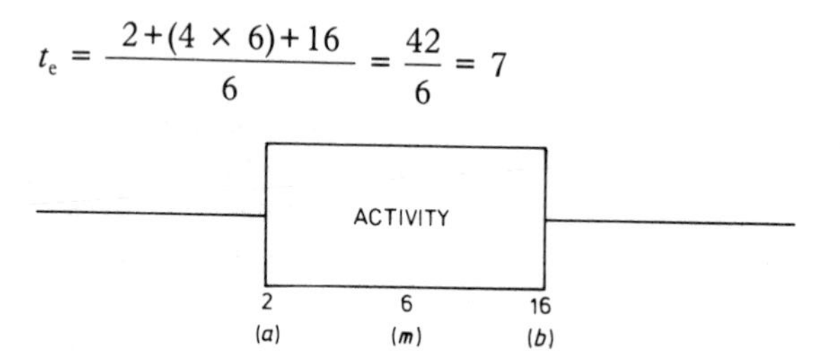

Fig. 5.33 Activities with three time estimates

In Fig. 5.34 t_e can be seen to be on the right of the most likely time. However, if the three estimates had been 2, 12 and 16 there would be a shift to the left of the most likely time (Fig. 5.35).

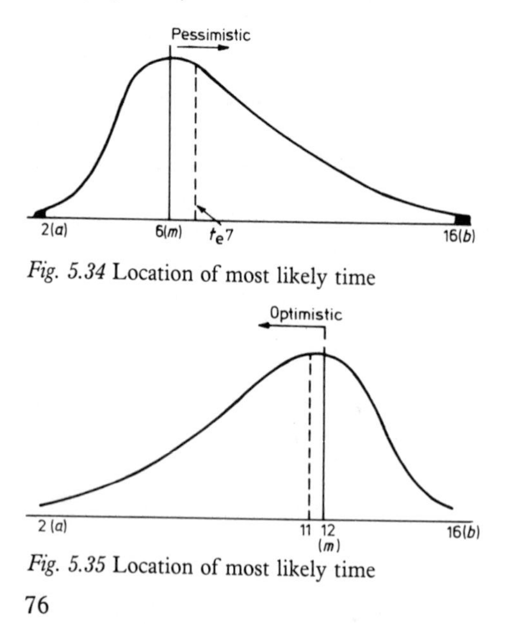

Fig. 5.34 Location of most likely time

Fig. 5.35 Location of most likely time

Using the weighted average formula:

$$t_e = \frac{2+(4 \times 12)+16}{6} = 11$$

Because the estimates a, m and b involve uncertainty there must then be some uncertainty in the final answer for t_e and some uncertainty about all the time values in the network.

Activity standard deviation

On a normal distribution curve we can plot estimates, for example 2 and 20 (Fig. 5.36). Using the range of 2 to 20, one standard deviation for the activity is calculated as:

$$\frac{20-2}{6} = 3 \text{ weeks}$$

The value of the most likely time (m) does not affect the calculation of standard deviation. It is only affected by the range from optimistic to pessimistic estimates.

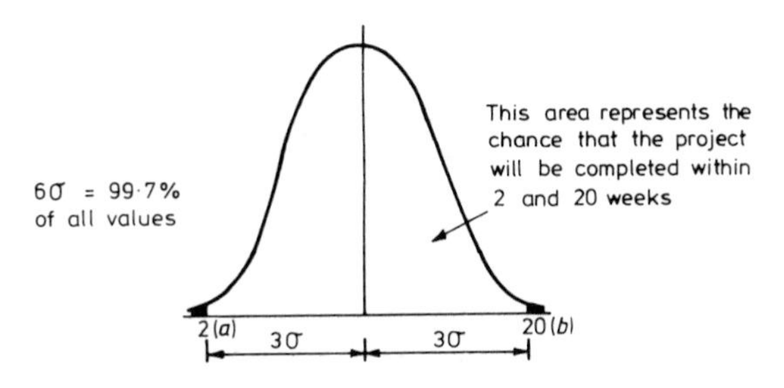

Fig. 5.36 Activity standard deviation

Consider the activities in Fig. 5.37 together with three time estimates. Calculate the individual standard deviations for each activity (Table 5.8). This gives a measure of spread of the activities around their most likely times. The expected elapsed time is now calculated for each of the activities (Table 5.9). The earliest expected date T_E for the network-ending is found by adding together the expected elapsed times for the activities t_e.

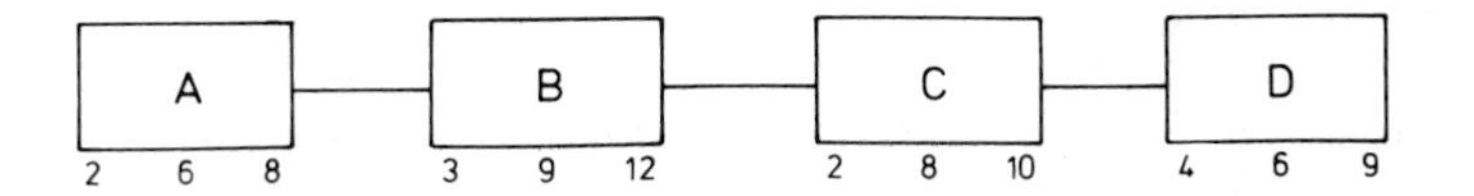

Fig. 5.37 Activities in series

Table 5.8

Activity	*a*	*b*	*b* − *a*	*Activity standard deviation* $\frac{b-a}{6}$ (from normal distribution)
A	2	8	6	1
B	3	12	9	1.5
C	2	10	8	1.33
D	4	9	5	0.833

Table 5.9

Activity	*a*	*m*	*b*	$\frac{a+4m+b}{6}$	*Expected elapsed time* (from beta distribution) t_e
A	2	6	8	$\frac{2+24+8}{6}$	5.66
B	3	9	12	$\frac{3+36+12}{6}$	8.50
C	2	8	10	$\frac{2+32+10}{6}$	7.33
D	4	6	9	$\frac{4+24+9}{6}$	6.16

$T_E = 5.66 + 8.50 + 7.33 + 6.13 = 27.65$

In order to help us determine the chances of finishing the project on time we must calculate the probability measure of the complete network. This is found by calculating the standard (Std) deviation.

$$\text{Standard deviation} = \sqrt{\left[\frac{(\text{Std deviation})^2}{\text{Activity A}} + \frac{(\text{Std deviation})^2}{\text{Activity B}} + \frac{(\text{Std deviation})^2}{\text{Activity C}} + \frac{(\text{Std deviation})^2}{\text{Activity D}}\right]}$$

$$\sigma T_E = \sqrt{[(1.0)^2 + (1.5)^2 + (1.33)^2 + (0.833)^2]} = 2.39$$

There are now two measures of the network, i.e. its T_E (earliest expected date for the network-ending) and σT_E (the standard deviation of the network-ending).

The T_E and standard deviation of an activity form a probability curve. If the standard deviation is large compared to its T_E variance is wide. If the standard deviation is small compared to its T_E then the estimators are confident that the actual time will not be far from the T_E.

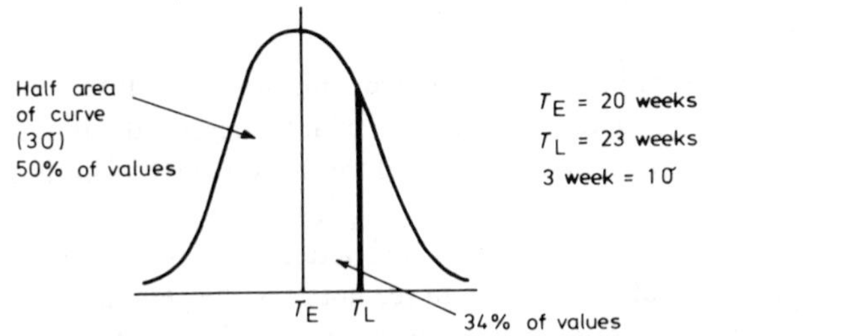

Fig. 5.38 Chances of project requiring more or less than 20 weeks are 50 per cent

A manager may wish to know the probability of completing a project by say its latest allowable date T_L. For this we use a probability curve or tables, the T_E value and its σT_E. For example working from the data in Fig. 5.38, the latest allowable date T_L is 23 weeks.

We know that approximately 68 per cent of all values in a normal shaped distribution are within $\pm$ 1σ from the average, thus since the T_L point is exactly 1σ to the right about 34 per cent of the values must be between T_E and T_L. Because 50 per cent of the values under the curve lie to the left of T_E and because 34 per cent lie between T_E and T_L, a total of 84 per cent of the values lie between the left-hand tail and T_L. There is, therefore, a probability of 84 per cent of finishing before the latest allowable date. An alternative is to use statistical tables, e.g.:

$$\frac{T_L - T_E}{\sigma} = \frac{23-20}{3} = 1$$

Using this figure look up normal distribution tables (areas under curve). This gives 0.84. Therefore as before, percentage probability of finishing before latest allowable date = 84 per cent.

Referring back to the previous example, in Fig. 5.37 it may be useful to determine the chances of finishing on time if the latest allowable time (T_L) is 25 weeks.

$$\frac{T_L - T_E}{\sigma} = \frac{25-27.65}{2.39} = -1.10$$

From normal distribution tables (areas under the curve) the chances of finishing on time are 0.14 or 14 per cent (Fig. 5.39).

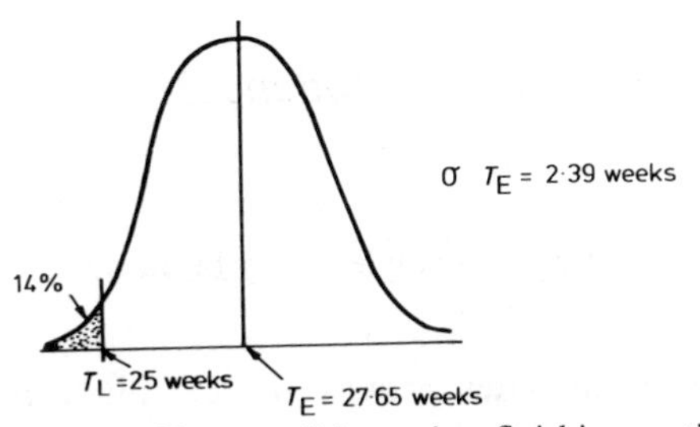

Fig. 5.39 Chances of the project finishing on time

5.8 Advantages of network planning

1. Separation of planning from scheduling: time periods are not taken into consideration when drafting the logic diagram. Sequence of activities are the initial criteria. This overcomes the simultaneous and therefore less effective planning and scheduling of bar charting.
2. The interdependence of activities is clearly shown (and determined) on networks. It enables those concerned, e.g. subcontractors, to see not only the general plan, but the ways in which their own activities depend upon or influence those of others.
3. Major subcontractors and specialists can contribute to the planning function by examining their relevant sections of the network, assessing its soundness and thus avoiding the possibility of unrealistic or superficial planning.
4. It permits resource smoothing to be carried out between early and latest start times, thus keeping within the network logic.
5. It permits examination of alternative methods or of individual job estimates on the project as a whole at the outset, before any particular plan is implemented.
6. Total resources can easily be calculated and time/cost/resource techniques can be utilised.
7. The network is simply a statement of logic which remains constant whether the activities take a longer or shorter time than estimated.
8. Identification of the critical path enables management to concentrate on a relatively few influential activities.
9. Bar charts based on true logic can be derived from a network.
10. It facilitates the use of a computer to deal with calculations.

Disadvantages

1. The preparation of networks is a tedious and exacting task.
2. Because of uncertainties, delays of materials, labour shortages, etc. frequent reviews are necessary.
3. There is often a communication barrier between the planner and site manager/foreman/operative, resulting in lack of commitment.
4. A network indicates one construction sequence. There may often be a number of satisfactory alternative forms of construction.

5.9 Worked examples: Networks (CPA and precedence)

Q.5.1.

In Fig. 5.40 is shown a network for the construction of a single-storey building.

(a) Answer the following questions:
   (i) Which activities can start immediately after activity 7–10, tiling, is completed?

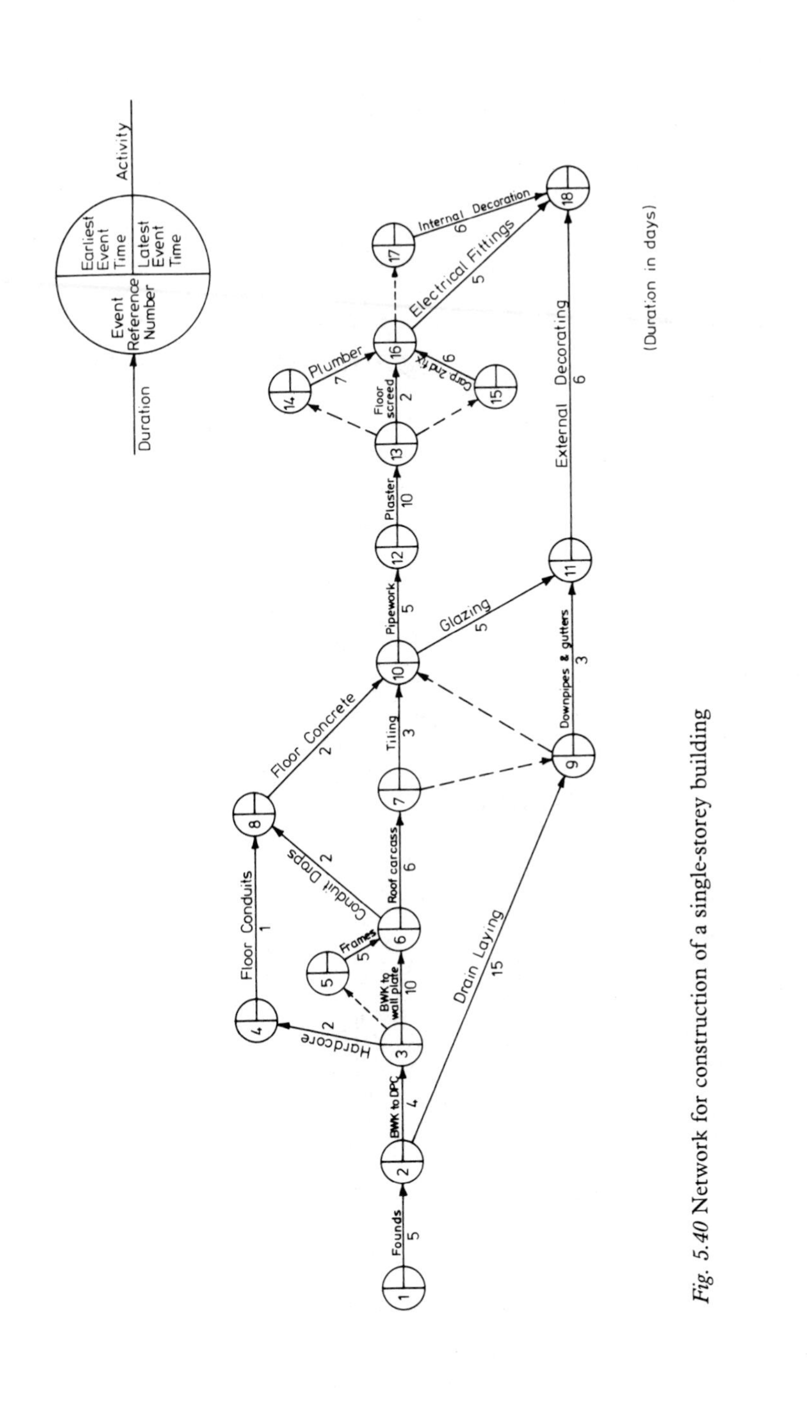

Fig. 5.40 Network for construction of a single-storey building

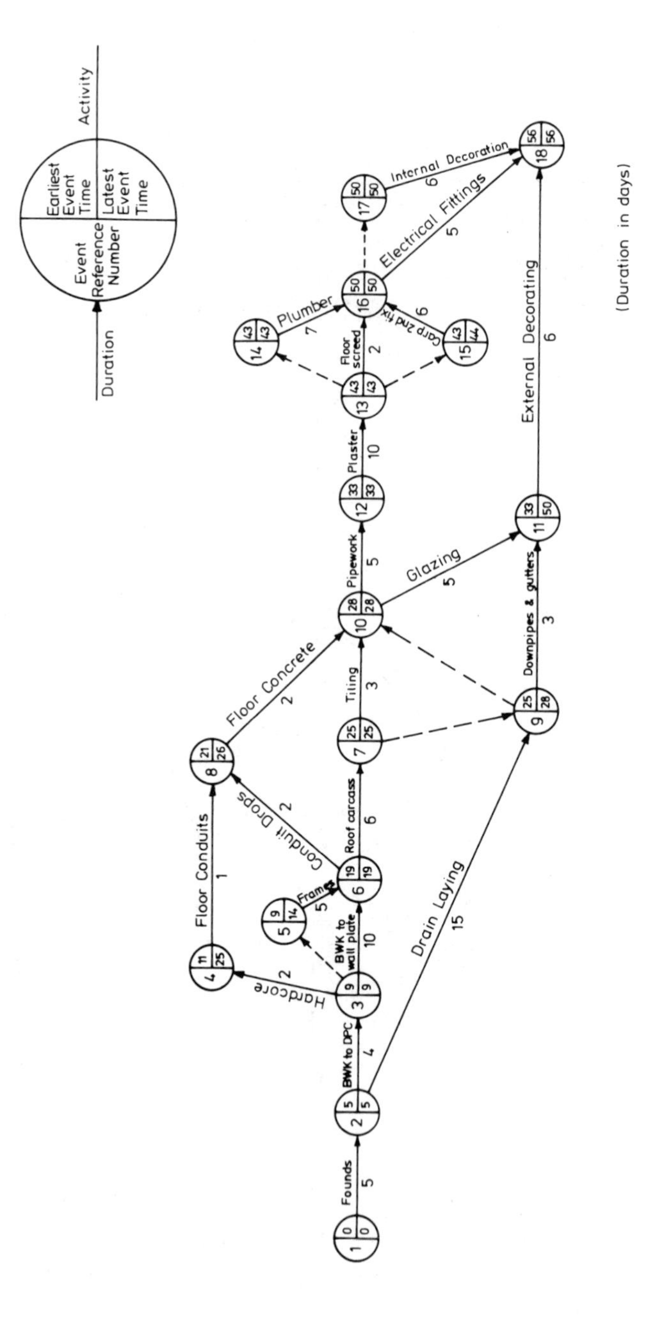

Fig. 5.41 Solution

(ii) What is the earliest starting time in days for activity 9–11, downpipes and gutters, and activity 8–10, floor concrete?
(iii) What is the earliest finishing time in days for activity 12–13, ceilings and plastering?
(iv) What is the earliest finishing time for event 18, completion of project?
(v) In order that the project should be completed on time, what is the latest starting time of activity 10–11, glazing?
(vi) What is the latest finishing time of activity 2–9, drainlaying?
(vii) How much additional time could activity 9–11, downpipes and gutters, take, without extending the total time of the project?

(b) Indicate the critical path of the network.
(c) If activity 15–16, carpenters' second fixings, takes 4 days longer than planned, how many days would the durations of activities 16–18, electrical fittings, and 17–18, internal decorating, need to be so that the project would be completed on time?

Q.5.1. Solution (Fig. 5.41)

(a) (i) Activity 7–10, tiling is completed after 28 days. After this is completed work can commence on activities 10–12, pipework, and 10–11, glazing.
(ii) Earliest starting time (EST) for activity 9–11, downpipes and gutters:
EST (9–11) = after 25 days;
EST (8–10) = after 21 days;
(iii) EFT (12–13) = 56 days;
(iv) EFT (18) = 56 days;
(v) LST (10–11) = 28 days;
(vi) LFT (2–9) = 28 days;
(vii) 9–11 could take an additional $(50 - 33 + 5) = 22$ days.

(b) Critical path indicated by solid line.
(c) Activity 16–18, 3 days; activity 17–18, 3 days.

Q.5.2.

Analyse the precedence diagram shown in Fig. 5.42.
See Fig. 5.43 for solution.

Q.5.3.

The engineer of a drainage authority has planned a sewerage scheme incorporating a sewage pumping station (see Fig. 5.44) to which external sewers connect via a main inlet. His operation listing and precedence network for the overall construction programme is shown in Table 5.10 and Fig. 5.45. He intends to carry out the project using three contractors A, B and C to do work as follows:

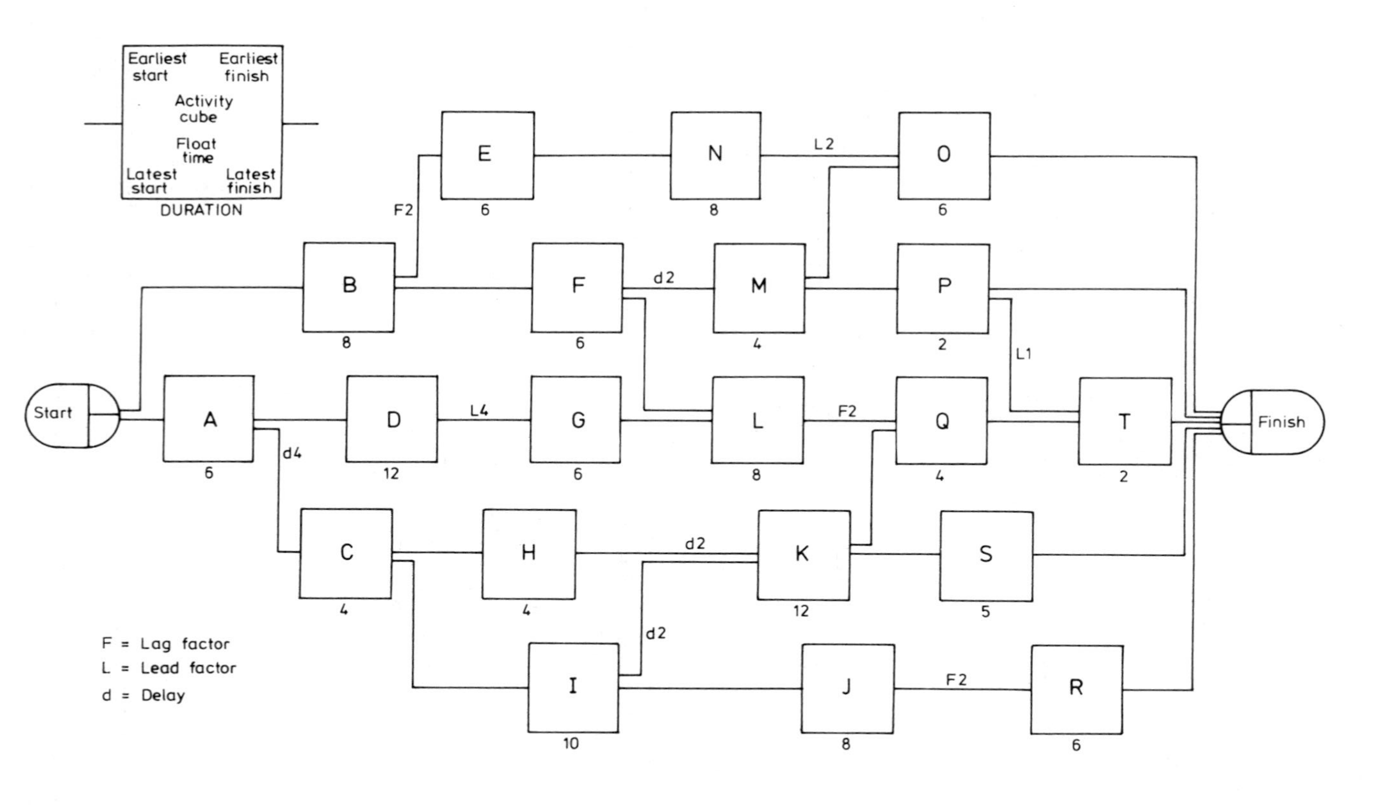

Fig. 5.42 Precedence network

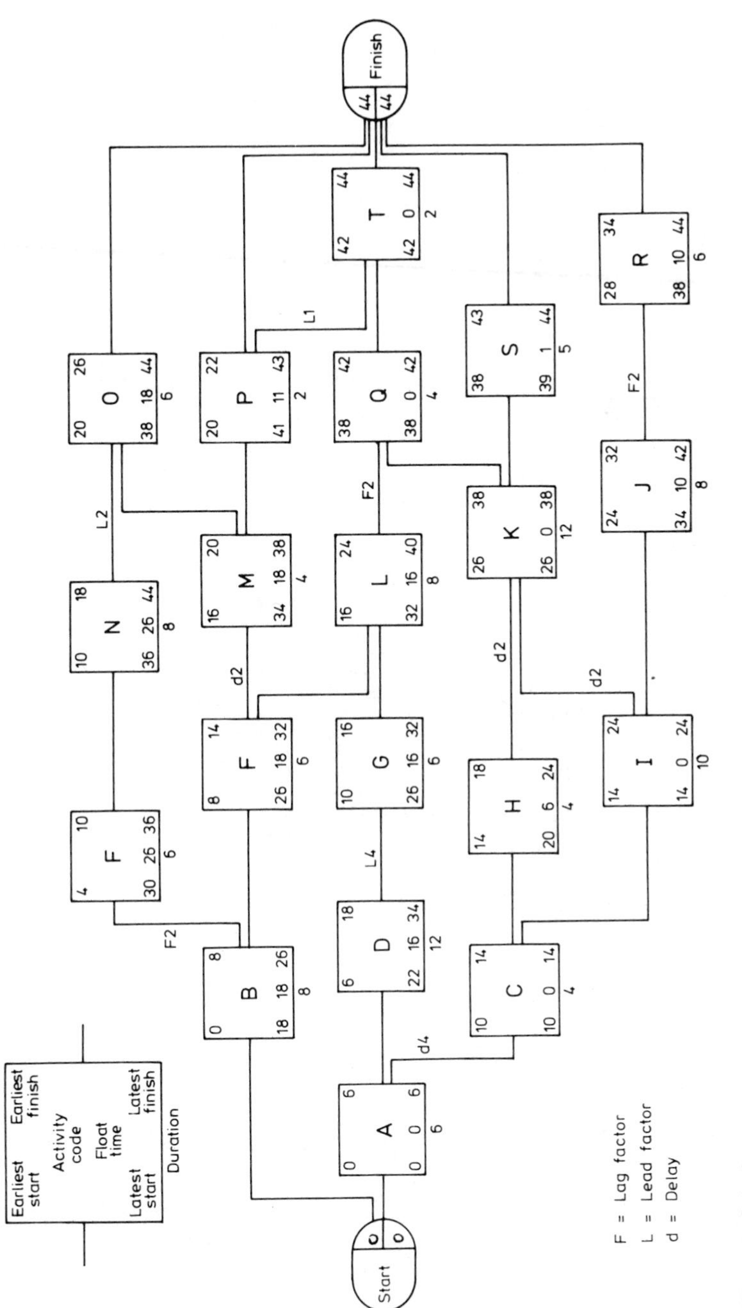

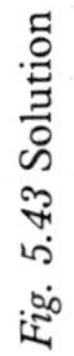
Fig. 5.43 Solution

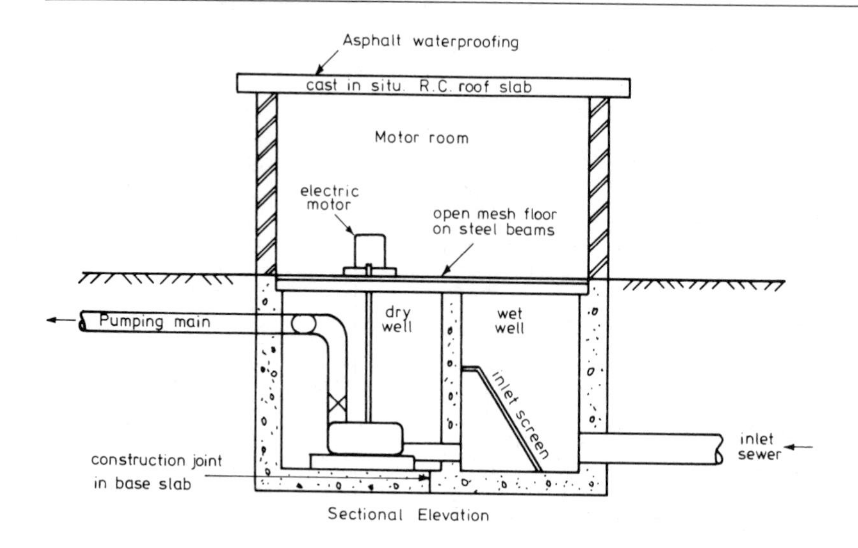

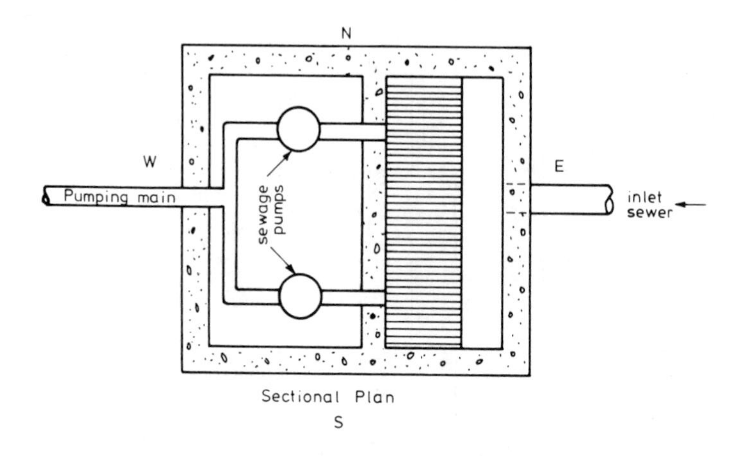

Fig. 5.44 Sketch of sewage pumping station

A: Contractor A will perform activity 5 (trenching and laying sewers upstream of the inlet).
B: Contractor B will be the local electricity undertaking which will lay all external cabling (activity 4) for the power supply of the pumphouse.
C: Contractor C will carry out all other activities.

Table 5.10 Operation listing for sewage pumping station

Code	*Operation description*	*Duration*	*Contractor*
1	Excavations	3	C
2	Prefabricate and deliver to site steel beams and open mesh flooring for motor room floor	48	C
3	Prefabricate and deliver wet well screens	36	C
4	External cabling by electricity authority	56	B
5	Trench and lay main sewers upstream of pumphouse	62	A
6	Wet well base slab	7	C
7	Dry well base slab	8	C
8	Basement wall (east) – (external)	6	C
9	Basement wall (north) – (external)	5	C
10	Basement wall (south) – (external)	4	C
11	Inlet sewer connection	4	C
12	Basement wall (west) – (external)	6	C
13	Backfill excavations	2	C
14	Basement internal walls	7	C
15	Pump foundations	4	C
16	Connect pump suction pipes	2	C
17	Install pumps	8	C
18	Install screens to wet well	4	C
19	Trench and lay pumping main	24	C
20	Brickwork motor room walls	9	C
21	Shutter, reinforcement and cast *in situ* RC motor room roof	14	C
22	Connect pump and pipe	3	C
23	Install motor room floor	2	C
24	Waterproof motor room roof slab (asphalt)	4	C
25	Fix windows and doors	3	C
26	Install switch gear and wiring	4	C
27	Internal finishes to motor room	10	C
28	Test run motors and pumps – Hand over to drainage authority	2	C

Answer the following questions:

(i) How long will it take to complete the project?

(ii) Which contractors, if any, can delay starting their work without affecting the overall construction period and by how much time in each case?

(iii) If under the conditions of contract, the engineer can impose liquidated damages for delay of £100/day for any delay in completion caused by Contractor C and if Contractor C takes 18 days longer than planned to prefabricate the steel beams and open mesh flooring for the motor room floor (activity 2) what damages, if any, will he be liable for?

The solution is given in Fig. 5.46.

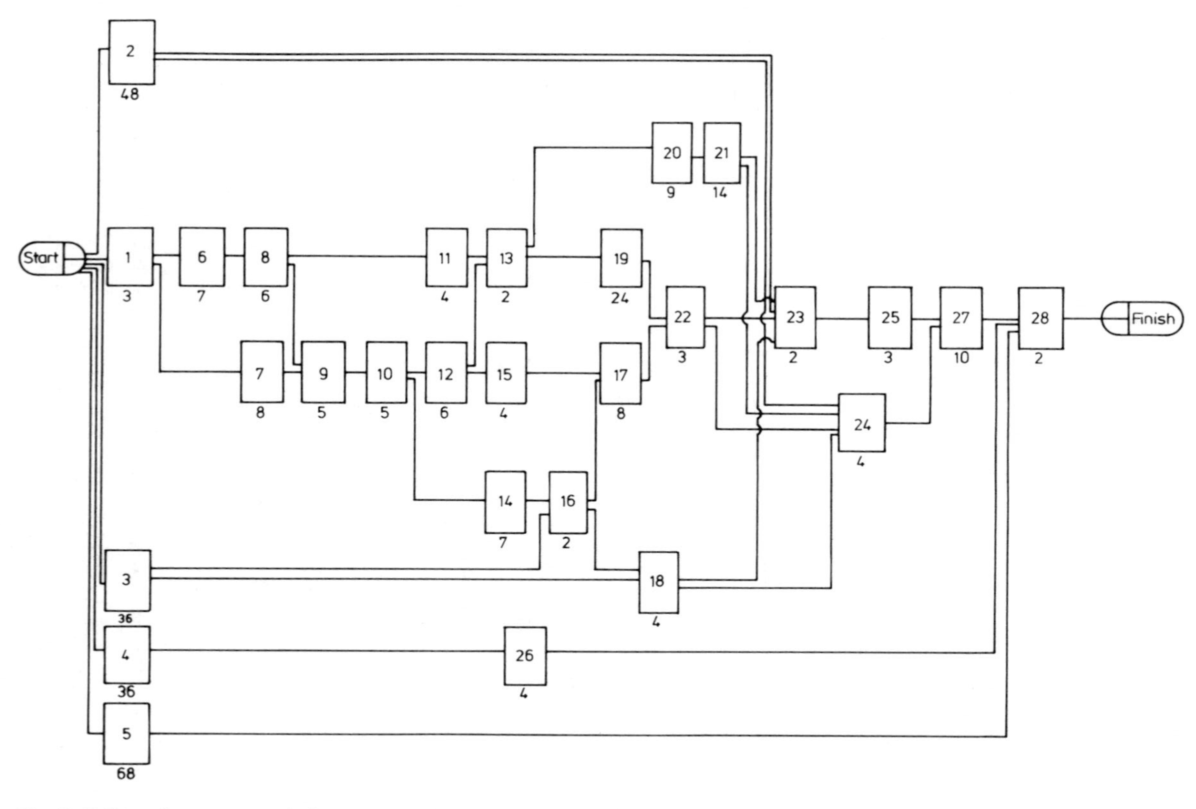

Fig. 5.45 Precedence network for construction of sewage pumping station

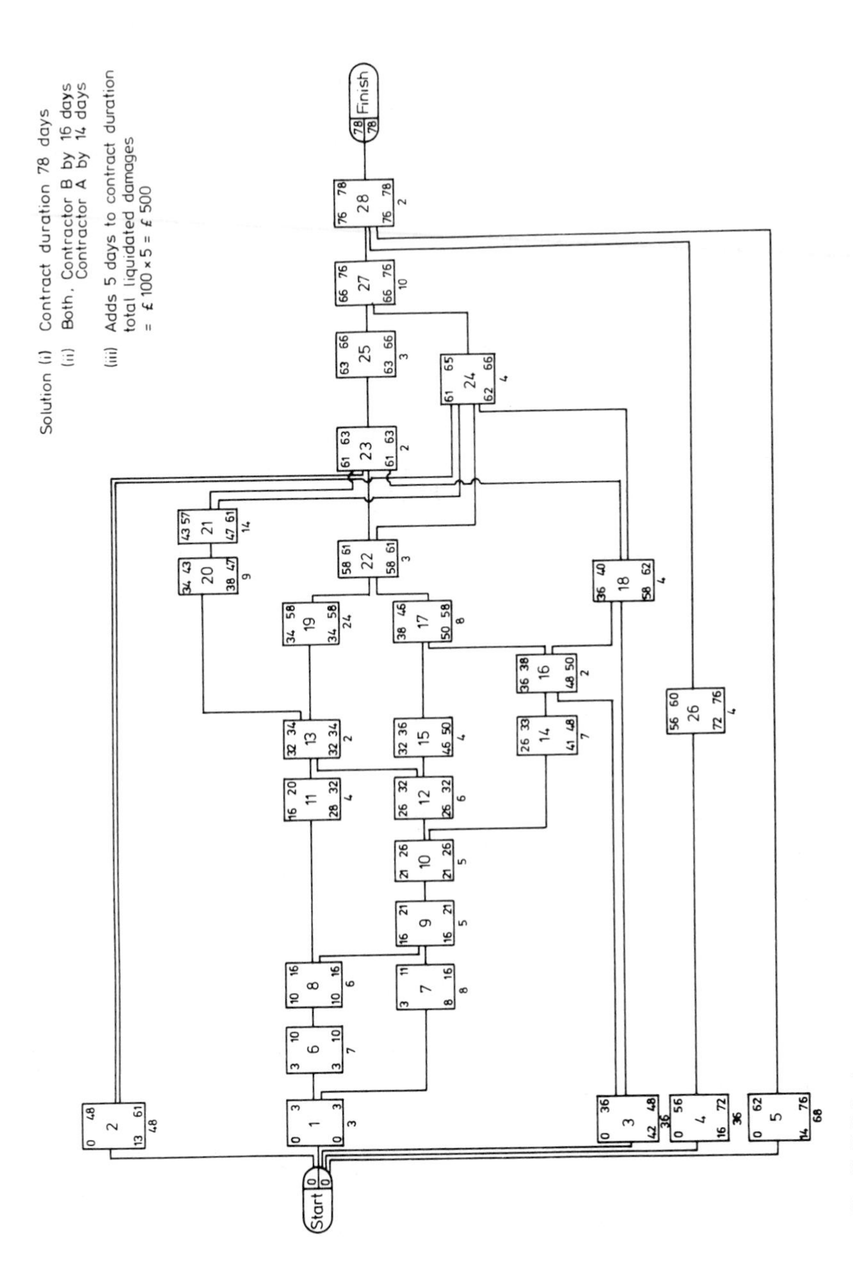

Solution (i) Contract duration 78 days
(ii) Both, Contractor B by 16 days
Contractor A by 14 days
(iii) Adds 5 days to contract duration
total liquidated damages
= £ 100 × 5 = £ 500
Start
Finish

Fig. 5.46 Solution

Q.5.4.

An ore company is proposing to construct a new mineral railway between a mine at A and a smelting plant at B, 9 km away (Fig. 5.47). The track-laying contractor can commence operations only at either A or B and could complete his operations working forward on a single heading in 9 weeks, if he did not need to wait for completion of earthworks. The earthworks contractor can start in any cutting and his plant has an output of 30,000 m^3/week on a haul of 400 m. Output is reduced by 5 per cent for each 100 m haul in excess of 400 m. (On a haul of 1 km the output would be 21,000 m^3/week.)

Use a network analysis, allowing for delays to track-laying until sections of cuttings and fills are completed, to determine:

(a) the minimum time required to complete the project;
(b) whether track-laying materials should be delivered to A or B.

Assume that possession of a cutting or fill, or part thereof, cannot be given to the track-laying contractor until all earth-moving operations in the cutting or fill have ceased. There is no loss of earthworks volume due to compaction.

The solution is given in Table 5.11 and Fig. 5.48.

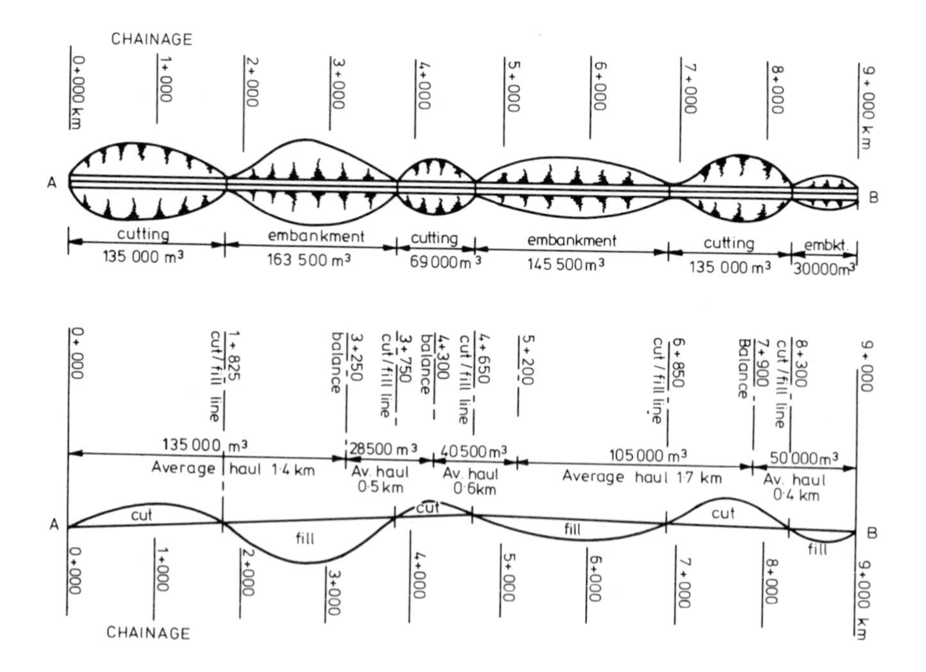

Fig. 5.47 Mineral railway scheme

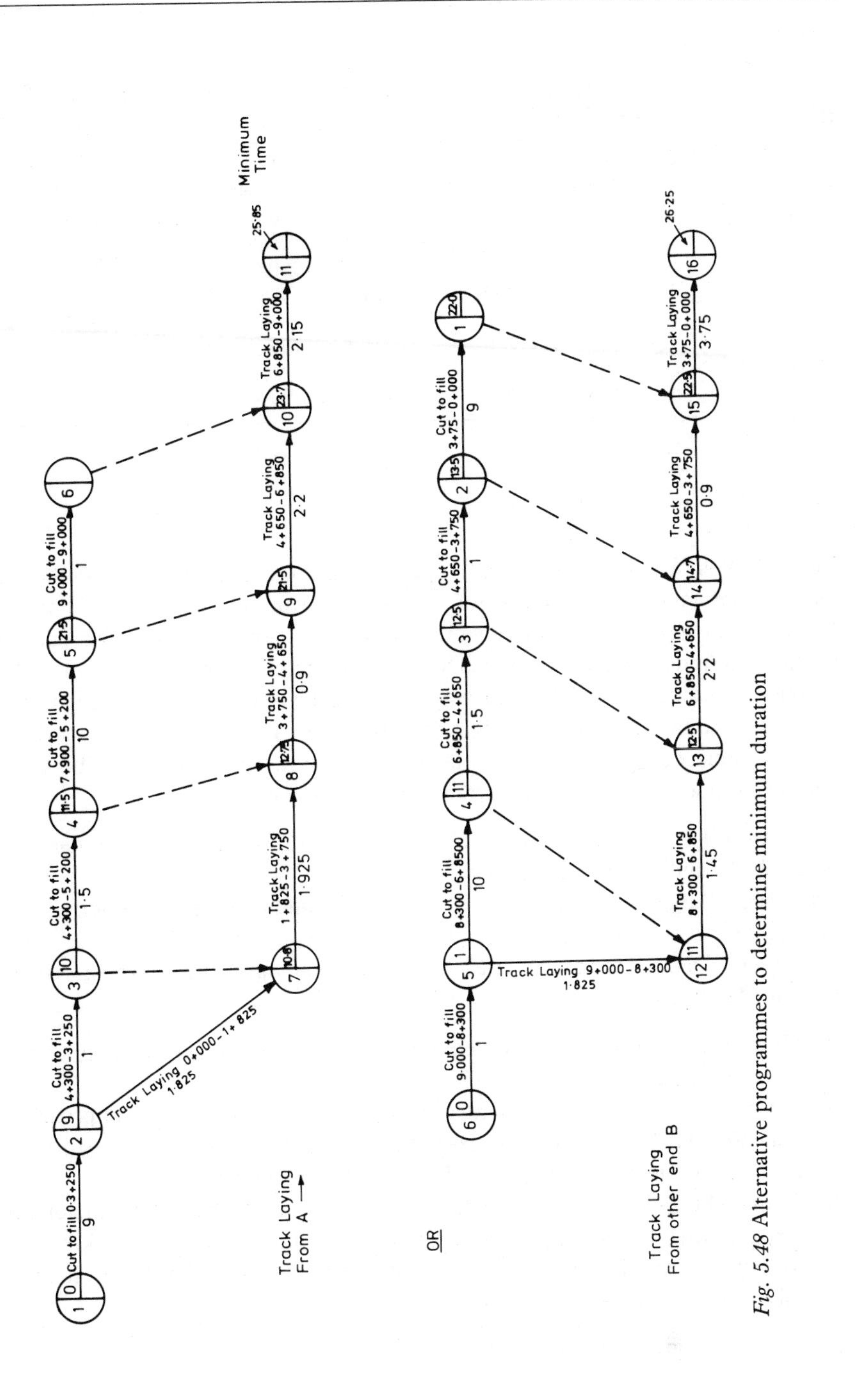

Fig. 5.48 Alternative programmes to determine minimum duration

Table 5.11 Solution

Activity	*Description*	*(m³) Vol.* + *(km) Dist.*	*(Weeks) Time*
1–2	Cut to fill 0 + 000 → 3 + 250	135,000 1.4 (15,000 m^3/week)	9
2–3	Cut to fill 4 + 300 → 3 + 250	28,500 0.5 (28,500 m^3/week)	1
3–4	Cut to fill 4 + 300 → 5 + 200	40,500 0.6 (27,000 m^3/week)	1.5
4–5	Cut to fill 7 + 900 → 5 + 200	105,000 1.7 (10,500 m^3/week)	10
5–6	Cut to fill 7 + 900 → 9 + 000	30,000 0.4 (30,000 m^3/week)	1.0

Q.5.5.

Figure 5.49 shows the network for programming the construction of a tanker berth which a port authority will use for the reception of North Sea oil. Figure 5.50 illustrates the proposed development. The port authority has its own barge and ancillary equipment which it can use to carry out the piling, reinforced concrete construction and fendering of the breasting and mooring dolphins at the jetty head. The barge can also be adapted for dredging the ship's berth.

The authority's engineer decides, as he is unable to undertake all of the work, to employ contractors to carry out certain elements as follows:

Contractor x To construct the approach jetty EF and the oil pumphouse – network activities 1–16 and 15–20 respectively.
Contractor y To prefabricate and erect the catwalks and to install the moorings on the dolphins – network activities 1–17, 17–18, 18–20.
Contractor z To manufacture and install the navigation lighting equipment – network activities 1–19, 19–20.

(i) How long will it take to complete the project based upon the activity durations given on the network?
(ii) If Contractor x has an accident during the construction of the approach jetty which causes him to incur 18 days of additional time on activity 1–16 will this delay the project as a whole, and if so, by how many days?
(iii) If, in addition to Contractor x having an accident in (ii) above, Contractor y has a strike lasting 14 days while erecting catwalks, and Contractor z takes 12 days longer than estimated to install the navigation lighting equipment how long will it take to complete the project?

The solution is given in Fig. 5.51.

(iv) Convert the CPA network (Fig. 5.51) into precedence diagram format.

The solution is given in Fig. 5.52.

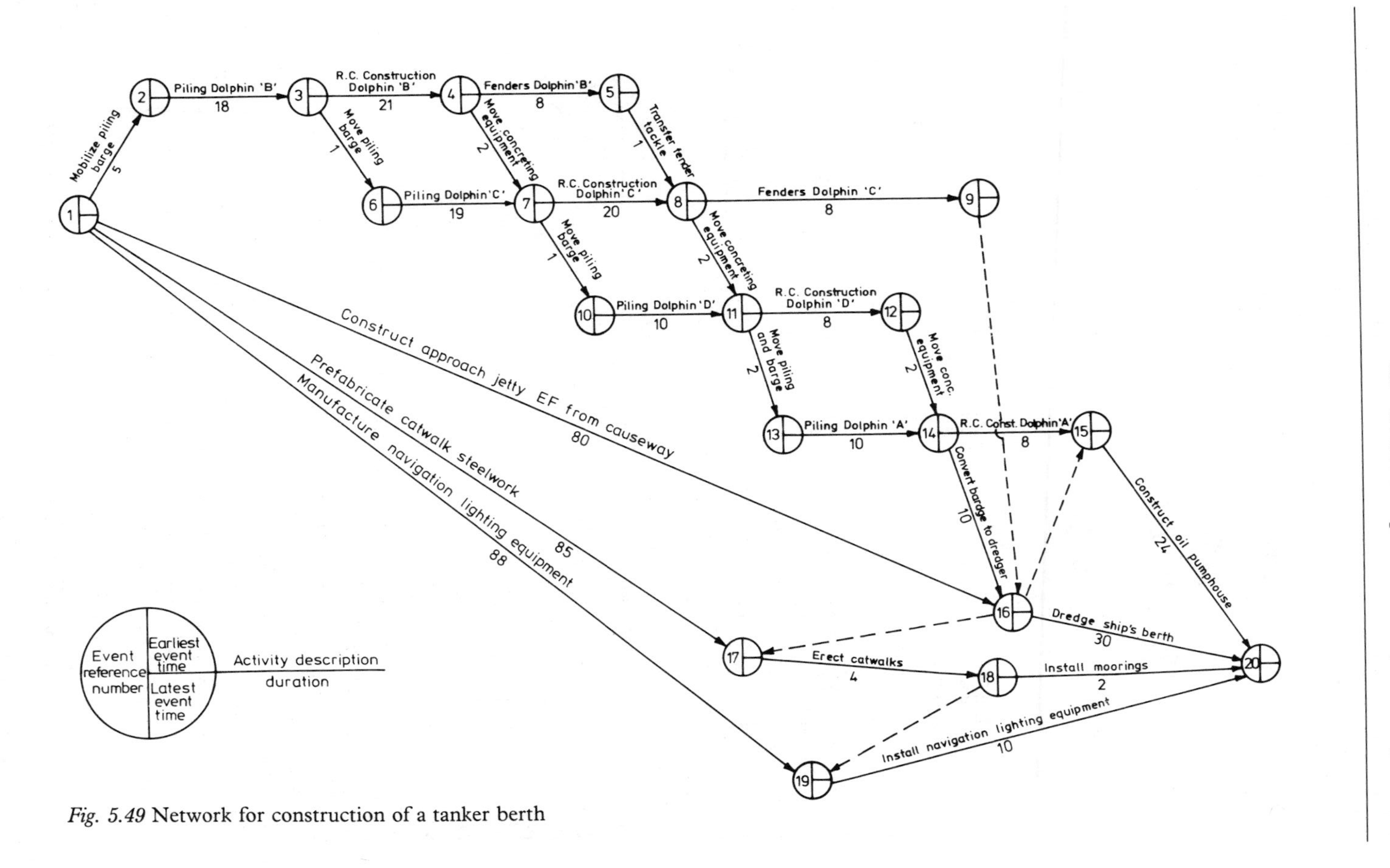

Fig. 5.49 Network for construction of a tanker berth

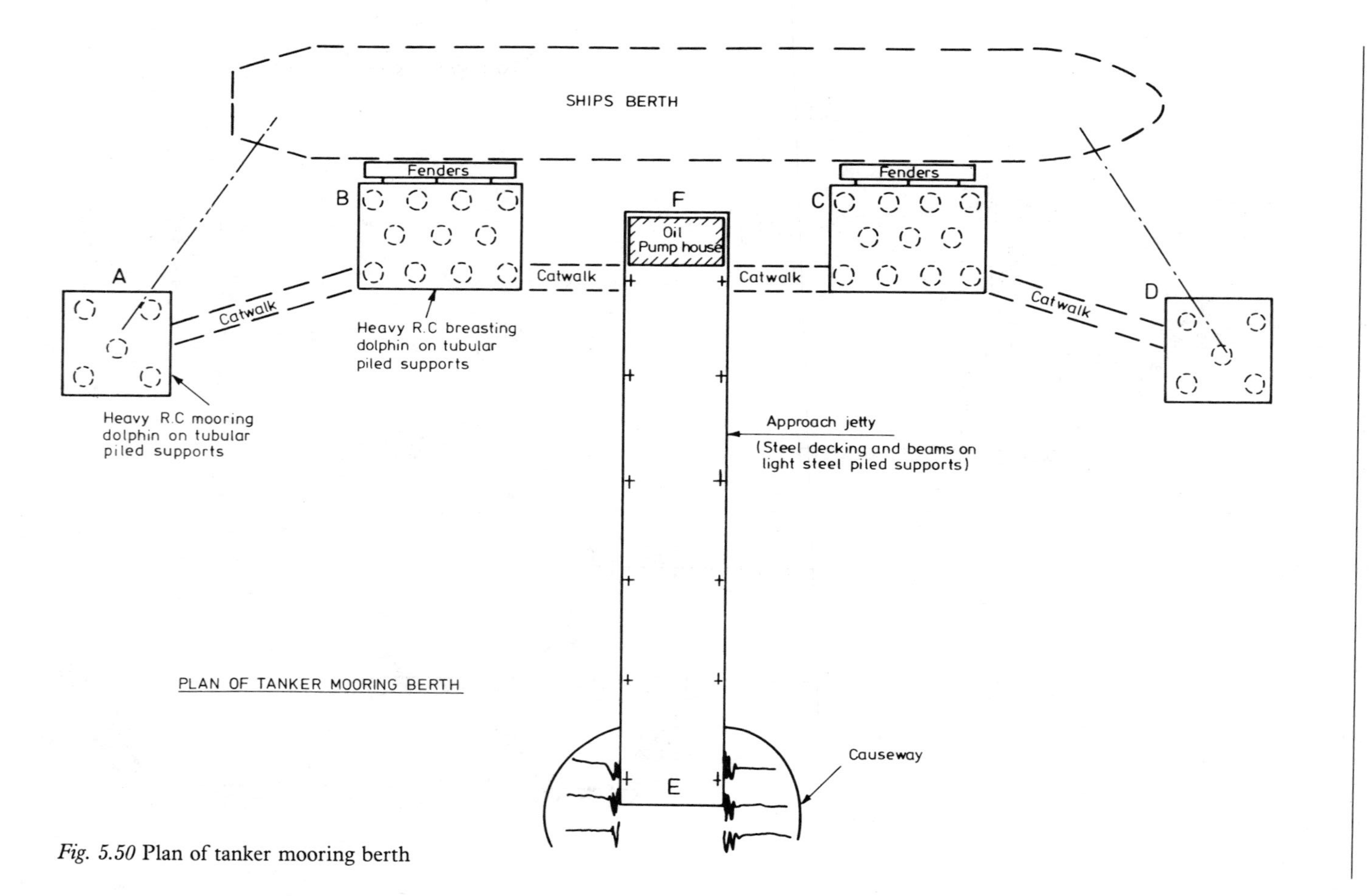

Fig. 5.50 Plan of tanker mooring berth

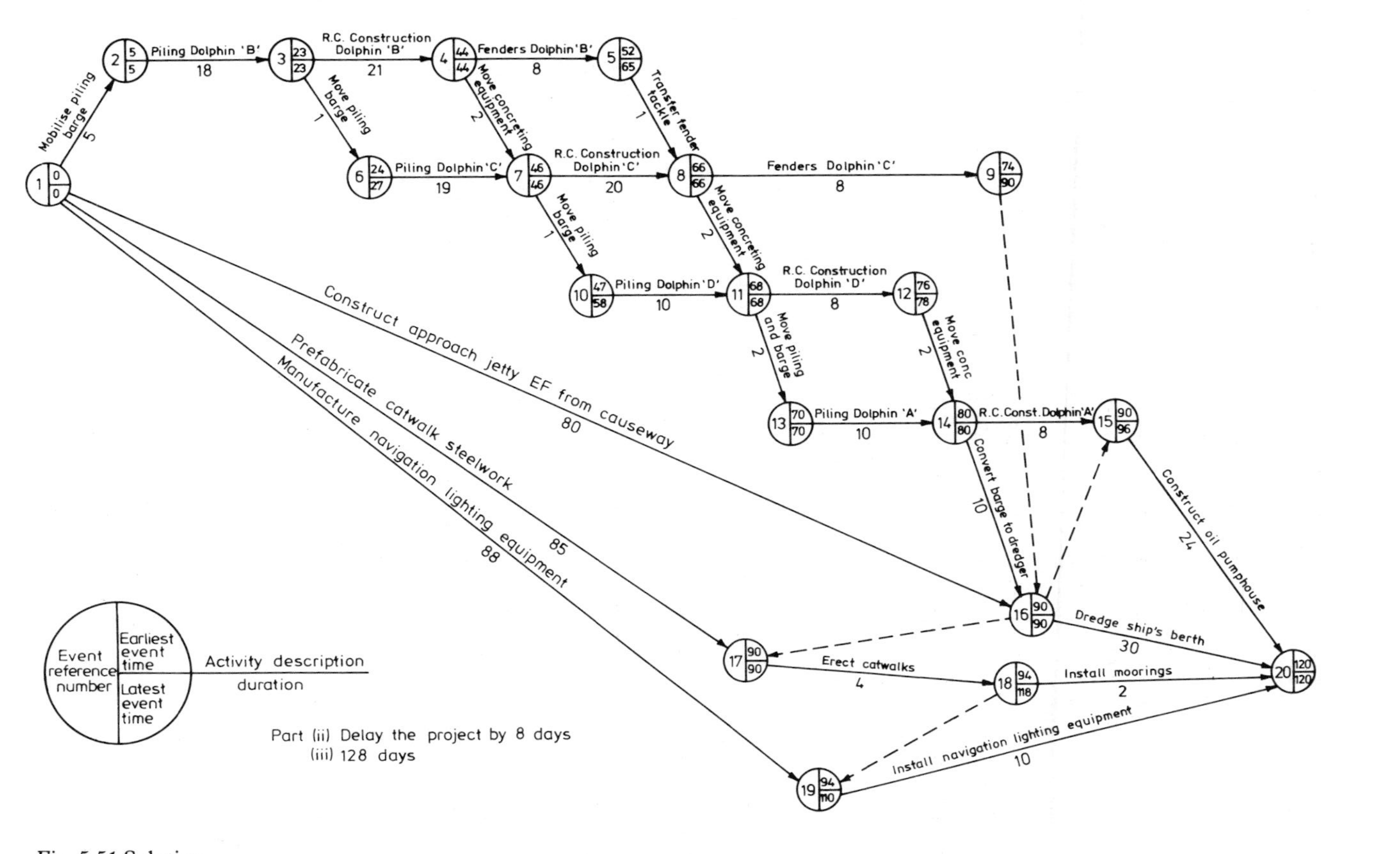

Fig. 5.51 Solution

Q.5.6.

Table 5.12 relates to a small construction project including costs of executing the work for both normal and crash durations.

(i) Using the i–j numbers and activity codes draw the networks in both critical path (CPM) and precedence diagram formats.
(ii) Determine the optimised direct cost of the project executed in crash duration. (Assume that each activity duration must be reduced by full amount.)
(iii) Determine the optimum contract time when indirect costs are estimated to be £18,000 at normal contract duration reducing by £1,000/week for each week of contract duration saved.
(iv) Draw the total cost curve for the contract.
(v) Check solution by utilising a time/cost optimisation computer package.

The solutions are given in Figs. 5.52 to 5.56.

Table 5.12

		Time (weeks)		*Cost (£)*	
Activity code	*i–j No.*	*Normal*	*Crash*	*Normal*	*Crash*
A	1–2	7	3	5,000	13,000
B	1–3	9	5	4,000	8,000
C	2–4	8	5	2,800	4,300
D	3–4	14	9	1,000	4,000
E	4–6	4	2	12,000	13,800
F	3–5	9	5	5,000	8,500
G	5–6	7	3	2,000	5,200
H	5–7	11	6	7,500	9,500
I	6–7	12	7	5,000	7,750

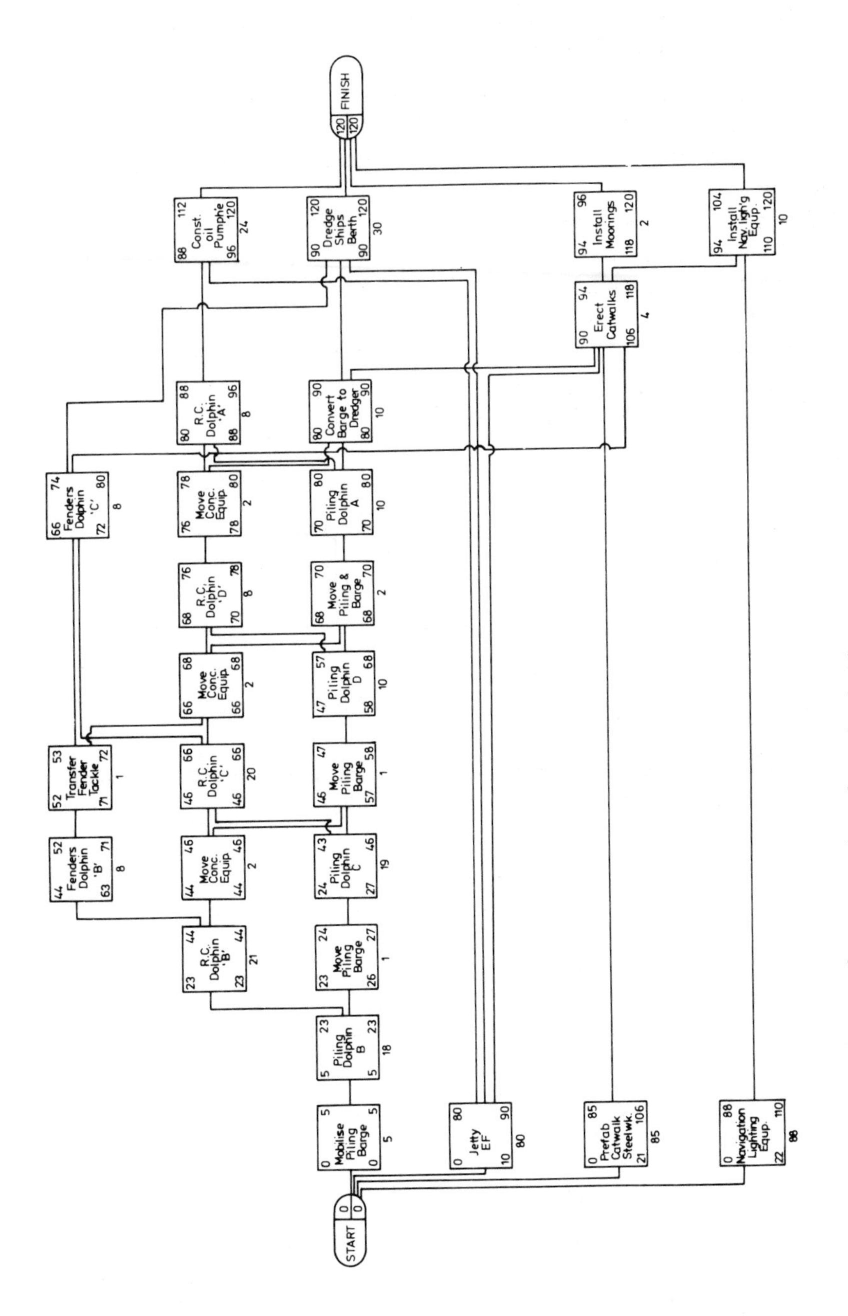

Fig. 5.52 Precedence diagram for construction of a tanker berth

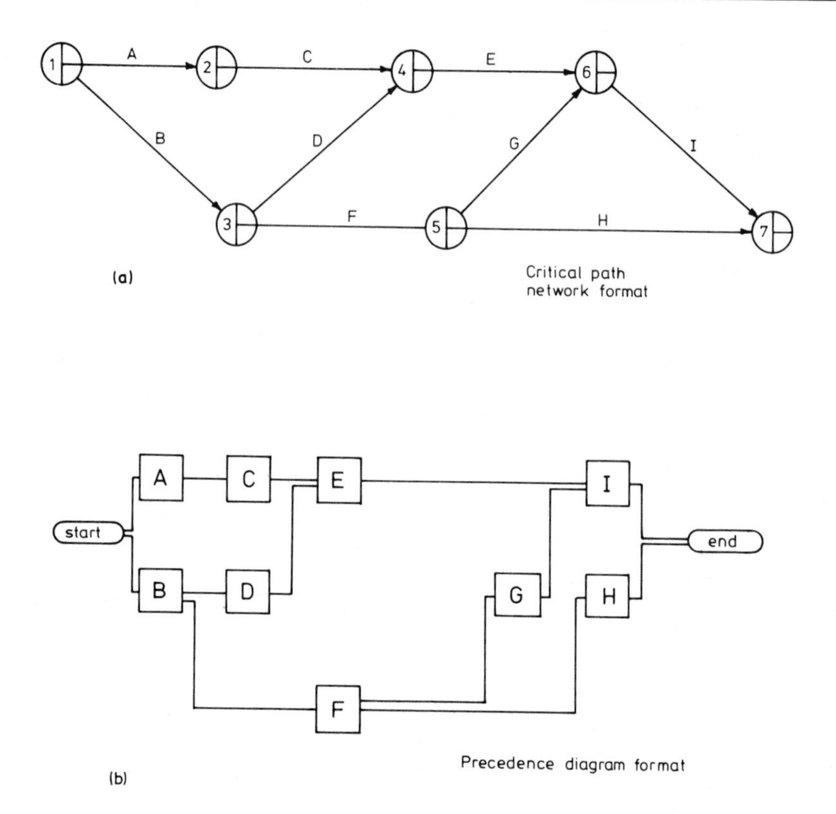

Fig. 5.53 (a) Critical path network format; (b) precedence diagram format (solution to Q.5.6(i))

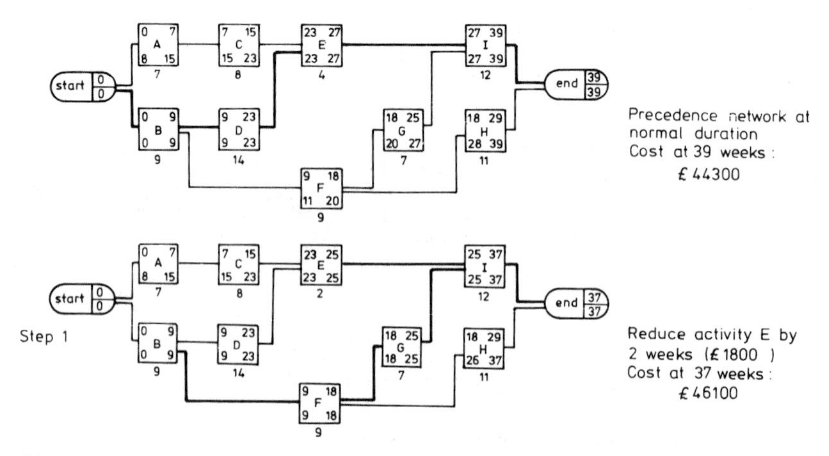

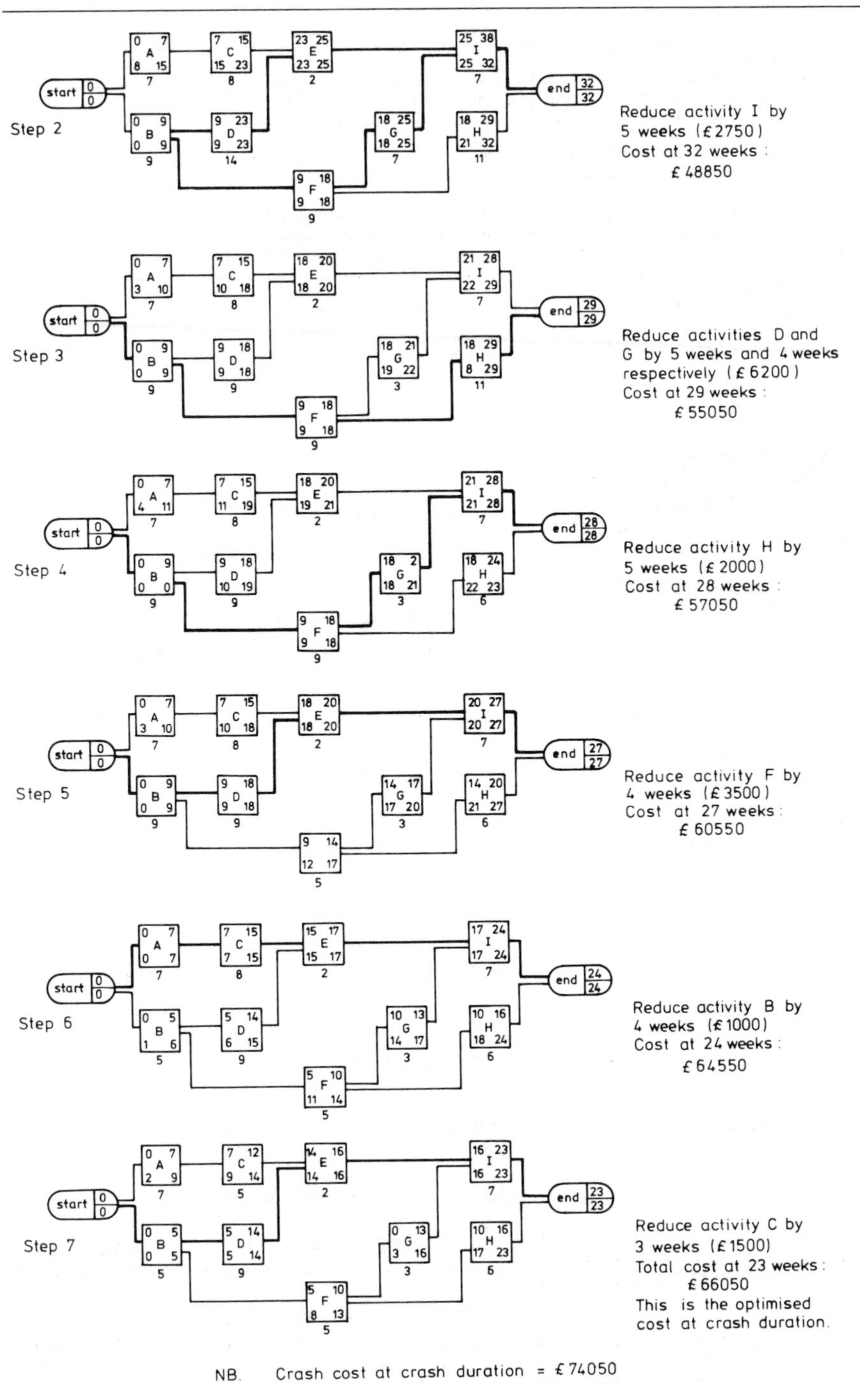

Fig. 5.54 Time–cost optimisation – solution to Q.5.6(ii)

Week No.								
Contract duration	39	37	32	29	28	27	24	23
Normal contract sum	44300	44300	44300	44300	44300	44300	44300	44300
Increase in direct cost		1800	4550	10750	12750	16250	20250	21750
Total of normal & direct costs	44300	46100	48850	55050	57050	60550	64550	66050
Indirect costs	18000	16000	11000	8000	7000	6000	300	2000
Contract costs according to duration	62300	62100	59850	63050	64050	66550	67550	68550

Optimised duration of 30 weeks when overheads are included
Least cost

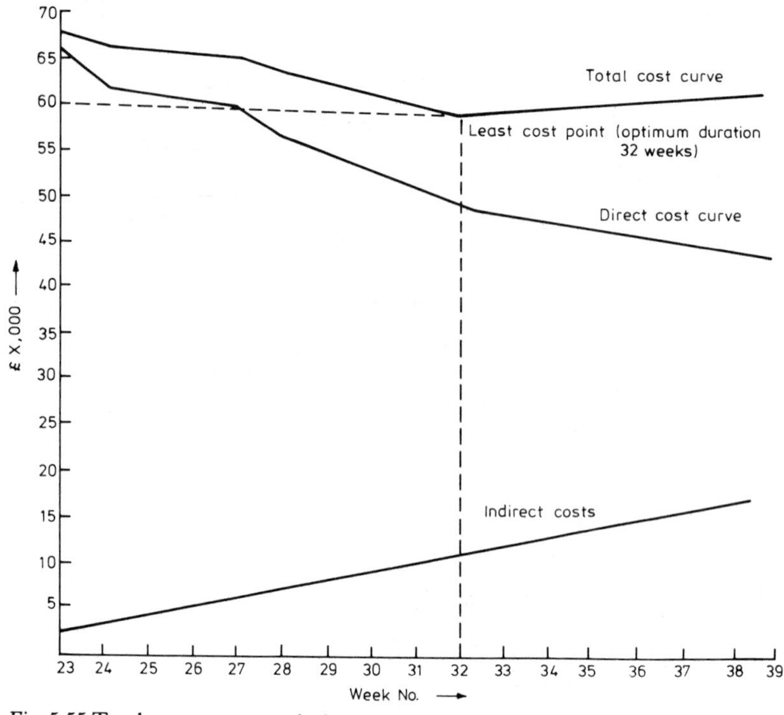

Fig. 5.55 Total cost curve – solution to Q.5.6(iii) and (iv)

Example of print out utilising a time/cost optimisation computer package.

ACTIVITY DESCRIPTION	PREC EVENT	SUCC EVENT	DURA	EARL START	LATE START	EARL FINISH	LATE FINISH	TOTAL FLOAT	FREE FLOAT
B	1	3	9.0	0.0	0.0	9.0	9.0	0.0	0.0
D	3	4	14.0	9.0	9.0	23.0	23.0	0.0	0.0
E	4	6	2.0	23.0	23.0	25.0	25.0	0.0	0.0
F	3	5	9.0	9.0	9.0	18.0	18.0	0.0	0.0
G	5	6	7.0	18.0	18.0	25.0	25.0	0.0	0.0
I	6	7	12.0	25.0	25.0	37.0	37.0	0.0	0.0
A	1	2	7.0	0.0	8.0	7.0	15.0	8.0	0.0
H	5	7	11.0	18.0	26.0	29.0	37.0	8.0	8.0
C	2	4	8.0	7.0	15.0	15.0	23.0	8.0	8.0

```
The project duration is 37.0 and the total cost is £46100.

There is more than 1 critical path in the network.
The next change may not alter the total duration.

Do you require further analysis using amended data?
Type Yes or No
:Yes
Is amended set of activities to be printed?
Type Yes or No
:No
The amended activity is:
I            6 - 7

The project duration is 32.0 and the total cost is £48850.

There is more than 1 critical path in the network.
The next change may not alter the total duration.

The amended activity is:
D            3 - 4

The project duration is 32.0 and the total cost is £51850.

Do you require further analysis using amended data?

:Yes

The amended activity is:
G            5 - 6

The project duration is 29.0 and the total cost is £55050.

Do you require further analysis using amended data?

:Yes

The amended activity is:
H            5 - 7

The project duration is 28.0 and the total cost is £57050.

Do you require a further analysis using amended data?

:Yes

The amended activity is:
F            3 - 5

The project duration is 27.0 and the total cost is £60550.

Do you require a further analysis using amended data?

:Yes

The amended activity is:
B            1 - 3

The project duration is 24.0 and the total cost is £64550.

Do you require further analysis using amended data?

:Yes

The amended activity is:
C            2 - 4

The project duration is 23.0 and the total cost is £66050.

Do you require further analysis using amended data?

:Yes

The last result cannot be improved since each critical activity has only
one duration/cost combination.
```

Fig. 5.56 Computer print out – solution to Q.5.6(v)

Q.5.7.

The following data list the optimistic, most likely and pessimistic durations of the activities in weeks of a construction project. Draw the network in both CPM and precedence diagram format and find the critical path assuming the most likely values for the durations.

Activity code	Activity i–j no.	Optimistic	Durations (weeks) most likely	Pessimistic
A	0–1	6	11	17
B	0–2	4	5	9
C	1–3	4	6	11
D	2–4	5	8	12
E	2–5	6	12	16
F	3–4	2	4	6
G	4–5	2	3	5
H	4–6	4	7	9
I	5–6	1	2	4

Calculate the standard deviation of the length of the critical path and find the probability of it taking longer than 30 weeks (see Figs. 5.57(a) and (b)).

Critical path

Code	i–j no.	a	m	b	$\sigma = \frac{b-a}{6}$
A	0–1	6	11	17	1.83
C	1–3	4	6	11	1.16
F	3–4	2	4	6	0.67
H	4–6	4	7	9	0.83

Standard deviation (σ) of network-ending E_T is

$$\sqrt{1.83^2 + 1.16^2 + 0.67^2 + 0.83^2} = 2.41$$

Calculation of expected elapsed time and earliest expected date for network-ending using formula $(\alpha + 4m + b)/6$,

$$\text{A}\ \frac{6 + (4 \times 11) + 17}{6} = 11.1 \qquad \text{C}\ \frac{4 + (4 \times 6) + 11}{6} = 6.5$$

$$\text{F}\ \frac{2 + (4 \times 4) + 6}{6} = 4 \qquad \text{H}\ \frac{4 + (4 \times 7) + 9}{6} = 6.8$$

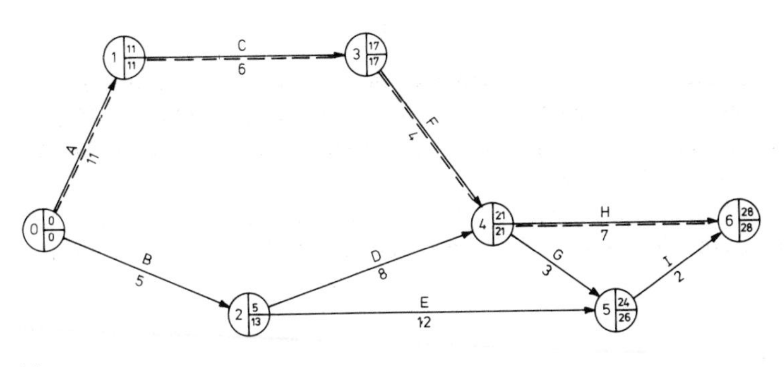

Fig. 5.57(a) Network format

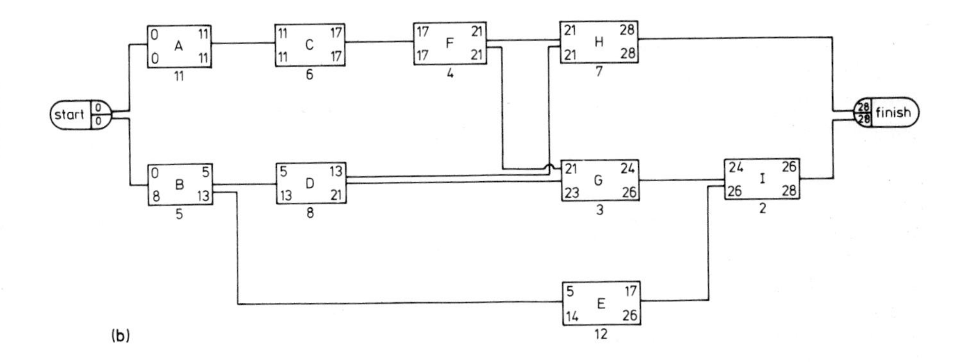

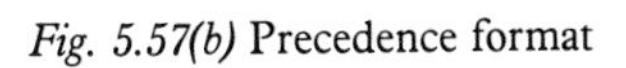

Fig. 5.57(b) Precedence format

Expected date for network-ending:

$$T_E = 11.1 + 6.5 + 4 + 6.8 = 28.4$$

$$\frac{30 - 28.4}{2.41} = +0.664$$

See statistical tables: area under normal curve = 0.75, say 75 per cent chance of finishing before week 30.

Q.5.8.

Table 5.13 lists the activities and durations for the construction of a single-span bridge shown in Fig. 5.58. The estimated overhead costs for the project are £120/day. Delivery times of materials can be reduced by purchase from alternative suppliers but costs will increase, as in Table 5.14.

Prepare a precedence diagram of the construction sequence and determine whether an acceleration of supplies will result in a reduction of the cost of the project (Fig. 5.59).

Table 5.13

Activity	*Duration (days)*
Clear site	7
Deliver reinforcing steel	32
Excavate abutment footings	5
Erect site offices	4
Shutter and concrete north footings	4
Shutter and concrete south footings	5
Place north approach fill	14
Place south approach fill	14
Deliver handrail	38
Erect steel deck beams	2
Shutter and concrete south abutments	10
Shutter and concrete north abutments	10
Deliver structural steel	65
Cure north abutment	7
Cure south abutment	7
Shutter and reinforced concrete deck slab	14
Pave approaches	6
Fix handrail	9
Cure deck slab	14
Place rip-rap	8
Align and trim channel	10
Clear site on completion	3

N.B. Activities are not listed in construction sequence

Table 5.14

Material/Component	*Cost (£/day)*
Handrail	40
Reinforcing steel	65
Structural steel	95

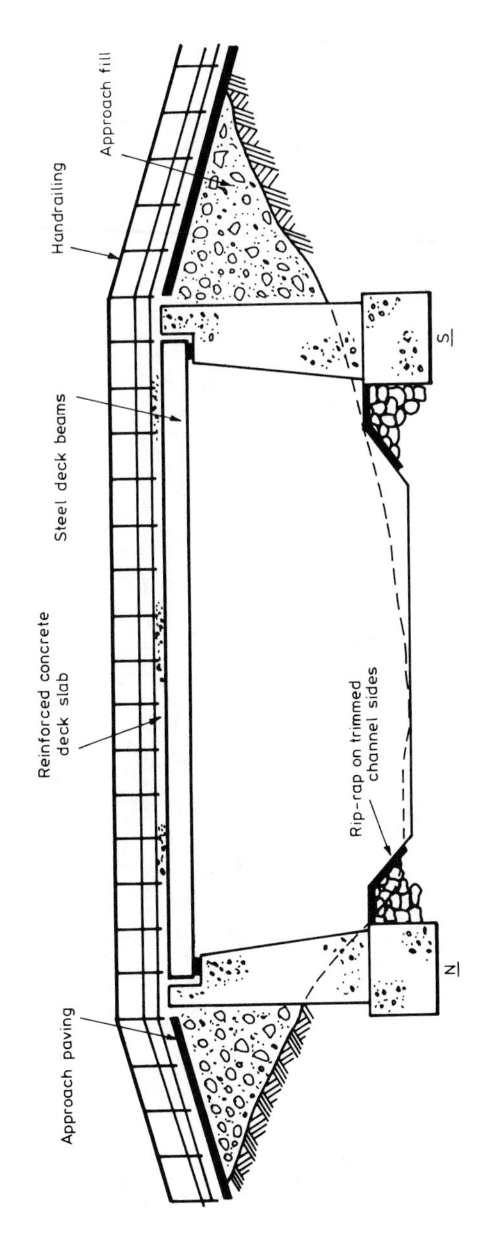

Fig. 5.58 Single-span bridge

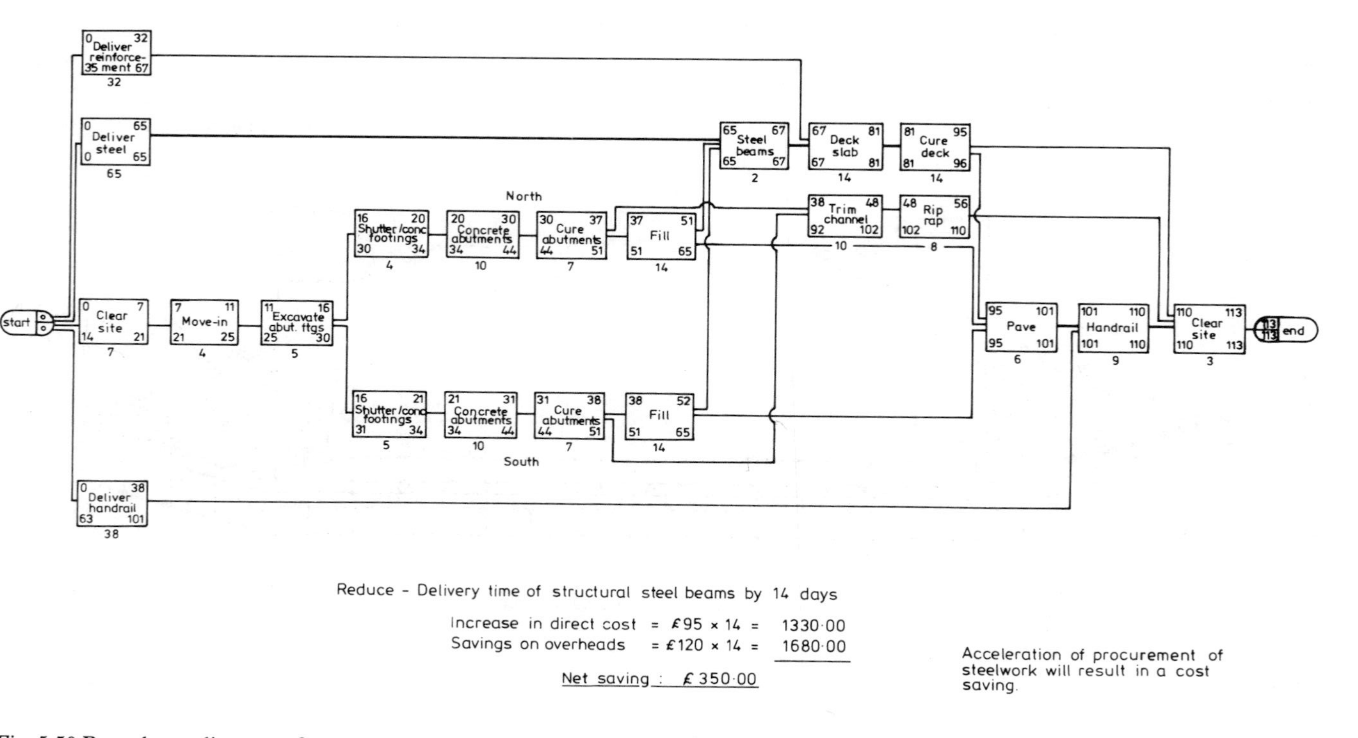

Fig. 5.59 Precedence diagram of construction sequence: single-span bridge – solution

Q.5.9.

Figure 5.60 illustrates a road realignment scheme, 8,000 m long, to replace the existing roadway and three-span masonry bridge. The masonry bridge is to be demolished on completion of the new works and the designer has decided to adopt a cast *in situ* reinforced concrete two-span bridge with central pier to carry the realigned road over the river. In order that the new bridge is constructed in the 'dry' a temporary channel and cofferdams will be constructed to divert the river.

By utilising the existing masonry bridge material can be transported to fill without waiting for construction of the new two-span bridge.

The minimum breakdown of operations required in order to integrate the bridge programme with the road contract is given below:

	Duration
Excavate diversion channel	8
Install cofferdams	4
Dewater old river course	1
Construct foundations, west abutment (bridge)	15
Construct foundations, east abutment	15
Construct pier foundation	12
Superstructure west abutment	20
Pier superstructure	16
East abutment superstructure	20
Excavate cutting chainage 8+00 to chainage 7+00 and fill embankment chainage 0.5 to chainage 3+00	18
Construct west deck span	24
Construct east deck span	24
Remove cofferdams	2
Backfill diversion channels	8
Excavate cutting chainage 6+00 to chainage 7+00 and fill embankment chainage 5+00 to chainage 6+20	10
Construct roadworks	15

Constraints

14-day delay after completion of dewatering before pier foundation can be commenced.

7-day delay after completion of dewatering before concreting to east abutment foundation can commence.

Prepare a network programme (either CPA or precedence) for this contract based on the above data and convert to bar-chart format.

The solution is given in Figs. 5.61 and 5.62.

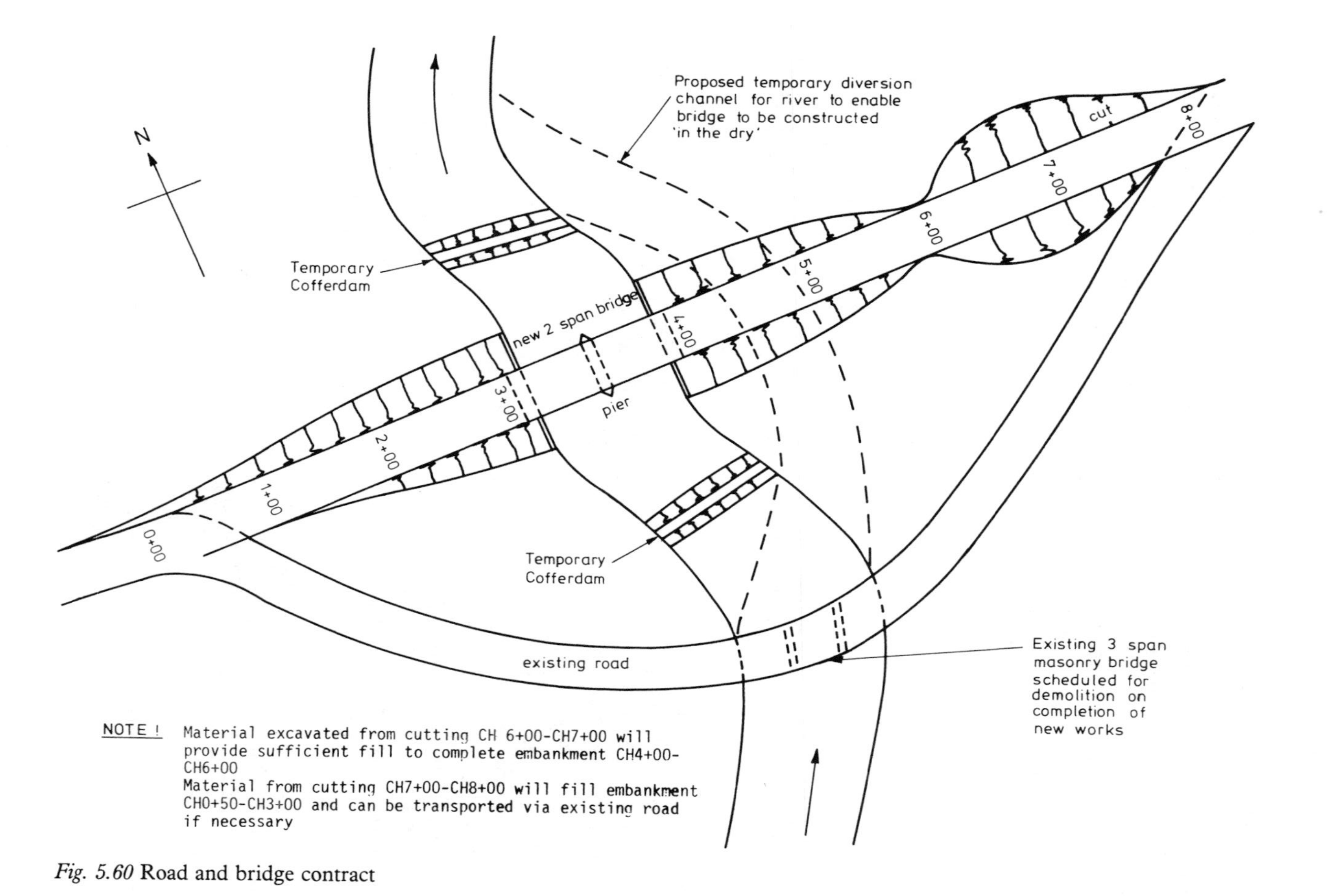

Fig. 5.60 Road and bridge contract

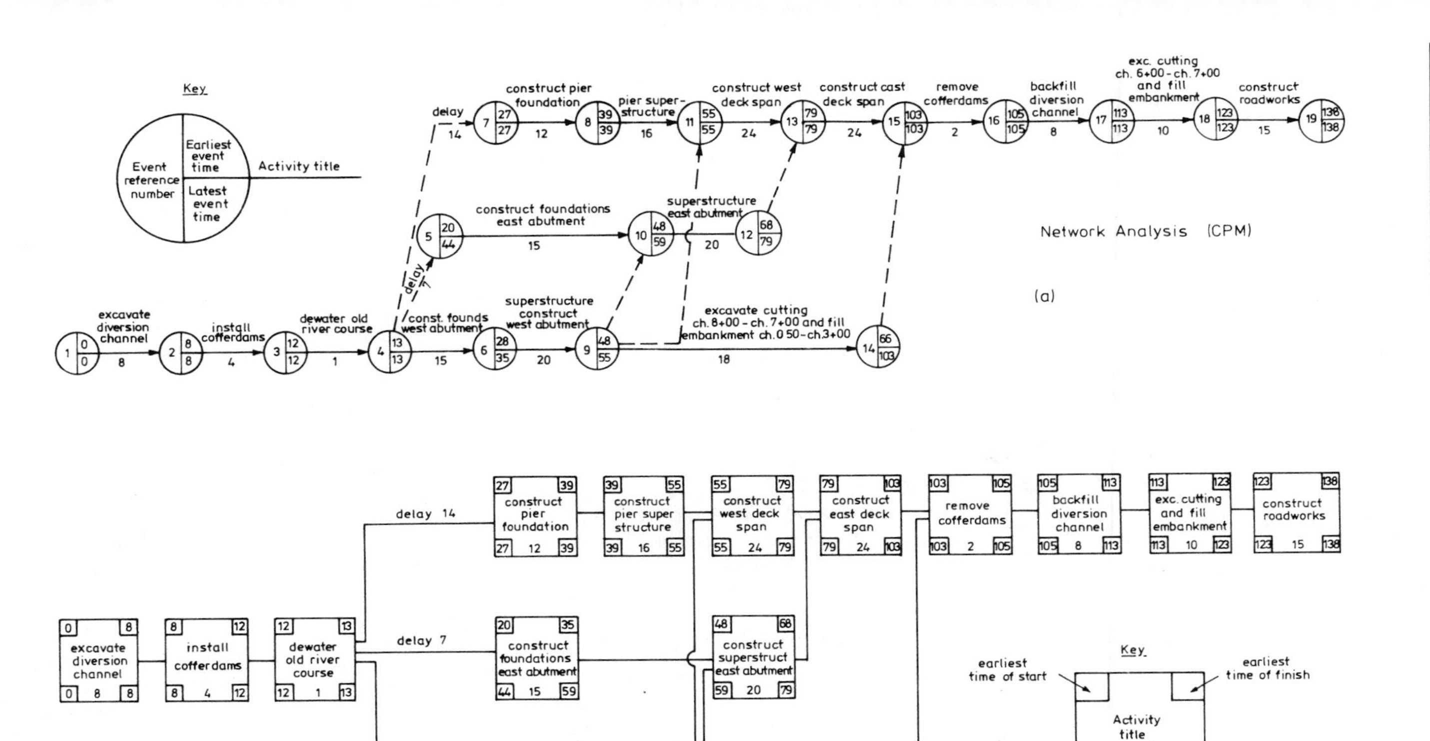

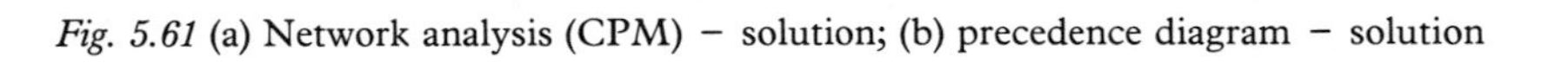

Fig. 5.61 (a) Network analysis (CPM) – solution; (b) precedence diagram – solution

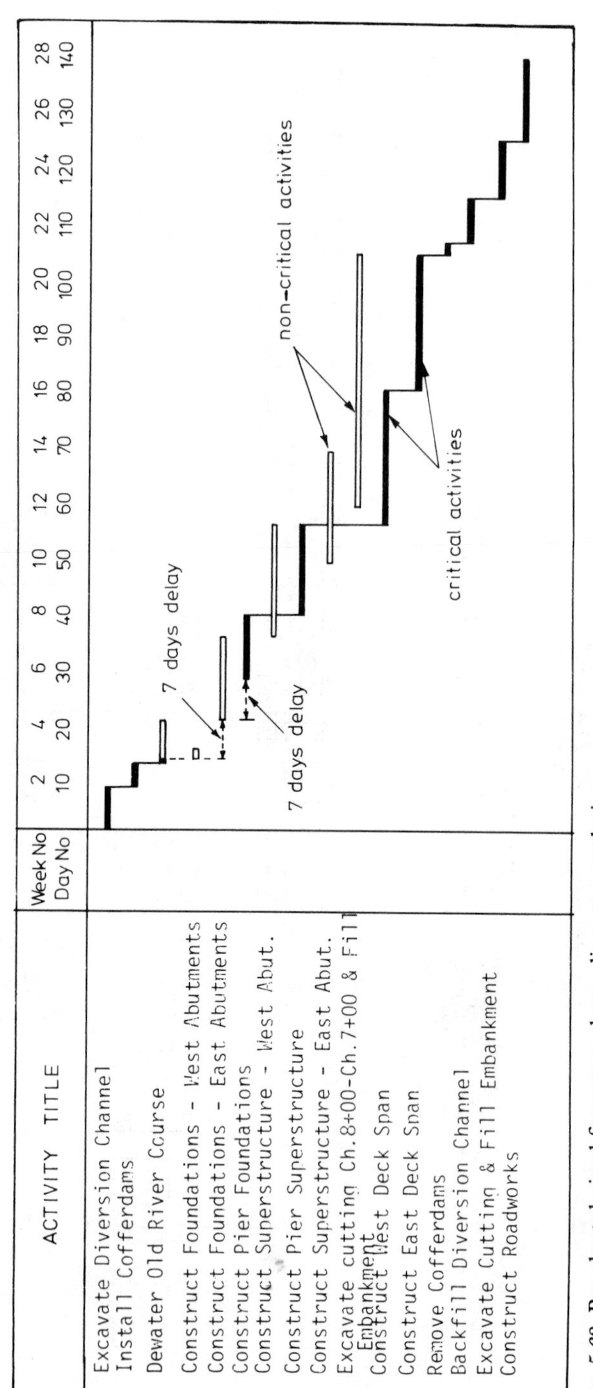

Fig. 5.62 Bar chart derived from precedence diagram – solution

Chapter 6

The line of balance technique (LOB)

6.1 Uses and advantages

The line of balance technique is a planning technique used to programme production of repetitive construction units. It was originally developed during the Second World War by the US Navy for planning and controlling mass production processes in factories. However, initial acceptance by industry in the UK has been generally slow. It was later due to much research effort by Philip Lumsden who contributed greatly to the development and successful application of this method for programming repetitive house building in the National Building Agency (NBA) that greater public interest in its effectiveness (as a powerful management tool) has been reawakened. It can be used to advantage on any construction project incorporating repetitive activities, i.e. large housing sites, blocks of flats, high-rise office blocks, pile-driving and capping, pre-cast concrete production and system building.

Advantages claimed for this technique are:

1. Combining the logic of network analysis with the principles of line of balance provides a very detailed picture of any repetitive project.
2. Reduces the amount of network planning and scheduling since only one network is used for each type of unit.
3. Provides a simple and effective tool for programming the ordering and delivery of materials and subsequently their incorporation into the construction.
4. By monitoring progress, individual jobs which are falling behind schedule can be easily identified and early corrective action taken.

There are two main approaches to the formulation of line of balance programmes.

6.2 Method 1: Using optimum gang sizes – after NBA

A hypothetical housing contract will be considered in order to illustrate the principles of this method.

The contract is for forty houses with a planned hand-over rate of three per week. The logic network (Fig. 6.1) and man-hours per activity (Table 6.1) are given.

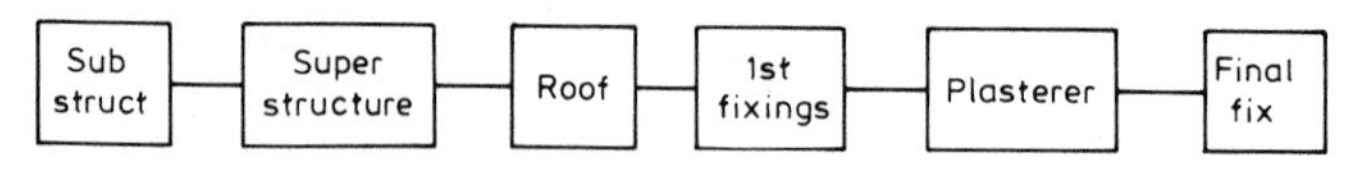

Fig. 6.1 Simplified network

Table 6.1

Activity	*Man-hours per activity (M)*
Substructure	100
Superstructure	200
Roof	60
1st fixings	175
Plasterer	120
Final fixing	350

Procedure

(a) Optimum resources are put against each activity.

(b) Produce a table and include man-hours per house (M) and the men per house (Q) (Table 6.2).

(c) Calculate required gang sizes to give approximate rate of three per week, e.g. gang size $G = 3M/40$.

(d) Convert the fractional numbers produced into whole numbers which must be a multiple of number of men per house. This may alter the hand-over rate slightly.

(e) The time allowed per activity for the construction of one house is now calculated as follows:

$$T \text{ days} = \frac{M}{8 \times Q}$$

(f) Calculate the horizontal component of the activity lines. This gives the slope. It is found by

$$S = \frac{(N-1)5}{U} \text{ days}$$

where N is the number of houses in the contract. Assume 5-day working week.

For practical consideration, time buffers should be allowed between the completion of each operation and the start of the succeeding operations.

It may be necessary to change the manning of an operation part of the way through the contract. For example, if certain tradesmen are in short supply in a particular area, larger buffers may be included to provide flexibility. If the rate of hand-over achieved with a particular gang varies too much, the gang sizes may be altered and the schedule replotted.

Table 6.2

Activity	Man hours per house	Gang size at 3 houses per 40-hour week	Men per house	Actual gang size	Actual rate per week	Time in days for 1 house	Horizontal component of activity lines
	M	$G = \frac{3M}{40}$	Q	g	$U = \frac{g}{G} \times 3$	$T = \frac{M}{8 \times Q}$	$S = \frac{(n-1)5}{U}$
Substructure	100	7.5	4	8	3.19	3.12	61.1
Superstructure	200	15.0	8	16	3.19	3.12	61.1
Roof	60	4.5	4	4	2.7	1.9	72.2
1st fixings	175	13.1	12	12	2.7	1.8	72.2
Plasterer	120	9.0	4	8	2.7	3.75	72.2
Final fixings	350	26.2	12	24	2.7	3.6	72.2

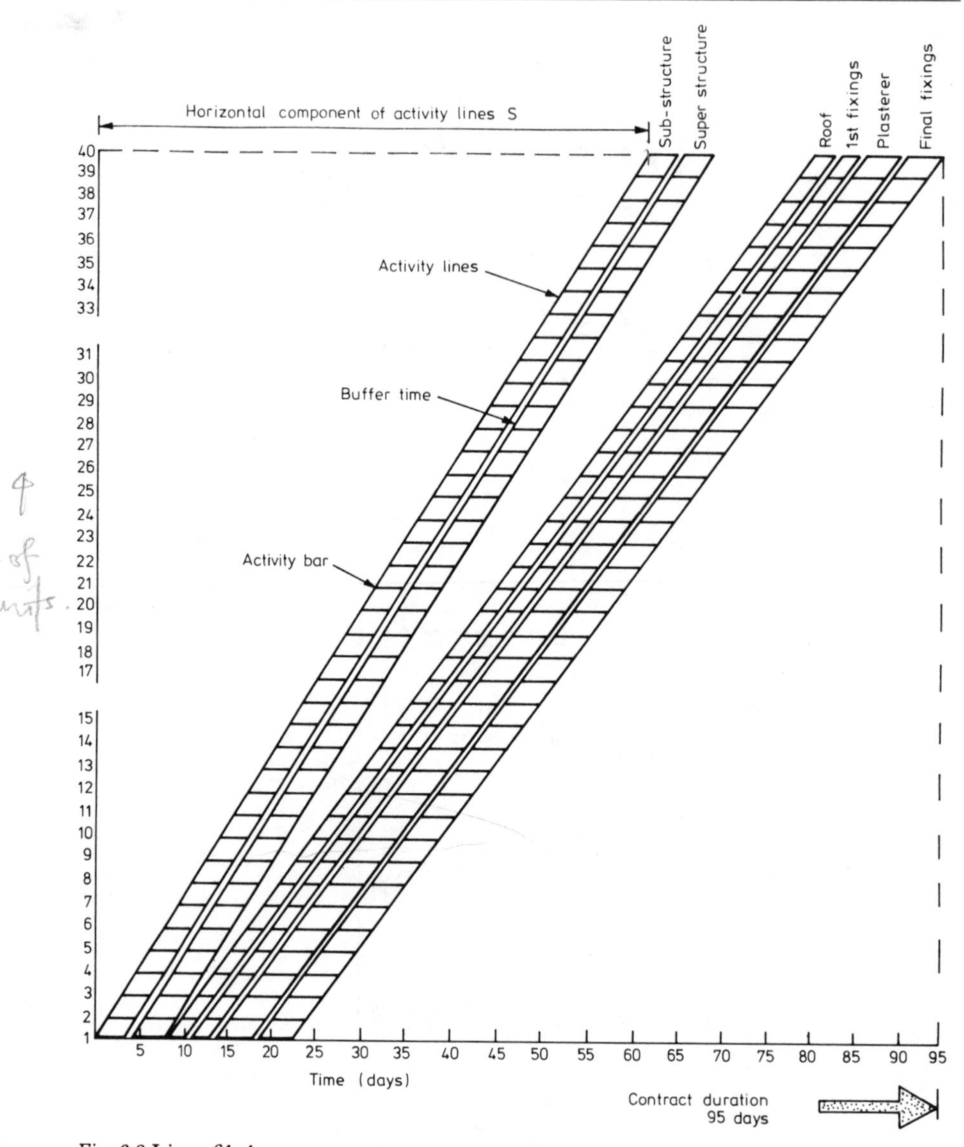

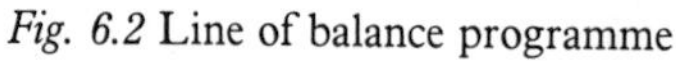
Fig. 6.2 Line of balance programme

The line of balance diagram (Fig. 6.2) for this simplified contract is easy to read. If a more complex network is used to display the construction logic of one housing unit in greater detail things will become slightly more difficult. The network in Fig. 6.3 is similar to the previous network, but now highlights activities that can be carried out concurrently. A table is produced as before (Table 6.3), but the line of balance diagram (Fig. 6.4) will differ in that certain activity lines will appear to be running through others. This is perfectly acceptable providing it is in accordance with the logic of the precedence or CPM diagram.

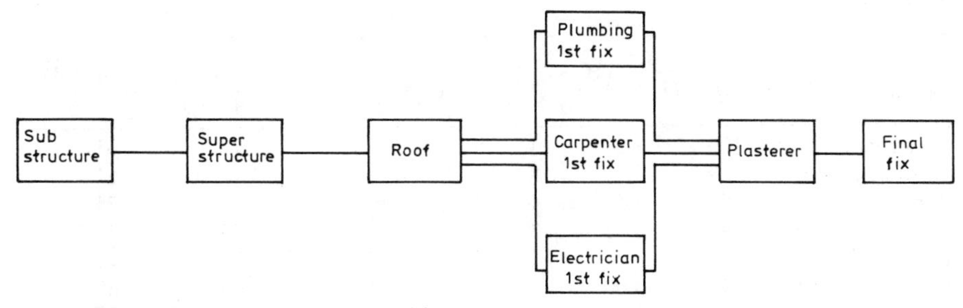

Fig. 6.3 Network with concurrent activities

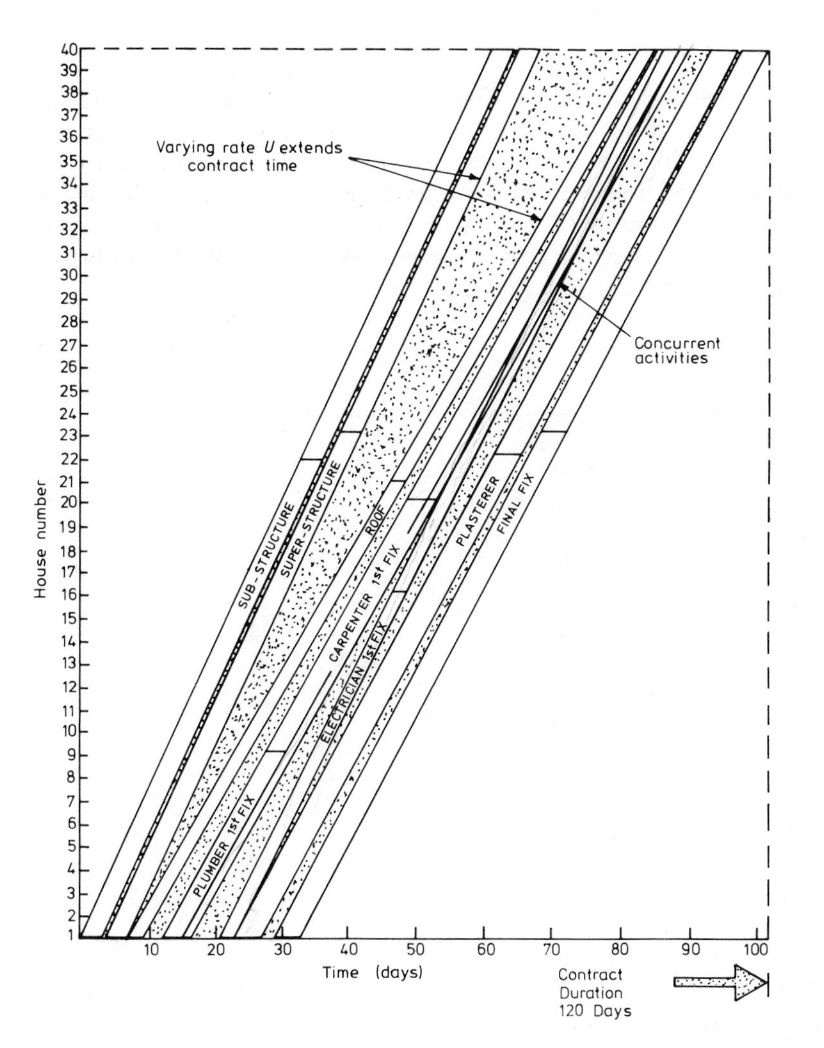

Fig. 6.4 Line of balance programme

Table 6.3

Activity	M	$G = \frac{3M}{40}$	Q	g	$U = \frac{g}{G} \times 3$	$T = \frac{M}{8 \times Q}$	$S = \frac{(n-1)5}{U}$
Substructure	100	7.5	4	8	3.19	3.12	61
Superstructure	200	15.0	8	16	3.19	3.12	61
Roof	60	4.5	4	4	2.6	1.87	75
Carpenter 1st fix	90	6.75	3	6	2.66	3.75	73.3
Plumber 1st fix	45	3.37	3	3	2.67	1.88	73
Electrician 1st fix	40	3.00	3	3	3.00	1.66	65
Plasterer	120	9.00	4	8	2.66	3.75	73.3
Final fix	350	26.25	12	24	2.74	3.64	71

A realistic example will now be considered. Assume the contract is for 100 semi-detached houses (50 blocks) of similar design and area. The planned hand-over rate is five blocks per week. From the drawings and specification the planner has derived a precedence diagram and man-hour requirements as shown in Fig. 6.5 and Table 6.4. A table is completed as before (Table 6.5). Because of large variations in initially planned hand-over rates for certain activities, gang sizes have been altered to achieve a more uniform rate of working (Fig. 6.6.).

Table 6.4

Activity	*Man-hours per house*
Substructure	110
Bwk to 1st floor	70
1st flr joists	10
Bwk 3rd, 4th lifts	130
Roof carcass	40
Roof tiling	20
Carpenter 1st fix	50
Glazing	20
Flashings	6
Partitions	30
Gas installation	30
Plumbing 1st fix	20
Electrician 1st fix	25
Plasterer	120
Carpenter 2nd fix	200
Plumbing 2nd fix	90
Electrician 2nd fix	20
Floor finish	15
External paintw'k	40
Clean out	3

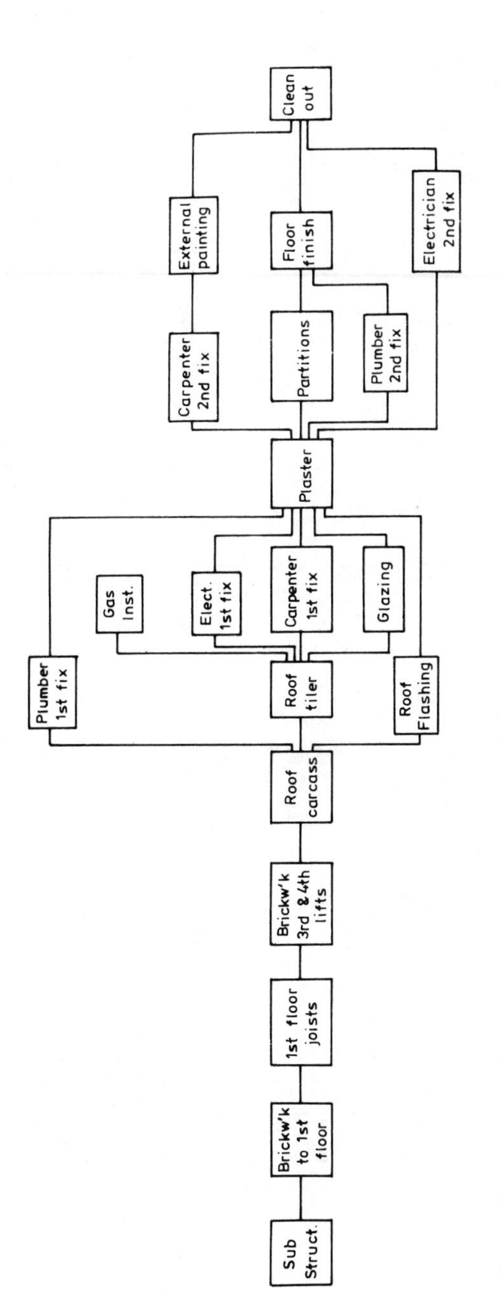

Fig. 6.5 Precedence/logic diagram for one house unit

Table 6.5

Activity description	*Man-hours per house*	*Gang size at 5 houses per 40 hr week*	*Men per house*	*Actual gang size*	*Actual rate per week*	*Time in days for one house*	*Horiz. comp. of activity lines*
	M	$G = \frac{5M}{40}$	Q	g	$U = \frac{g}{G} \times 5$	$T = \frac{M}{8 \times Q}$	$S = \frac{(n-1)5}{U}$
Substructure	110	13.75	5	15	5.45	2.75	45
Bwk to first floor	70	8.75	4	8	4.57	2.18	53.6
1st floor joists	10	1.25	2	1 & 2	4 & 6	0.50	30 & 20
Bwk 3rd and 4th lifts	130	16.25	4	16	4.9	4.06	50
Roof carcass	40	5.00	2	4	4	2.50	61.25
Roof tiling	20	2.50	2	2	4	1.25	61.25
Carpentry 1st fix	50	6.25	2	6	4.8	3.13	51
Glazing	20	2.50	2	2	4	1.25	61.25
Flashings	8	1.00	1	1	5	1.00	49
Partitions	30	3.75	2	4	5.33	1.87	46
Gas installation	30	3.75	2	4	5.33	1.87	46
Plumbing 1st fix	20	2.50	2	2	4	1.25	61.25
Electrician 1st fix	25	3.13	2	2 & 4	3 & 6	1.56	40 & 20
Plasterer	120	15.00	4	16	5.3	3.75	46.20
Carpentry 2nd fix	200	25.00	3	24	4.8	8.33	51
Plumbing 2nd fix	90	11.25	2	12	5.33	5.60	46
Electrician 2nd fix	20	2.50	2	4 & 2	3 & 6	1.25	40 & 20
Floor finish	15	1.88	2	2	5.33	0.94	46
External painting	40	5.00	3	6	6	1.48	41
Clean out	3	0.38	1	1	5	0.37	49

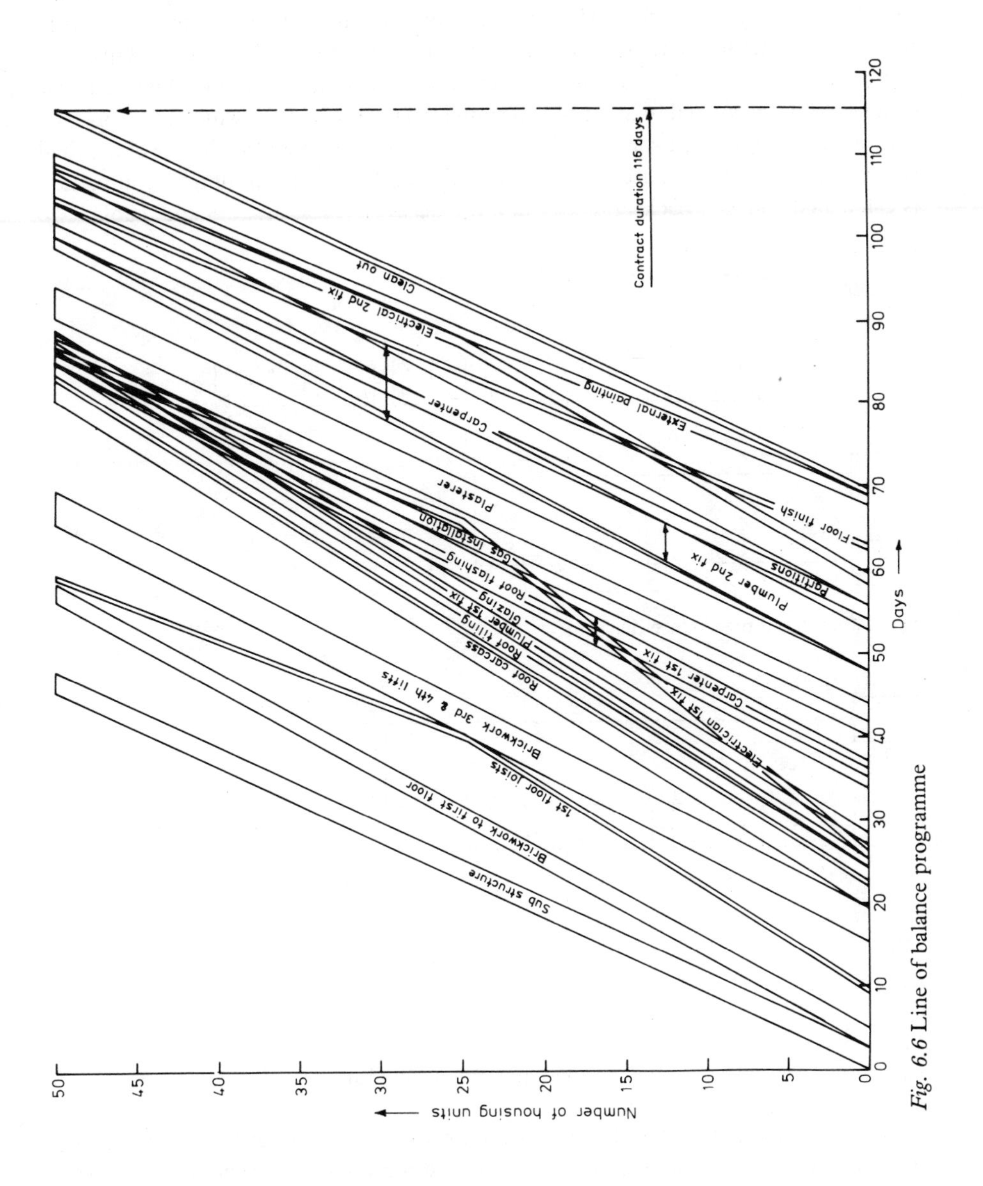

Fig. 6.6 Line of balance programme

6.3 Method 2: Natural rhythm or minimum loss working

In this method the line of balance is geared to the delivery of completed units, therefore the logic network must be analysed in relation to its completion time. As in the previous method, one network for a typical unit is used. The rate of delivery of completed components has a major influence on unit completion time. The lead time of activities is calculated in order to determine the timing of work which must precede completion. The earliest completion date is given by the earliest finish time of the final activity. This is shown to occur at time zero. Each activity time to contract completion is calculated working backwards through the network (Figs. 6.7, 6.8 and Table 6.6).

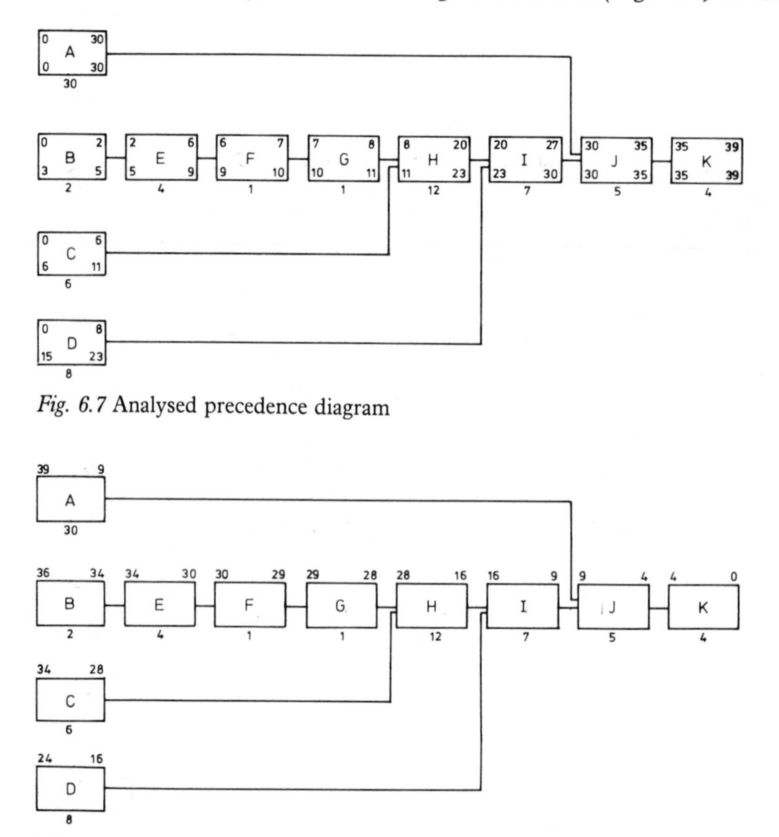

Fig. 6.7 Analysed precedence diagram

Fig. 6.8 Precedence diagram with activity lead times included

Commencement of activity K with a duration of 4 days must lead the contract completion date by 4 days. Similarly activity J has a duration of 5 days and its earliest start must precede contract completion by $5+4 = 9$ days.

By working through the network in this manner the delivery lead times can be calculated. For example the delivery of cabinets must be in hand 39 days before the house unit is due to be finished. Plumbing fittings must be ordered 24 days before planned completion and roof trusses 34 days.

Table 6.6

Activity code	*Activity description*
A	Order and deliver cabinets
B	Substructure
C	Order and deliver roof trusses
D	Order and deliver plumbing fixtures
E	Footings
F	Joists
G	Roof
H	1st fixings
I	Plaster
J	2nd fixings
K	Painting

Assuming a contract comprising 50 housing units with a planned hand-over rate of 5 per week and completion of unit 1 at week 20, the completion schedule would appear as in Fig. 6.9. From the network, ordering and delivery of components must start 39 days before unit completion. We can add this line to the programme (Fig. 6.10).

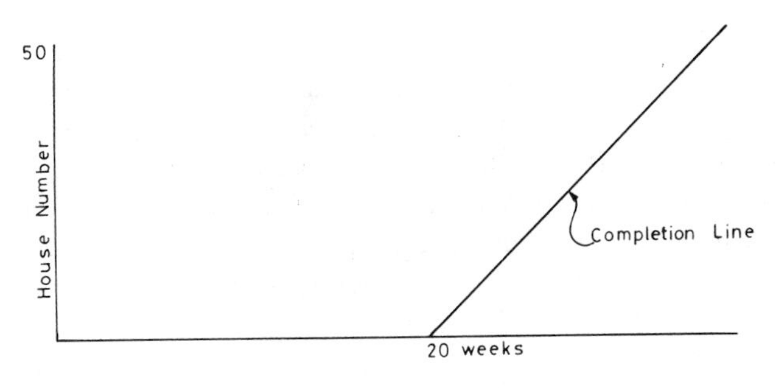

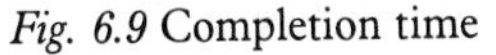
Fig. 6.9 Completion time

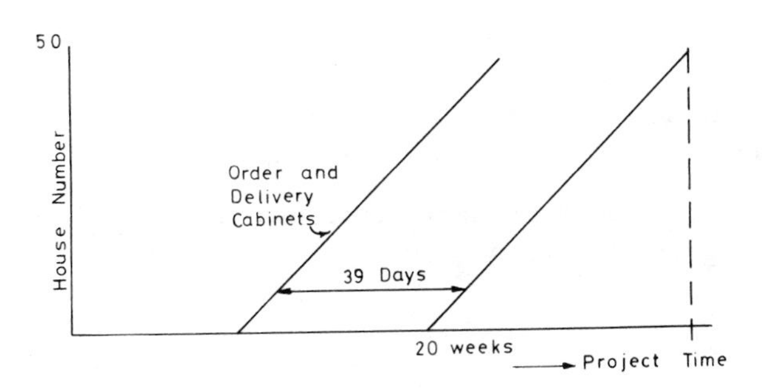

Fig. 6.10 Contract durations

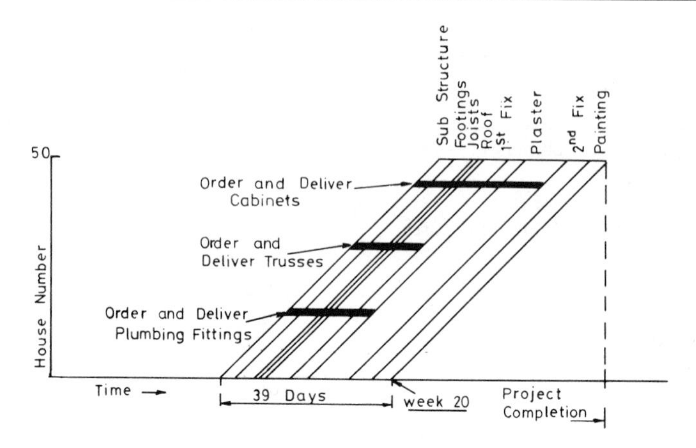

Fig. 6.11 Completed LOB

Each activity can now be drawn in parallel to the completion line (all same rate of working) using the lead times from the network. The planned completion date of each activity can be read directly from the line of balance diagram (Fig. 6.11).

Gang number

It is necessary to determine the number of gangs required for each activity in order to maintain the planned rate of progress. (In Fig. 6.12 this is one housing unit per week.) This is easily obtained from examination of the line of balance programme. The histogram in Fig. 6.12 shows the build-up and run-down of labour for a project. Assuming that the gang size for activity A is four and from the diagram it can be seen that a total of three gangs are required, the peak labour requirement will therefore be $4 \times 3 = 12$.

The total man-weeks for each activity can be determined by multiplying the peak labour requirement by the equivalent base line of the histogram. In Fig. 6.12 this is $12 \times 10 = 120$ man-weeks. The equivalent base of the histogram is the time from the start of labour build-up to the end of the period where the labour force has been constant and is drawn as follows:

- Set up a labour resource scale on the vertical axis.
- Draw the labour resource histograms for each activity.
- Drop a perpendicular from commencement of labour run-down.

The equivalent base histogram can now be scaled off the diagram. It is possible to calculate this base line by using the following formula:

$$Q \frac{-\text{ No. of gangs}}{M} + A$$

where Q is the total number of times an activity is repeated, M the planned hand-over rate (units/week) and A the activity duration.

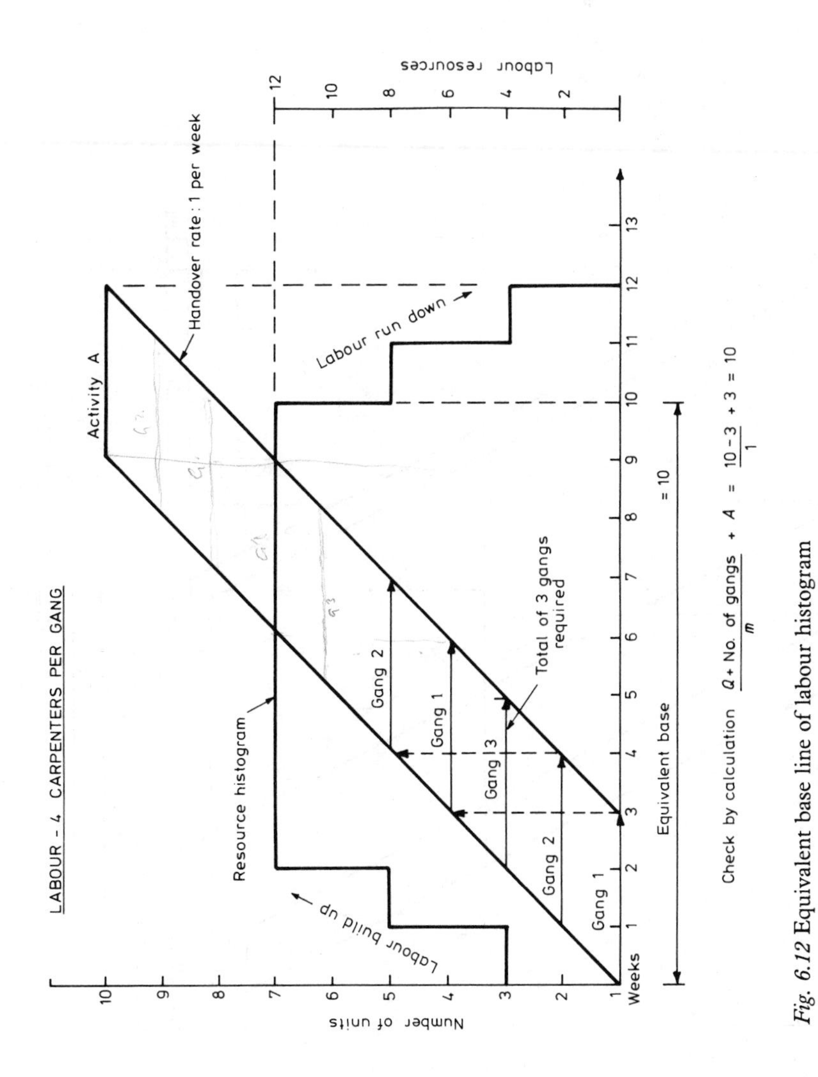

Fig. 6.12 Equivalent base line of labour histogram

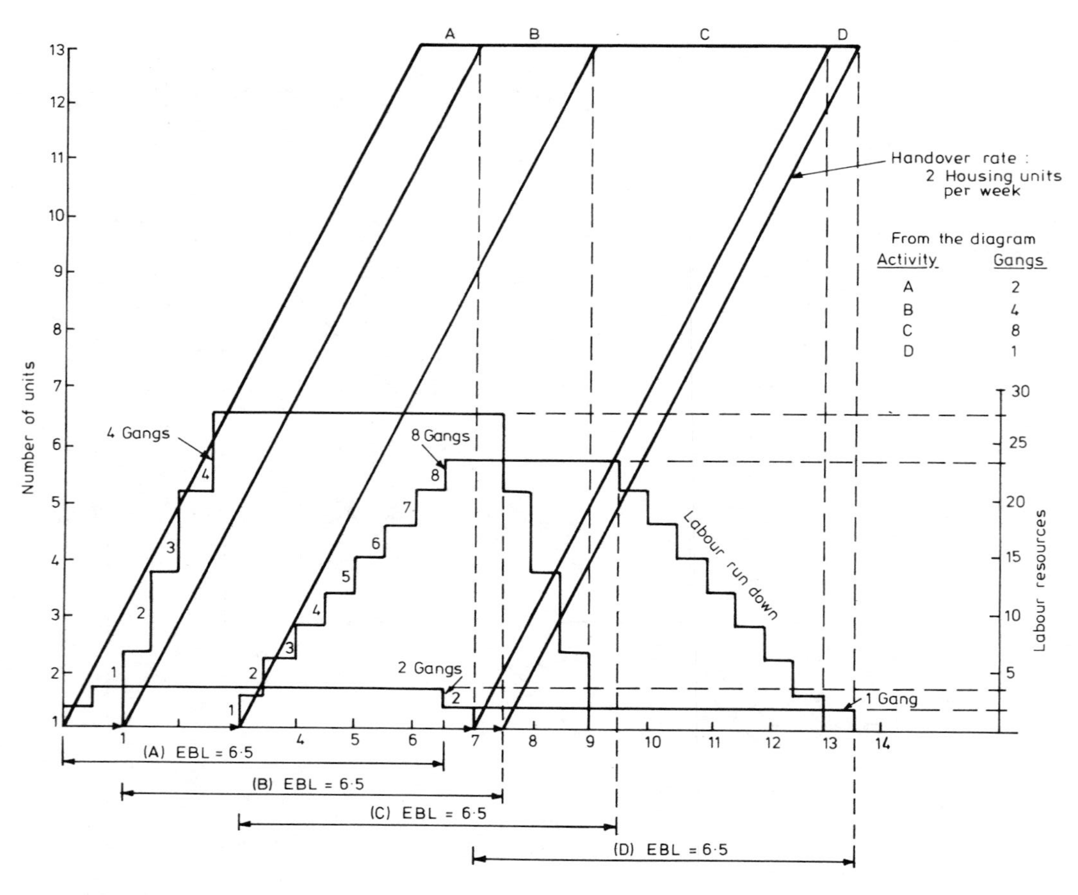

Fig. 6.13 Total labour requirements

The line of balance programme in Fig. 6.13 has been analysed using both methods. For example, for activity C the equivalent base line is

$$\frac{13-8}{2} + 4 = 6.5 \text{ weeks}$$

The total labour requirement for the line of balance programme in Fig. 6.13 can now be determined assuming the following gang sizes:

Activity	Gang size (assumed)	No. of gangs (from diagram)
A	2	2
B	7	4
C	3	8
D	2	1

Total man-weeks are obtained from Table 6.7.

Table 6.7

Activity	*1* *Duration (weeks)*	*2* *No. of men per gang*	*3* *Activity man-weeks (1 × 2)*	*4* *No. of gangs*	*5* *Maximum no. of men (2 × 4)*	*6* *Equivalent base of histogram*	*7* *Man-weeks per activity (5 × 6)*
	1	2	2	2	4	6.5	26
	2	7	14	4	28	6.5	182
	4	3	12	8	24	6.5	156
	0.5	2	1	1	2	6.5	13
						Total no. of man weeks	377

Balancing gangs

The importance of balancing gangs to achieve optimum productivity has been discussed in Chapter 3 dealing with bar charts. The same principles can be used here. This provides a great deal of flexibility in obtaining a figure corresponding to the natural rhythm of the majority of operations.

There are two basic methods as outlined by P. Lumsden:

1. By taking the work content of an activity expressed in man-hours and by adjusting the gange size, achieve an activity duration which is equal to or a multiple of the greatest common activity duration of the construction network.
2. By grouping activities which are the responsibility of particular trades thereby increasing the duration of the gang activity such that a low and hence flexible natural rhythm is achieved.

A combination of both methods is usual for best results in practice (Fig. 6.14).

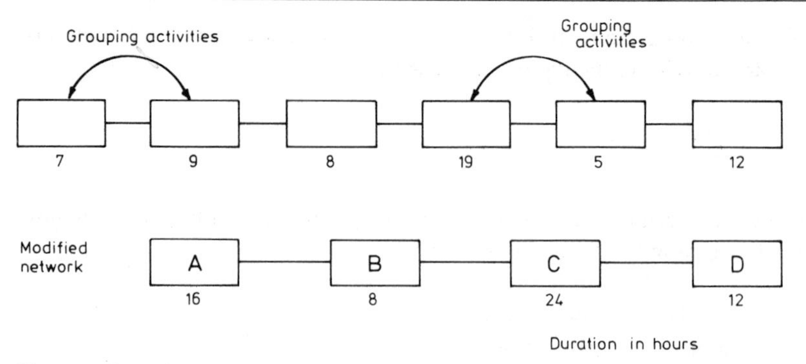

Fig. 6.14 Grouping activities

Example 1

Assuming a contract for six housing units, a line of balance diagram can be prepared with each activity drawn at its natural rhythm (Fig. 6.15).

Balancing of all activities to a common repetition is only possible if their durations are divisible by a common factor. The line of balance diagram in Fig. 6.15 is capable of being balanced to a rate of 2 units/day.

Natural rhythm	A	0.5 unit/day × 4 gangs gives 2 units/day
	B	1.0 unit/day × 2 gangs gives 2 units/day
	C	0.33 unit/day × 6 gangs gives 2 units/day
	D	0.66 unit/day × 3 gangs gives 2 units/day

This is shown in Fig. 6.16.

Example 2

Consider the simplified network in Fig. 6.17 for the construction of ten identical garages. The client requires a hand-over rate of at least six per week. Activity durations are calculated on the resource allocations given in Table 6.8. The total man-hours therefore for 10 garages would be 1,560 if all labour was fully employed at all times. However, examination of the line of balance diagram (Fig. 6.18) based on a hand-over rate of six garages per week clearly highlights non-productive time.

It can be seen that:

(a) Resources allocated to foundations on the first garage completes work in 4 hours. They are not compelled by the programme to start work on the second garage until 7 hours have elapsed, i.e. 3 hours' 'waiting' time.

(b) Resources allocated to brickwork on the first garage will complete work after 20 hours have elapsed, i.e. they will not be released until after the brickwork on the second and third garage should commence. Two further gangs of bricklayers must be employed on the second and third garages if the programme is to be met and there will be a delay of 4 hours between the time the gang allocated to brickwork on the first garage can start work on the fourth garage (unless foundations are completed ahead of programme).

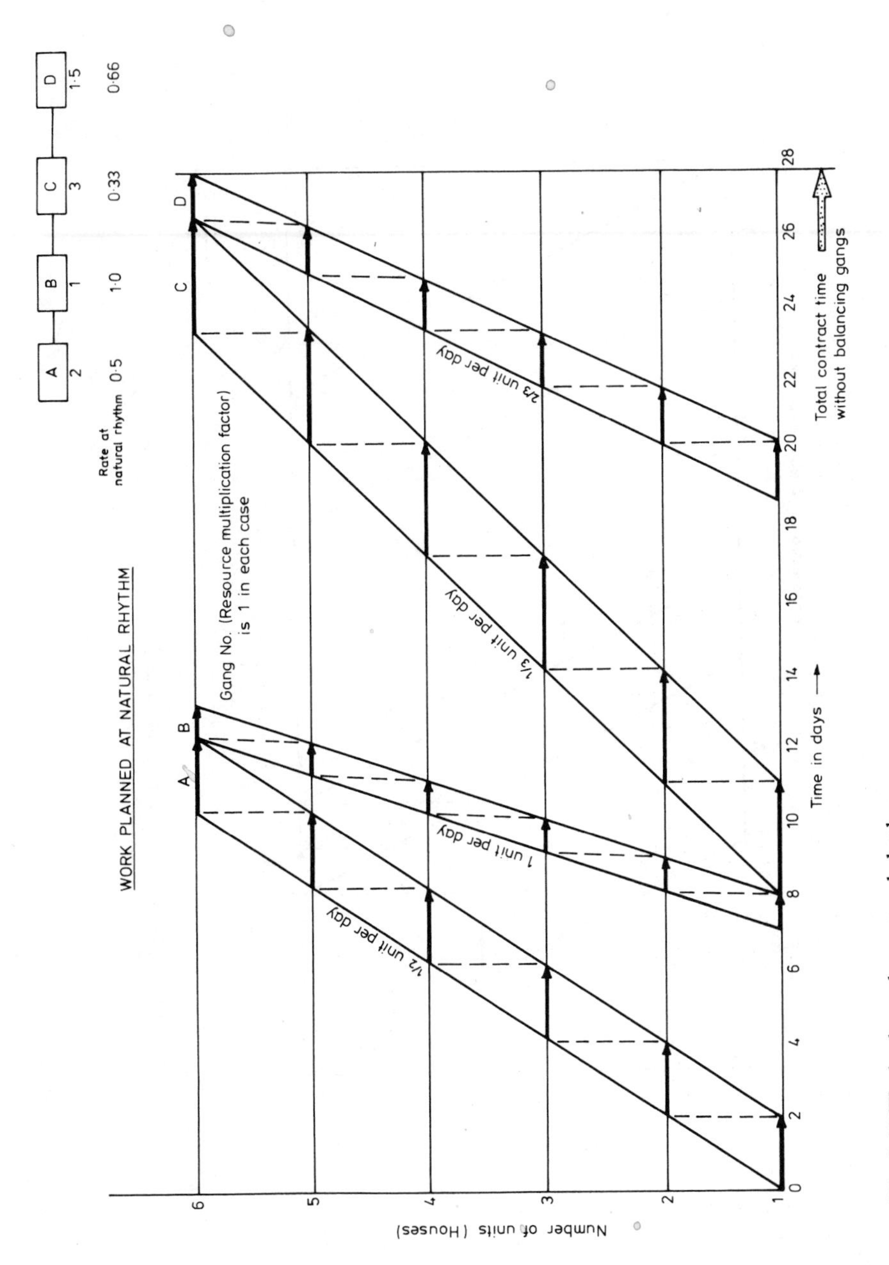

Fig. 6.15 Work planned at natural rhythm

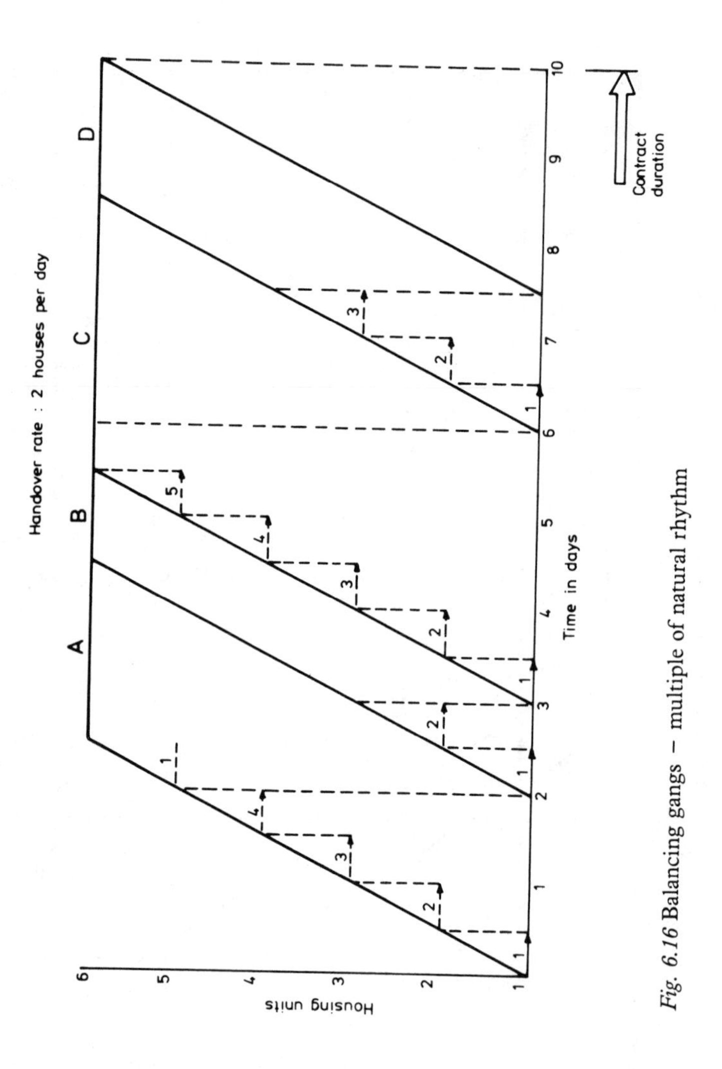

Fig. 6.16 Balancing gangs – multiple of natural rhythm

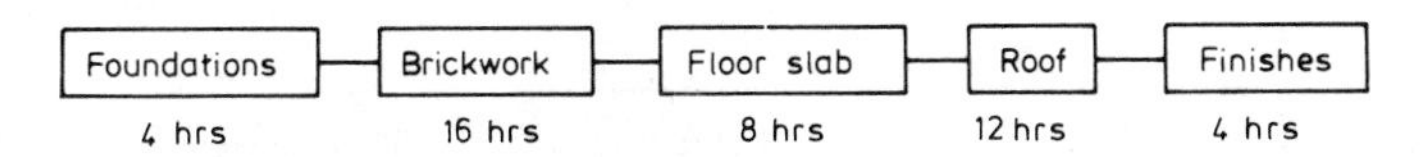

Fig. 6.17 Network for one garage

Table 6.8 Activity man-hours

Activity	*Duration (hr)*	*Resources (labour)*	*Activity man-hours*
Foundations	4	3	12
Brickwork	16	5	80
Floor slab	8	2	16
Roof	12	3	36
Finishes	4	3	12
		Man-hours for one garage	156

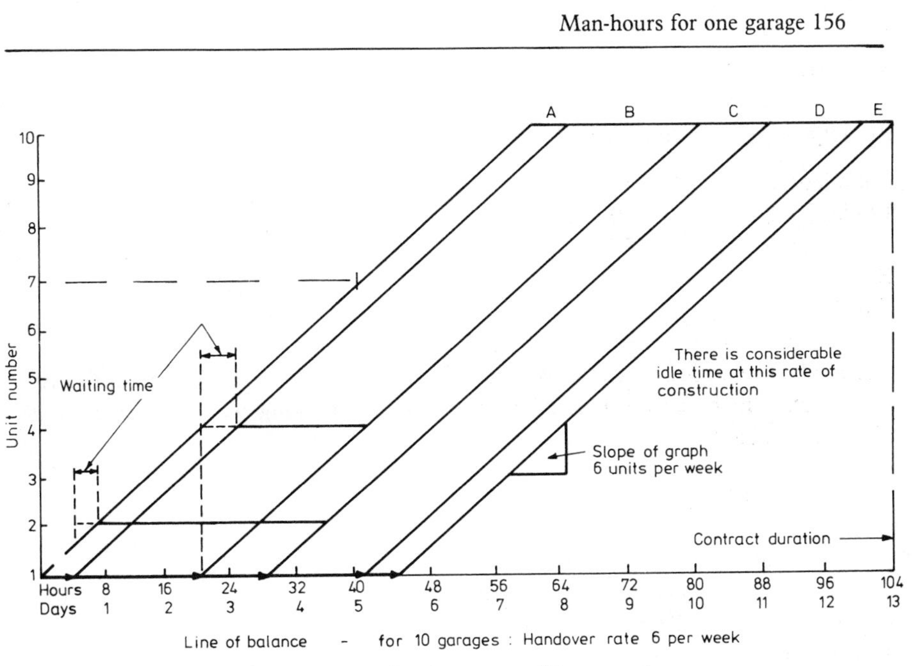

Fig. 6.18 Line of balance for 10 garages: handover rate of 6 per week

(c) Similarly, other activities will require extra allocations of resources. Floor slab: one additional gang; roof: one additional gang; and waiting time will be involved.

Hence from the line of balance diagram (Fig. 6.18), Table 6.9 is produced highlighting the number of gangs per activity (resource multiplication factor) and waiting times. Planned duration = 16, waiting time = 4. Total elapsed time for one activity is 20 hours (i.e. time from start of one unit to start of next), utilisation is

Table 6.9

Activity	*Duration (hrs)*	*Resources (labour)*	*Number of gangs required*	*Waiting time (hrs)*
Foundation	4	3	1	3
Brickwork	16	5	3	4
Floor slab	8	2	2	6
Roof	12	3	2	1
Finishes	4	3	1	2

N.B. Labour utilisation is very poor, e.g. considering brickwork operations.

$$\frac{16}{20} = 80 \text{ per cent}$$

Similarly:

Activity	Utilisation (%)
Foundations	57
Brickwork	80
Floor slab	57
Roof	92
Finishes	67

A method of improving the utilisation of resources is to calculate each activity's natural rhythm based on activity durations and a working week of, for example, 40 hours (Table 6.10).

Table 6.10

Activity	*Duration (hrs)*	*Natural rhythm (per 40 hr week)*
Foundations	4	10
Brickwork	16	2.5
Floor slab	8	5
Roof	12	3.33
Finishes	4	10

The next step is to calculate the number of gangs required to build at a multiple of each activity's natural rhythm. From Table 6.11 it can be seen that the modified common building rate is ten.

The modified line of balance diagram is shown in Fig. 6.19.

Table 6.11

Activity	*Duration (hrs)*	*Natural rhythm (per 40 hr week)*	*Common building rate*	*No. of gangs required*
Foundations	4	10	10	1
Brickwork	16	2.5	10	4
Floor slab	8	5	10	2
Roof	12	3.33	10	3
Finishes	4	10	10	1

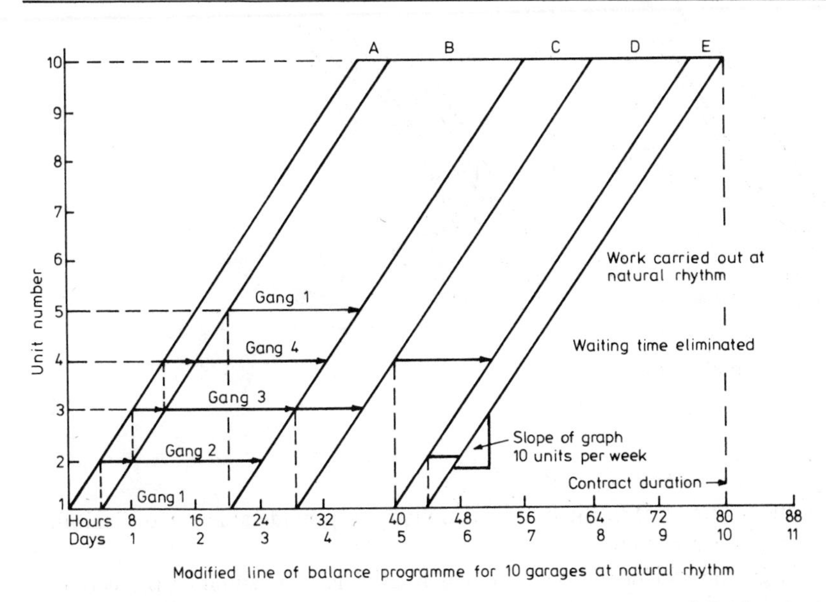

Fig. 6.19 Modified line of balance programme for 10 garages at natural rhythm

6.4 Buffer times

In the examples shown there has been no allowance for the many unforeseen circumstances that may occur in construction operations, e.g. poor weather, fluctuating labour forces, poor site conditions, erratic delivery of components. Obviously, some of these factors are impossible to predict! What is needed is a degree of flexibility built into the programme, based on:

(a) sound experience of previous contracts;
(b) analysis of future trends.

There are two basic ways of providing this intuitive flexibility:

1. An activity buffer.
2. A stage buffer.

These buffers will increase the contract time and must always be consistent with securing the contractor's planned profit.

Activity buffers

This is an allowance included in each activity time estimate to cater for random differences in productivity. The function is to delay the start of subsequent activities to an extent that will minimise the effects of delays (Fig. 6.20).

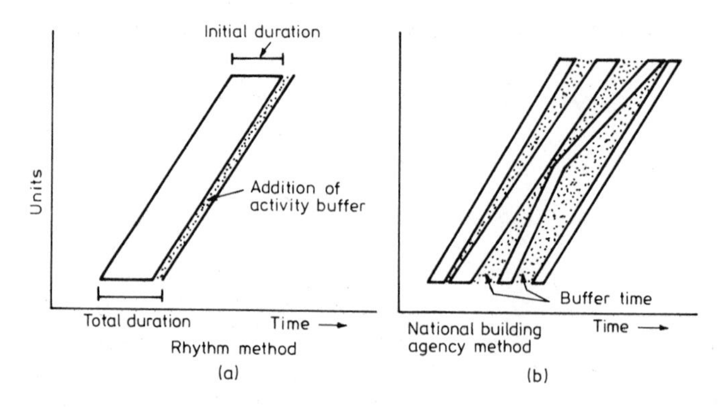

Fig. 6.20 Activity buffer times

Advantages of activity buffer

(a) It is standard practice in the building industry to provide a degree of flexibility (allowance) for each activity.

(b) For a large contract of the type for which LOB is most suitable; protection is automatically incorporated at each activity level.

Disadvantages

(a) There is an increase in overall contract duration. This should be allowed for at estimating stage.

(b) There is always a temptation to dispense with the additional time allowances on each activity if the job is at least on schedule.

(c) As with all building work, the hazards of working in the open are unpredictable and therefore at best, the activity buffer is only an approximate figure.

Stage buffers

This is a time allowance between main stages of the work, e.g. substructure and superstructure, superstructure and internal works, etc. It facilitates the programming of these stages as a multiple of their respective natural rhythms (Fig. 6.21).

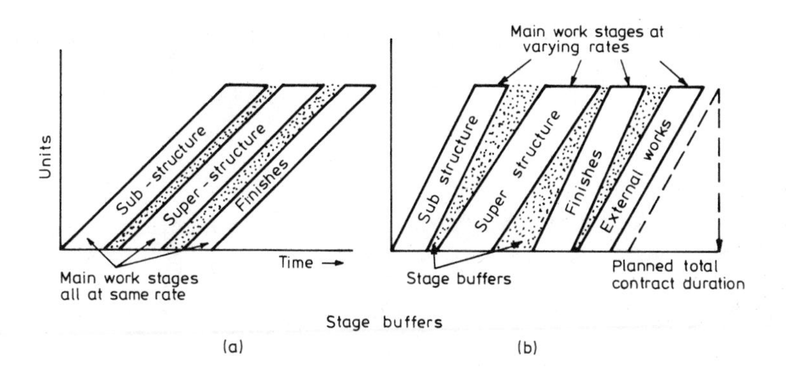

Fig. 6.21 Stage buffers

Advantages

It is possible to programme major construction stages at different rates of rhythms.

The essential function then of the time buffer, whether activity or stage, is to minimise the effect of disturbances to enable all operations to progress smoothly. The presence of any type of time buffer, does much to minimise the effects of irregular material deliveries and of fluctuating labour forces.

6.5 Controlling progress

The line of balance technique, like any other planning technique, will be useless (however well prepared) if it is not made use of for control purposes.

Progress is constantly monitored and recorded on the line of balance programme chart by shading in activities (Fig. 6.22) or main work stages (Fig. 6.23). In Fig. 6.22(a), progress appears quite satisfactory, i.e.:

(a) Excavation work is on schedule.
(b) Concrete to foundations is falling behind and action will be taken to speed up this activity.
(c) Brickwork is just about on schedule.
(d) Floor joists are ahead of schedule and manning levels will have to be checked.

As an alternative to the above it is feasible to plot the cumulative starts and finishes for each activity (Fig. 6.22(b)). Delays in housing units can easily be highlighted, as in Fig. 6.22(c).

Figure 6.23 displays similar means of recording progress on line of balance programmes based on the natural rhythm method. Main stages of work are depicted, e.g. substructure, superstructure, finishes, external works. Progress per block can be monitored and early corrective action taken if necessary.

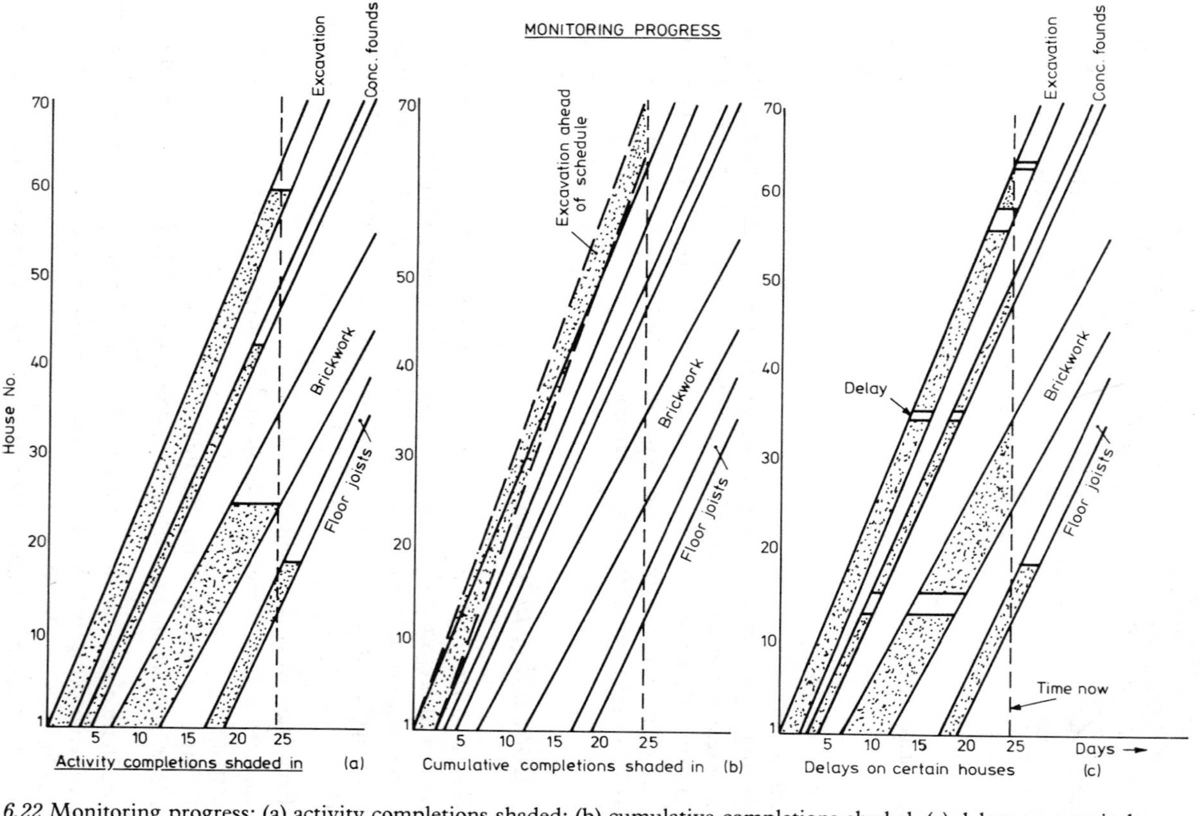

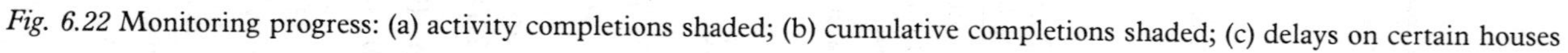
Fig. 6.22 Monitoring progress: (a) activity completions shaded; (b) cumulative completions shaded; (c) delays on certain houses

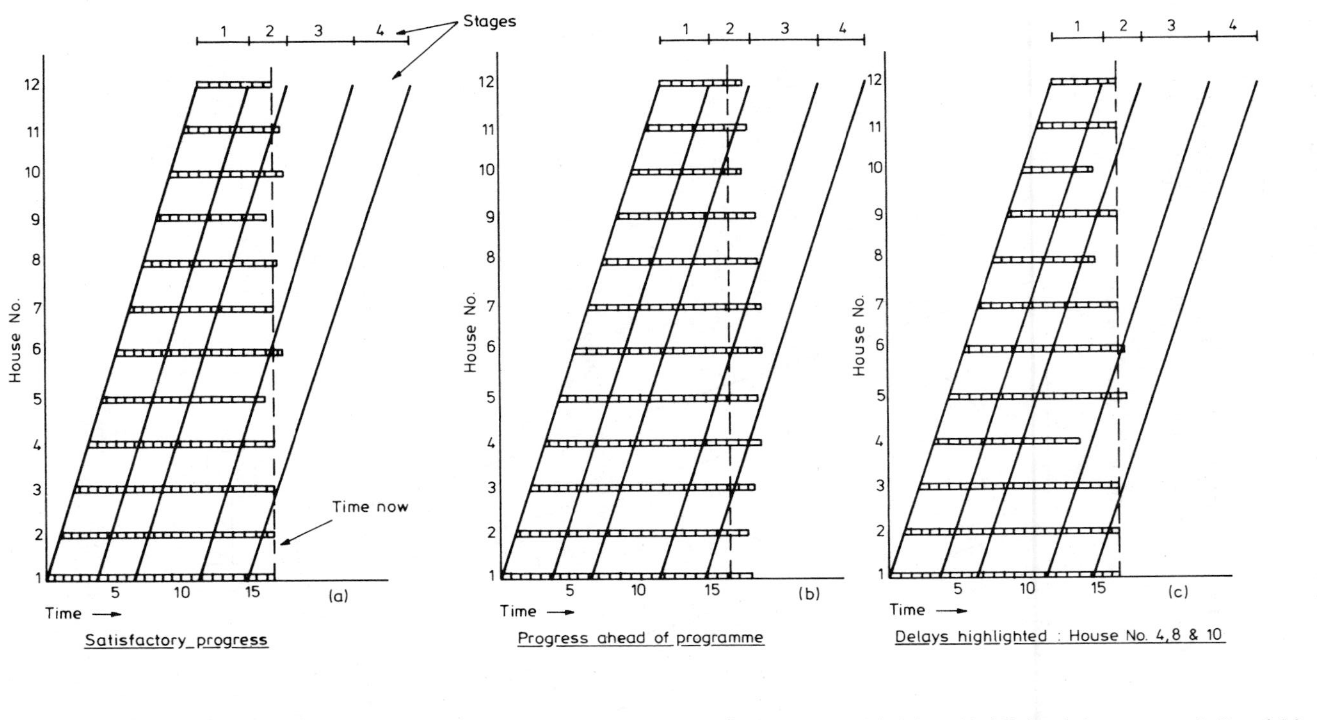

Fig. 6.23 Monitoring progress: (a) satisfactory progress; (b) progress ahead of programme; (c) delays highlighted: house nos. 4, 8 and 10

6.6 Worked examples: Use of line of balance technique

Q.6.1.

From the line of balance diagram (Fig. 6.24), precedence diagram (Fig. 6.25) and Table 6.12, determine the total number of man-weeks required to complete the project.

The solutions are given in Fig. 6.26 and Table 6.13.

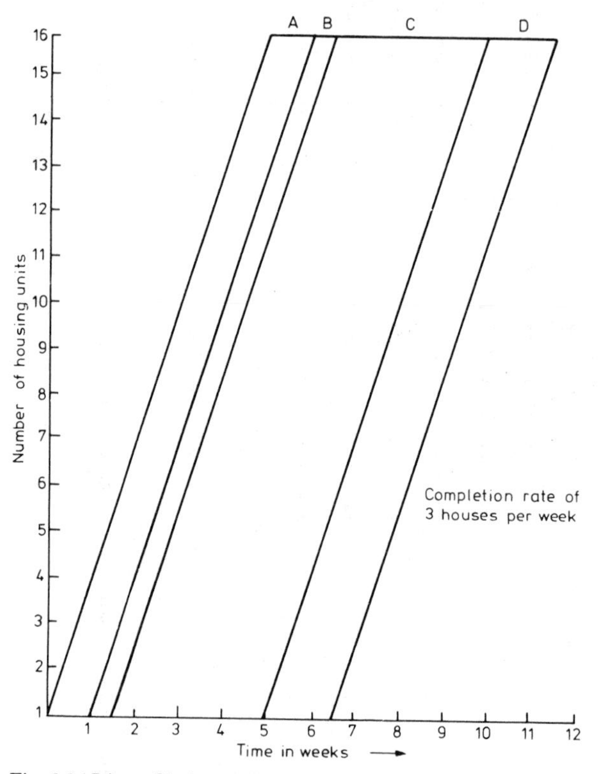

Fig. 6.24 Line of balance diagram

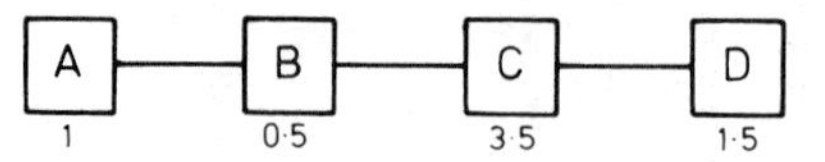

Fig. 6.25 Simplified network

Table 6.12

Activity	Duration	Activity manweeks
A	1 week	4
B	0·5	1·5
C	3·5	10·5
D	1·5	6·0

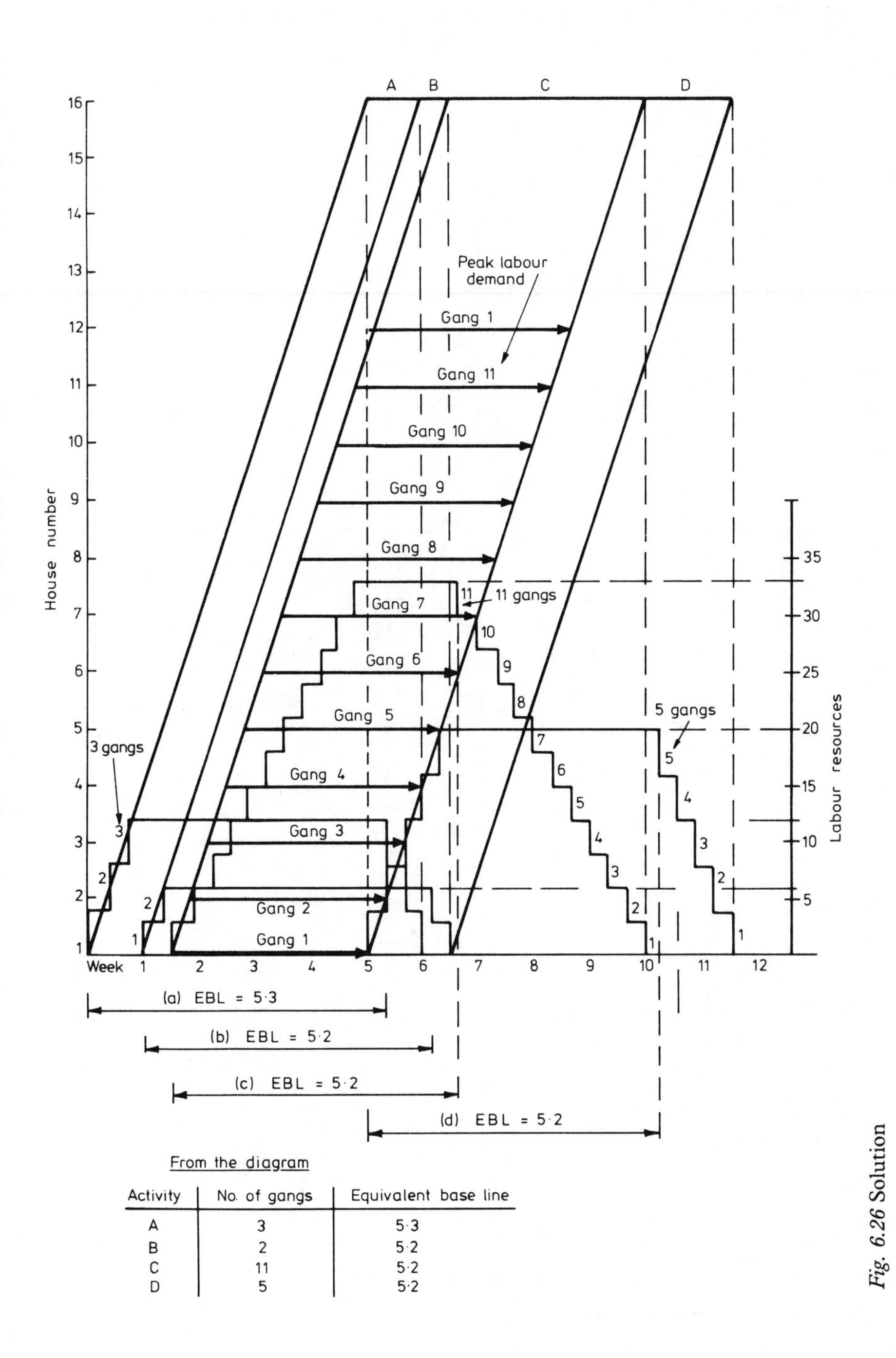

Activity	No. of gangs	Equivalent base line
A	3	5·3
B	2	5·2
C	11	5·2
D	5	5·2

Fig. 6.26 Solution

Table 6.13

Activity code	*Duration (weeks)*	*Activity man-weeks*	*No. of men per gang*	*No. of gangs*	*Maximum no. of men*	*Equivalent* base histogram (from diagram)*	*Man-weeks per activity (for 16 houses)*
A	1	4	4	3	12	5.3	63.6
B	0.5	1.5	3	2	6	5.2	31.2
C	3.5	10.5	3	11	33	5.2	171.6
D	1.5	4	4	5	20	5.2	104
						Total man-weeks 370.4	

Total man-weeks to carry out the work at a handover rate of 3 houses/week = 370.4

* Check equivalent histogram by calculation using $\frac{Q - \text{no. of gangs}}{m} + A$

Activity: A $\frac{16 - 3}{3} + 1 = 5.3$; B $\frac{16 - 2}{3} + 0.5 = 5.2$; C $\frac{16 - 11}{3} + 3.5 = 5.2$;

D $\frac{16 - 5}{3} + 1.5 = 5.2$.

Q.6.2.

What would be the maximum requirement of resource Y on a contract for fifteen houses if the first house was to be completed after 15 weeks and completions to be at the rate of one per week thereafter. The network for one house with resource requirements and durations shown on the activities is shown in Fig. 6.27.

The solution is given in Fig. 6.28.

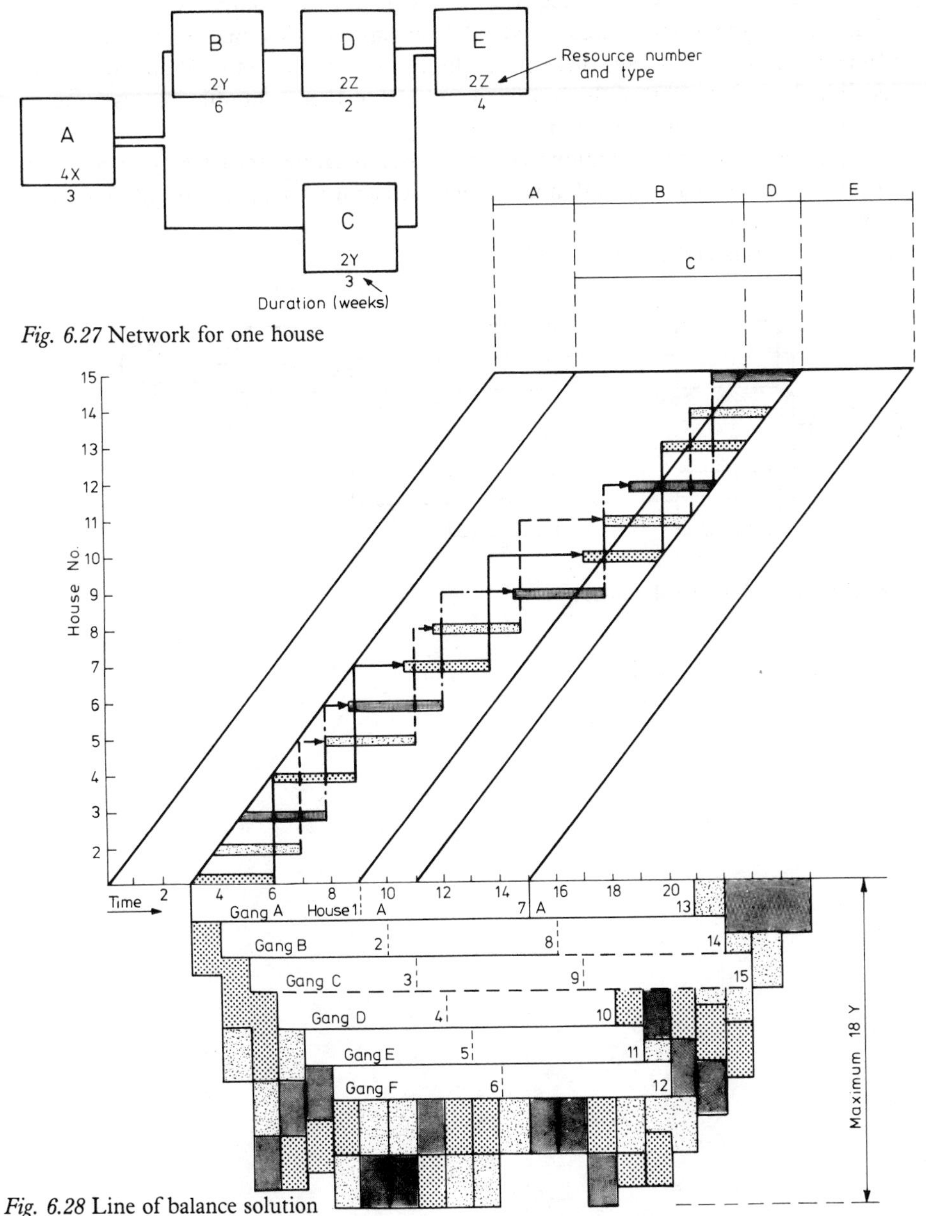

Fig. 6.27 Network for one house

Fig. 6.28 Line of balance solution

Q.6.3.

Figure 6.29 shows the logic of a network describing the production of a small building. Activity B requires 50 resource x day units of resource R and the activity is such that it is possible to vary the quantity of resource applied to it from 1R up to a maximum of 5R. All other activities have fixed durations as shown (in days) on the diagram. Activity E has a fixed resource requirement of two units of resource R and the remaining activities do not require allocations of this resource.

The first building is to be completed in 60 days and completions are to be at the rate of one building every 5 days thereafter. The contractor is able to call on ten units of resource R from the start of the contract and has a further eight units of this resource which can be called upon when necessary.

Using 'line-of-balance' techniques, or otherwise, determine the latest time by which the total allocation of resource R must be increased above ten units if contract output is to be maintained.

The solution is given in Fig. 6.30.

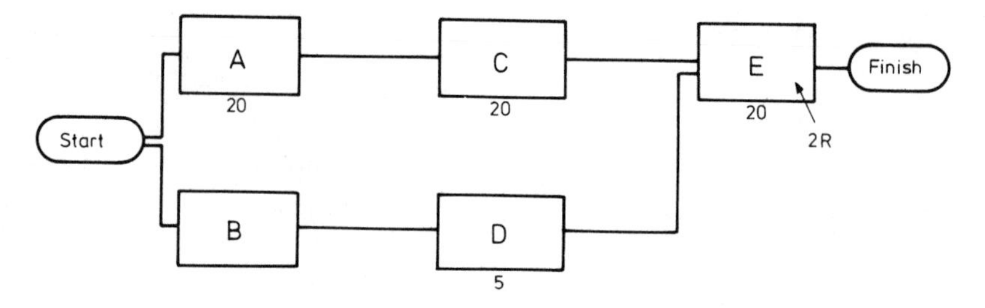

Fig. 6.29 Network for small building

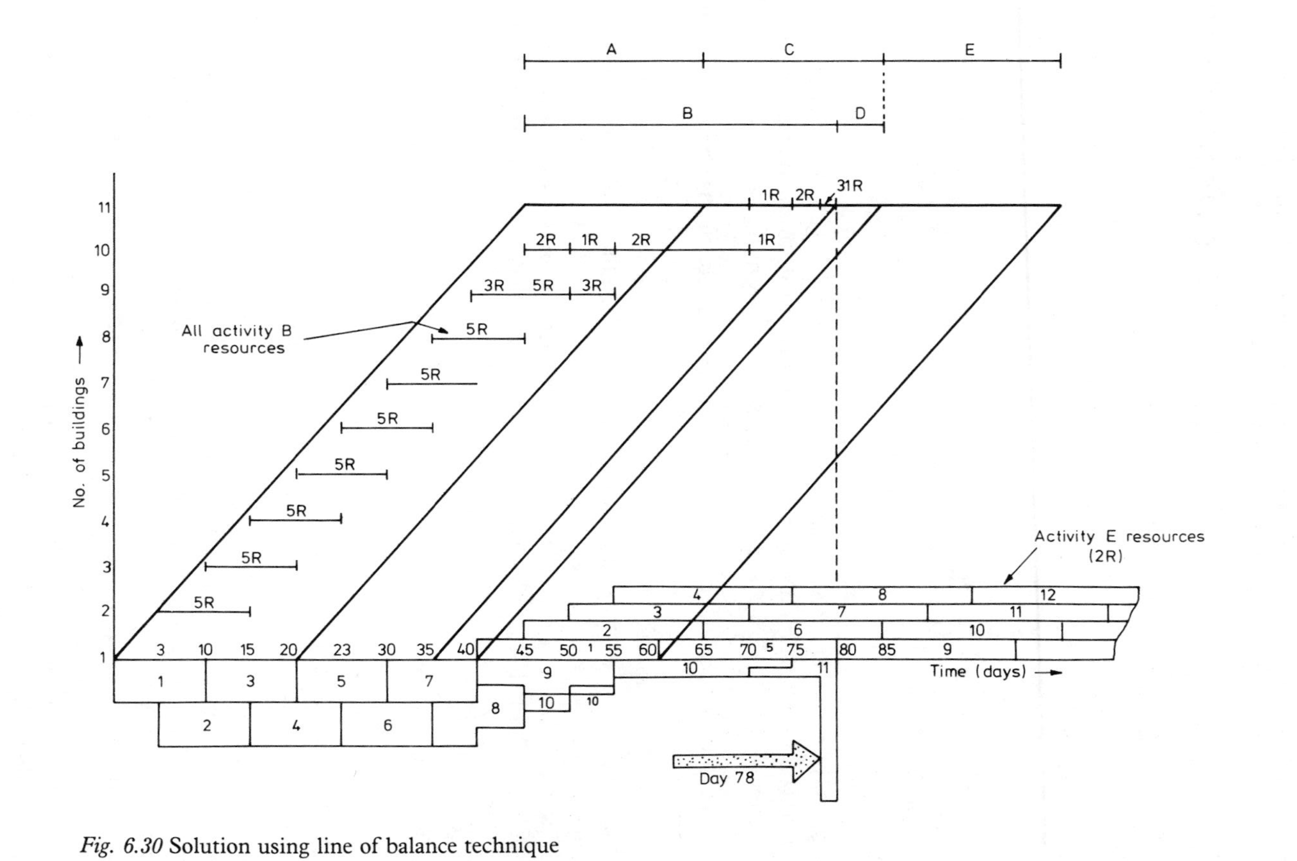

Fig. 6.30 Solution using line of balance technique

Chapter 7

Linear programming

7.1 Introduction

Linear programming is a mathematical technique for determination of the optimal allocation of resources, or an optimal plan chosen from a large number of possible alternatives. Construction management problems amenable to solution using linear programming are numerous and fall into three broad categories:

1. Optimising the use of scarce resources in order to obtain maximum profit or minimum costs.
2. Transportation models, e.g. the possibility of moving numbers of items (plant, materials, etc.) to stations (sites, workshops, etc.) where the numbers in both cases are not necessarily the same.
3. The allocation or assignment of men, machines and materials to production activities. There is a set of tasks and a set of methods, a one-to-one correspondence.

The procedure is to:

(a) formulate the problem;
(b) construct the mathematical model;
(c) derive a solution;
(d) test and evaluate the solution;
(e) implement and maintain the solution.

It must be appreciated that in practical construction problems, the use of mathematical decision-making techniques often does not give complete solutions and often only a part of the overall problem which can be formulated in mathematical terms can be solved. Frequently a combined mathematical–intuitive approach can be successfully used. Generally, the more complex the problem, the less likely it can be solved by intuition alone.

Some linear programming calculations, particularly the simplex method, are in practice tedious for non-mathematicians to handle. Fortunately many standard computer programs are available which considerably reduce the calculation. The main task is formulating the problem into a linear programming model and feeding the equations into the computer program.

7.2 Requirements of the LP model

The following are the requirements for the construction of a linear programming model.

1. *Objective function.* There must be an objective the company wants to achieve. This may be to maximise profit, minimise cost, minimise time, determine best product mix, etc.
2. *Constraints.* There must be alternative courses of action, one of which will be optimal, thus achieving the objective.
3. *Linear relationship.* The objective function and constraint sets must be linear. The procedure will be explained.

Example 1

A joinery manufacturer produces two types of timber frame housing packages, i.e. *chalet* and *Georgian.* The operations are highly mechanised and the estimated average time required from each machine for the manufacturer of each package is given below.

Package	Machines		
	A	B	C
Chalet	10	10	40
Georgian	30	60	20

In a given period there are 600 hours of machine A time available, 850 hours of machine B and 800 hours of machine C. The expected profit for the production of each package is:

Chalet	£2,000
Georgian	£1,900

and the demand for each type is very high. Assuming that the machines are available when required determine the optimum number of each house type package to be manufactured in order to make maximum profit and state the maximum profit figure.

Formulation of the problem

Let x_c be the number of chalet packages to be manufactured, and x_g the number of Georgian packages to be manufactured. Each of the chalet packages (x_c) and the Georgian packages (x_g) requires 10 and 30 hours respectively of machine A time, i.e. a total of $10x_c + 30x_g$ hours is required.

The constraint of 600 hours' availability of machine A time means that $10x_c + 30x_g$ must be less than or equal to 600. This is represented as $10x_c + 30x_g \leqslant 600$.

The constraints on the availability of the machine B and machine C can be shown as follows:

Machine B	$10x_c + 60x_g \leqslant 850$
Machine C	$40x_c + 20x_g \leqslant 800$

The purpose is to determine the number of each house type to be manufactured in order to achieve maximum profit. This profit objective can be stated as follows:

$2{,}000x_c + 1{,}900x_g$ (objective function)

Since it is impossible to produce less than nothing of each house package x_c and x_g cannot be negative, the problem can therefore be stated in mathematical terms as: maximise

$2{,}000x + 1{,}900x_g$
subject to
$x_c \geqslant 0$
$x_g \geqslant 0$

and
$10x_c + 30x_g \leqslant 600$ (machine A)
$10x_c + 60x_g \leqslant 850$ (machine B)
$40x_c + 20x_g \leqslant 800$ (machine C)

This problem consists of only two unknowns and can therefore be easily solved using a straight-line graphical method.

The technique is to draw the straight lines of the equations for machines A, B and C above in the plane x_c and x_g as shown below:

Graph

1. $10x_c + 30x_g = 600$
Let $x_c = 0$
$30x_g = 600$
$x_g = 600 \div 30 = 20$ (0,20)
Let $x_g = 0$
$10x_c = 600$
$x_c = 60$ (60,0)

2. $10x_c + 60x_g = 850$
Let $x_c = 0$
$60x_g = 850$
$x_g = 850 \div 60 = 14.17$ (0,14.17)
Let $x_g = 0$
$10x_c = 850$
$x_c = 85$ (85,0)

3. $40x_c + 20x_g = 800$
Let $x_c = 0$
$20x_g = 800$
$x_g = 800 \div 20 = 40$ (0,40)
$x_g = 0$
$40x_c = 800$
$x_c = 800 \div 40 = 20$ (20,0)

The shaded area contains all the pairs of values of x_c and x_g which are feasible solutions to the problem (Fig. 7.1).

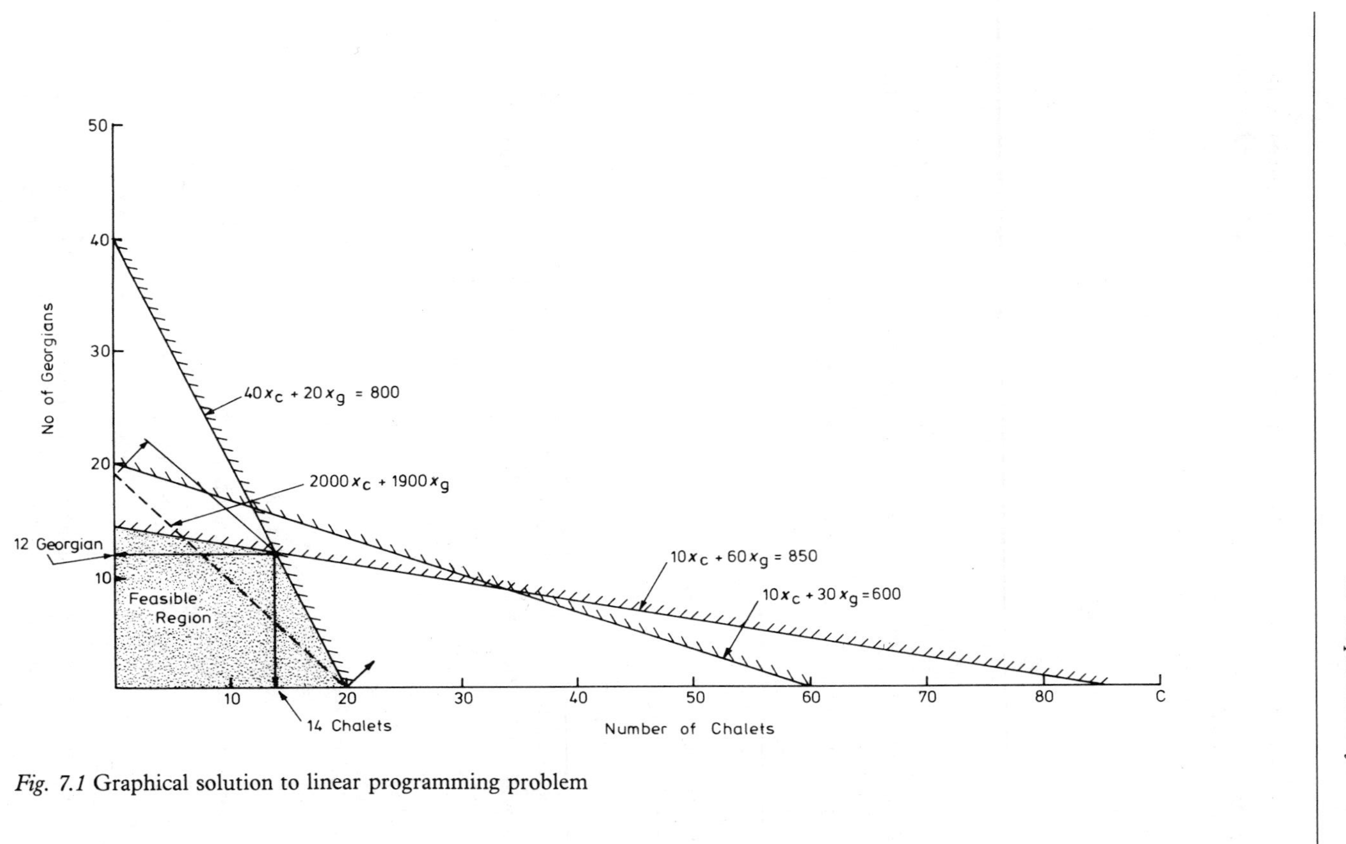

Fig. 7.1 Graphical solution to linear programming problem

With this feasible region lies the maximum profit point. The objective function line is used ($2{,}000x_c + 1{,}900x_g$) and is drawn parallel moving away from the origin until the maximum value is obtained – point Y the extreme point.

7.3 Simplex method

The graphical method of solution is of no practical value when the number of house types (variables) exceeds two or three. The simplex method is used when the number of variables exceeds two and consists of following a number of simple arithmetic steps. The previous example will be solved using this technique in order to demonstrate the factors involved. The aim as before is to determine the optimum number of house type packages to be manufactured in order to maximise profit, i.e. maximise

$2{,}000x_c + 1{,}900x_g$
subject to
$x_c \geqslant 0 \quad x_g \geqslant 0$
and
$10x_c + 30x_g \leqslant 600$
$10x_c + 60x_g \leqslant 850$
$40x_c + 20x_g \leqslant 800$

Since it is easier to handle equalities than inequalities convert to equations by introducing slack variables C_1, C_2 and C_3. These are the differences between the two sides of the inequalities, thus maximise

$2{,}000x_c + 1{,}900x_g$
subject to
$10x_c + 30x_g + C_1 = 600$
$10x_c + 60x_g + C_2 = 850$
$40x_c + 20x_g + C_3 = 800$
C_1, C_2 and C_3 are of course non-negative.

In a more general problem with say n variables and m inequalities constraints, we require m slack variables, and a solution that maximises has out of a total of $m+n$ variables including slacks, exactly m that are non-zero.

A set of variables satisfying the constraints with m non-zero and n zero is called a feasible solution.

The calculation starts with a feasible solution, tests to see if it is optimal; if not an improved solution is found, this iteration is continued until no further improvement is possible.

The calculations are written in tabular form. The rows correspond to the first set of non-zero variables. Columns are labelled x_c, x_g, C_1, C_2 and C_3. A final P is added to correspond to the constraints in the equations and a final row to show the coefficients (profits) Z is added.

	x_c	x_g	C_1	C_2	C_3	P
C_1	10	30	1			600
C_2	10	60		1		850
C_3	40	20			1	800
Z	2,000	1,900				

Procedure

1. Select the column which has the largest positive entry in row Z. In this case it is x_c with entry 2,000.
2. Divide the positive entries in the column selected in step 1 into the corresponding entries in P and select the smallest. In this case it is x_c. For $C_1 = 600$, $10 = 60$; $C_2 = 850$; $10 = 85$; $C_3 = 800$, $40 = 20$. Therefore we select row C_3.
3. Divide the row selected in step 2 by the entry in the column selected in step 1 and relabel the result with the heading of the column selected. In this case divide entries in row C_3 by 40 and relabel the row x_c. This results in

	x_c	x_g	C_1	C_2	C_3	P
x_c	1	0.5			0.025	20

4. Eliminate the entries in all rows other than that changed in step 3 (including the Z row) by subtracting suitable multiples of the relabelled row obtained in step 3. If all Z entries are negative or zero, the optimal solution has been found. If not, return to step 1. In this case we multiply row x_c by 10 and subtract it from C_1. Proceed similarly for row C_2 (multiplying x_c by 10) and row Z (multiplying x_c by 2,000). This gives Table 7.1. The constant obtained when eliminating from row Z is not required at subsequent steps; hence it is usually omitted.

Table 7.1

	x_c	x_g	C_1	C_2	C_3	P
x_c	1	0.5			0.025	20
C_2		25	1		− 0.25	400
C_3		55		1	− 0.25	650
Z		900			−50	

Because there still is a positive entry in row Z we return to step 1 and repeat the four steps. This gives Table 7.2. Because both entries in row Z of Table 7.2 are negative, we conclude that $x_g = 11.82$ and $x_c = 4.09$ maximise Z. See Fig. 7.1.

Table 7.2

	x_c	x_g	C_1	C_2	C_3	P
x_g		1		0.018	−0.0045	11.82
x_c	1			−0.009	0.00273	14.09
C_1			1	−0.45	−0.36	370.45
Z				− 16.2	−45.95	

We can, therefore, conclude that 12 Georgian packages and 14 chalet packages should be manufactured giving a maximum profit figure of

$$\begin{aligned} 12 \times £1{,}900 &= £22{,}800 \\ 14 \times £2{,}000 &= £28{,}000 \\ & \overline{£50{,}800} \end{aligned}$$

Note that step 2 in the procedure just described fails if none of the entries (other than Z) in the column selected at step 1 are positive. In this case the variable selected at step 1 can be increased indefinitely without violating a constraint of data supposed to represent a real problem; this usually means the real problem is non-linear or that a constraint has been omitted.

These arithmetical steps can become tedious with increased numbers of variables. By utilising a standard computer program package the arithmetics can be completely eliminated. All that is required is:

(a) Formulation of the problem in algebraic terms, i.e.:
$2{,}000x_c + 1{,}900x_g$
subject to

$$\begin{aligned} x_c &\geqslant 0 \quad \text{(constraint } C_1\text{)} \\ x_g &\geqslant 0 \quad \text{(constraint } C_2\text{)} \\ 10x_c + 30x_g &\leqslant 600 \quad \text{(constraint } C_3\text{)} \\ 10x_c + 60x_g &\leqslant 850 \quad \text{(constraint } C_4\text{)} \\ 40x_c + 20x_g &\leqslant 800 \quad \text{(constraint } C_5\text{)} \end{aligned}$$

(b) Answer two simple questions, i.e.:
  (i) How many variables? $(x_c + x_g) = 2$.
  (ii) How many constraints? $(C_1, C_2, C_3, C_4$ and $C_5) = 5$.

This example follows Table 7.3, having been run through a computer package on the Ulster Polytechnic's ICL 1902 T computer. The package was developed by Daellenbach and Bell.

In the worked example the number of iterations necessary to solve the problem was five. The computer batch version can handle up to 20 variables and 20 constraints, and up to 500 iterations to achieve the optimum solution.

Example 2

An an illustration, a more complex set of data relating to a similar type of problem is given below.

A joinery manufacturer has the capacity to produce five types of timber frame housing packages each offering a different profit figure but also requiring varying times from three machines as shown below.

Table 7.3 Solution of a linear programming problem

Title? – House types
No. of variables? – 2
Type variable names – XC XG
Type variable coefficients – 2,000 1,900
No. of constraints? – 5
Type constraints – C1 1 0 GE 0
–C2 0 1 GE 0
–C3 10 30 LE 600
–C4 10 60 LE 850
–C5 40 20 LE 800

Variable	Status	Value
XC	Basic	14.091
XG	Basic	11.818
C1	Basic	
C2	Basic	
C3	Basic	Maximal objective – 50636.364
C4		No. of iterations – 5
C5		

Is a full tableau print required? (Type yes or no) – Yes

1.000000	0.000000	0.000000	0.000000
0.000000	–0.009091	0.027873	0.000000
0.000000	0.000000	–0.009091	0.027273
14.090909			
0.000000	1.000000	0.060000	0.000000
0.000000	0.018182	–0.064545	0.000000
0.000000	0.000000	0.018182	–0.004545
11.818182			
0.000000	0.000000	0.000000	0.000000
1.000000	–0.454545	–0.136364	0.000000
0.000000	1.000000	–0.454545	–0.136364
104.545455			
0.000000	0.000000	0.000000	1.000000
0.000000	0.018182	–0.004545	0.000000
–1.000000	0.000000	0.018182	–0.004545
11.818182			
0.000000	0.000000	1.000000	0.000000
0.000000	–0.009091	0.027273	–1.000000
0.000000	0.000000	–0.009091	0.027273
14.090909			
0.000000	0.000000	0.000000	0.000000
0.000000	–16.363636	–45.909091	0.000000
0.000000	0.000000	–16.363636	–45.909091
–50636.363636			
0.000000	0.000000	0.000000	0.000000
0.000000	0.000000	–0.000000	–1.000000
–1.000000	–1.000000	–1.000000	–1.000000
–0.000000			

Timber frames package house type	*Estimated profit per package (£)*	*Time required from each machine (hr)*		
		A	B	C
Chalet	2,000	10	10	40
Georgian	2,900	20	40	20
Bungalow	1,750	15	5	5
Semi-detached	1,500	5	20	10
Tudor	2,600	30	25	20
Total available hours per machine		2,000	2,800	1,750

Due to the shortage of certain materials and components there is an upper limit to the number of each package that can be manufactured, i.e.:

Chalet	50
Georgian	100
Bungalow	70
Semi-detached	80
Tudor	50
Total of any combination	170

Assuming that there is a high demand for whatever package produced the problem is to determine the optimum number of each to be manufactured in order to obtain maximum profit.

Formulation of the problem:

Let x_c = chalets, x_g = Georgians, x_b = bungalows, x_s = semis, x_t = Tudors
Maximise

$$2{,}000x_c + 2{,}900x_g + 1{,}750x_b + 1{,}500x_s + 2{,}600x_t$$

subject to

$$x_c \geqslant 0, \quad x_g \geqslant 0, \quad x_b \geqslant 0, \quad x_s \geqslant 0, \quad x_t \geqslant 0$$

and

$$\begin{aligned}
10x_c + 20x_g + 15x_b + 5x_s + 30x_t &\leqslant 2{,}000 \text{ (machine A constraint)}\\
10x_c + 40x_g + 5x_b + 20x_s + 25x_t &\leqslant 2{,}800 \text{ (machine B constraint)}\\
40x_c + 20x_g + 5x_b + 10x_s + 20x_t &\leqslant 1{,}750 \text{ (machine C constraint)}\\
x_c &\leqslant 50 \text{ (package no. constraint)}\\
x_g &\leqslant 100 \text{ (package no. constraint)}\\
x_b &\leqslant 70 \text{ (package no. constraint)}\\
x_s &\leqslant 80 \text{ (package no. constraint)}\\
x_t &\leqslant 50 \text{ (package no. constraint)}\\
x_c + x_g + x_b + x_s + x_t &\leqslant 170 \text{ (total no. of packages)}
\end{aligned}$$

This complex problem can be solved in seconds by use of the standard computer program. An extract of the computer print out is shown in Table 7.4.

Table 7.4 Solution of a linear programming problem

Title?
- House types
No. of variables?
- 5
Type variable names
- XC XG XB XS XT
Type variable coefficients
- 2,000 2,900 1750 1500 2600
No. of constraints?
- 9
Type constraints
- C1 1 0 0 0 0 LE 50
- C2 0 1 0 0 0 LE 100
- C3 0 0 1 0 0 LE 70
- C4 0 0 0 1 0 LE 80
- C5 0 0 0 0 1 LE 50
- C6 1 1 1 1 1 LE 170
- C7 10 20 15 5 30 LE 2000
- C8 10 40 5 20 25 LE 2800
- C9 40 20 5 10 20 LE 1750

Maximal objective – 311326.923
No. of iterations – 13

Variable	Status	Value
XC	Basic	4.423
XG	Basic	24.038
XB	Basic	70.000
XS	Basic	68.846
XT	Basic	2.692
C1	Basic	
C2	Basic	
C3		
C4	Basic	
C5	Basic	
C6		
C7		
C8		
C9		

Note
Thirteen iterations were required in order to obtain the optimum solution. The optimum number of each house type package to be manufactured is given below.

Timber frame packages		*No. profit*	
Chalets	(4.423)	4 × 2,000	£8,000
Georgians	(24.038)	24 × 2,900	£69,600
Bungalows	(70.00)	70 × 1,750	£122,500
Semis	(68.846)	69 × 1,500	£103,500
Tudors	(2.692)	3 × 2,600	£7,800
		Profit	£311,400

In a real situation available machines and equipment can be increased or decreased, or new machines and methods introduced. Whatever the constraints, each problem can be formulated as above and an optimum solution found.

7.4 A transportation problem

This category of linear programming problem can easily be solved using an iterative process; as before, the procedure will be explained through worked examples.

Example 3

A large concrete manufacturer has fóur ready-mix concrete depots *a*, *b*, *c* and *d*. Each depot has the following quantities of concrete available: (*a*) 60 m^3; (*b*) 80 m^3; (*c*) 50 m^3; (*d*) 45 m^3; total 235 m^3.

The following quantities of concrete have to be delivered to three sites: (*x*) 85 m^3; (*y*) 100 m^3; (*z*) 50 m^3; total 235 m^3.

The transportation costs in £/m^3 from each depot to the three sites are shown in Table 7.5. The aim is to satisfy site requirements at minimal cost.

Table 7.5

SITES \ DEPOTS	a	b	c	d
X	6	10	6	5
Y	4	8	10	4
Z	3	9	8	7

Solution

A method of transporting the concrete which satisfies site requirements while not exceeding the depot's restricted available quantities is first determined. This solution is then tested to see if in fact it is the cheapest method. If not the process is repeated until no further cost reduction is possible.

An initial feasible solution is to allocate first as much as possible to that route having the least transportation cost per m^3 (Table 7.6).

Table 7.6

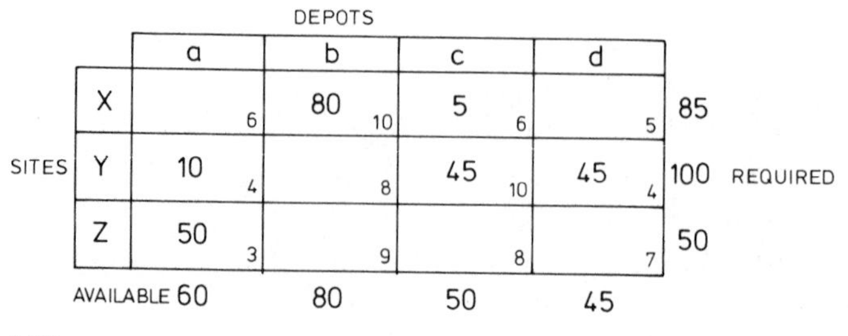

SITES \ DEPOTS	a	b	c	d	REQUIRED
X	(6)	80 (10)	5 (6)	(5)	85
Y	10 (4)	(8)	45 (10)	45 (4)	100
Z	50 (3)	(9)	(8)	(7)	50
AVAILABLE	60	80	50	45	

There is one route (az) which has a transportation cost of £3. Allocate as much as possible to this route. The maximum which can be allocated is the full 50 m^3, leaving 10 m^3 in depot a. The next cheapest routes are ay and dy. Allocate say 10 m^3 to ay and the available 45 m^3 to dy, 80 m^3 must be allocated to bx and 5 m^3 to cx. The requirements of each site has been satisfied – but is it the cheapest allocation?

Route	*Cost/m^3*	*No. of m^3*	*Cost/route (£)*
bx	10	80	800
cx	6	5	30
ay	4	10	40
cy	10	45	450
dy	4	45	180
az	3	50	150
		Total cost	£1,650

To determine whether or not there is a more economical routeing system each unused route is examined.

Table 7.7

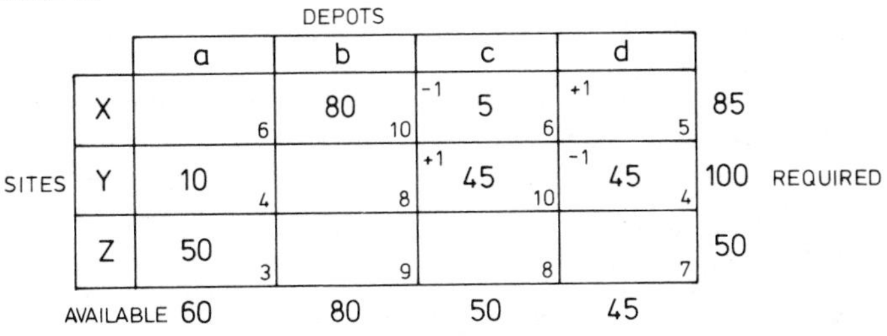

		DEPOTS				
		a	b	c	d	
	X	(6)	80 (10)	−1 5 (6)	+1 (5)	85
SITES	Y	10 (4)	(8)	+1 45 (10)	−1 45 (4)	100 REQUIRED
	Z	50 (3)	(9)	(8)	(7)	50
	AVAILABLE	60	80	50	45	

Try sending 1 m^3 along route dx (Table 7.7), which is the cheapest of the unused routes. A modification is now necessary to the existing allocations to ensure the quantity available at the depots is not exceeded. If 1 m^3 is sent along dx the 1 m^3 must be subtracted from dy, 1 m^3 added to cy and 1 m^3 subtracted from cx. These changes have no effect on the feasibility of the initial solution. The cost implication of this change is

Cost dx − cost dy + cost cy − cost cx
$= 5 - 4 + 10 - 6 = 5$

This represents an increase in cost. A negative figure represents a decrease in cost.

Use of shadow cost

If the previous inspection is repeated for each unused route in turn the procedure can become tedious. The technique of shadow costs is a short cut and is made up of two parts, i.e.:

A cost of dispatch from depots ($abcd$)
A cost of reception at sites (xyz)

The shadow cost of transporting 1 m^3 along route dx is $Cdx = d + x$. By assigning an arbitrary value to one of the shadow costs, say x, the remaining ones a, b, c, d, y, z may be determined provided the solution employs six routes, i.e. one route for each shadow cost required. Since the initial allocation uses six routes the shadow costs can be determined.

$b + x = 10, \quad c + x = 6, \quad a + y = 4, \quad c + y = 10$
$d + y = 4, \quad a + z = 3$

Let $x = 0$, therefore

$b = 10, \quad c = 6, \quad y = 4, \quad d = 0, \quad a = 0, \quad z = 3$

Table 7.8

		DEPOTS				
		a	b	c	d	
	Shadow Costs	0	10	6	0	
SITES X	0	(6)	80 (10)	5 (6)	(5)	85
Y	4	10 (4)	(8)	45 (10)	45 (6)	100 REQUIRED
Z	3	50 (3)	(9)	(8)	(7)	50
AVAILABLE		60	80	50	45	

Initial allocation and shadow costs are shown in Table 7.8. By introducing the new route dx and sending 1 m^3 along it altered the cost by:

$Cdx - Cdy + Ccy - Ccx$

In terms of shadow costs this is equivalent to:

$Cdx - (d + y) + (c + y) - (c + x) = Cdx - (d + x)$

This is the general result, i.e. sending 1 m^3 along a previously unoccupied route increases the total transportation cost by the unit transportation cost of the new route minus the sum of the shadow cost of that route. The difference for each unused route can now be evaluated:

$Cax - (a + x) = 6, \quad Cdx - (d + x) = 5, \quad Cby - (b + y) = -6$
$Cbz - (b+z) = -4, \quad Ccz - (c + z) = -1, \quad Cdz - (d + z) = 4$

Table 7.9

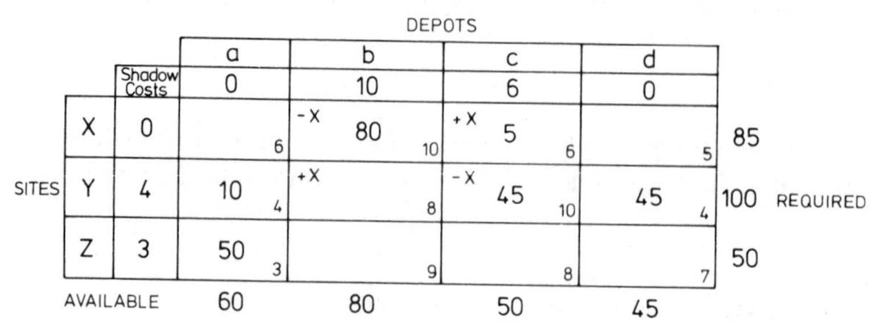

		DEPOTS				
		a	b	c	d	
	Shadow Costs	0	10	6	0	
SITES X	0	(6)	−X 80 (10)	+X 5 (6)	(5)	85
Y	4	10 (4)	+X (8)	−X 45 (10)	45 (4)	100 REQUIRED
Z	3	50 (3)	(9)	(8)	(7)	50
AVAILABLE		60	80	50	45	

Table 7.10

		DEPOTS				
		a	b	c	d	
	X	6	35 10	50 6	5	85
SITES	Y	10 4	45 8	10	45 4	100 REQUIRED
	Z	50 3	9	8	7	50
AVAILABLE		60	80	50	45	

Route *by* is lowest so reallocate along this route (Tables 7.9 and 7.10).

New cost

$35 \times £10 = £350$
$50 \times £6 = £300$
$10 \times £4 = £40$
$45 \times £8 = £360$
$45 \times £4 = £180$
$50 \times £3 = £150$

£1,380

A saving of £570.

A further iteration will be made to see if the optimum solution has been obtained.

Let $x = 0$, then

$x + b = 10, \quad x + c = 6, \quad y + a = 4, \quad y + b = 8, \quad y + d = 4, \quad z + a = 3$

therefore,

$b = 10, \quad c = 6, \quad y = -2, \quad a = 6, \quad z = -3, \quad d = 6$

Table 7.11

			DEPOTS				
			a	b	c	d	
		Shadow Costs	6	10	6	6	
	X	0	6	35 10	50 6	5	85
Sites	Y	-2	10 4	45 8	10	45 4	100 Required
	Z	-3	50 3	9	8	7	50
AVAILABLE			60	80	50	45	

The difference for each unused route is

$Cxa - (x + a) = 0, \quad Cxd - (x + d) = -1, \quad Cyc - (y + c) = 6$
$Czb - (z + b) = 2, \quad Czc - (z + c) = 5, \quad Czd - (z + d) = 4$

Route *xd* is negative, therefore reallocate along this route (Table 7.12).

Table 7.12

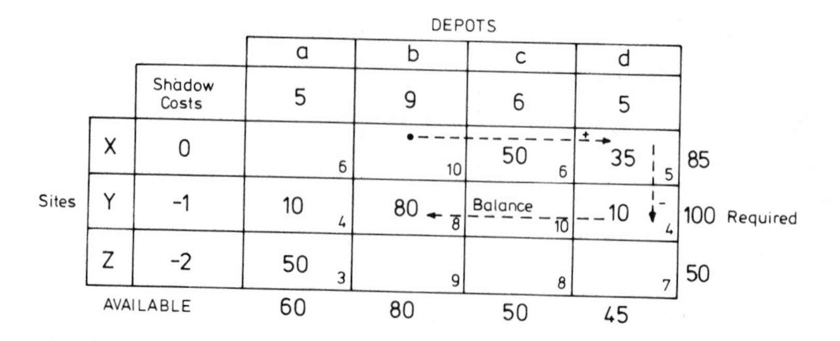

		DEPOTS				
		a	b	c	d	
	Shadow Costs	5	9	6	5	
Sites X	0	(6)	(10)	50 (6)	35 (5)	85
Sites Y	-1	10 (4)	80 (8)	Balance (10)	10 (4)	100 Required
Sites Z	-2	50 (3)	(9)	(8)	(7)	50
AVAILABLE		60	80	50	45	

New cost

50 m³ from depot *c* to site *x* at £6/m³	=	£300
35 m³ from depot *d* to site *x* at £5/m³	=	£175
10 m³ from depot *a* to site *y* at £4/m³	=	£40
80 m³ from depot *b* to site *y* at £8/m³	=	£640
10 m³ from depot *d* to site *y* at £4/m³	=	£40
50 m³ from depot *a* to site *z* at £3/m³	=	£150
		£1,345

Test for optimum solution

$x + c = 6, \quad x + d = 5, \quad y + a = 4, \quad y + b = 8, \quad y + d = 4, \quad z + a = 3$

Let $x = 0$, therefore

$c = 6, \quad d = 5, \quad y = -1, \quad b = 9, \quad a = 5, \quad z = -2$

The difference for each unused route is

$Cax - (a + x) = 1, \quad Cbx - (b + x) = 1, \quad Cyc - (y + c) = 5$
$Czb - (z + b) = 2, \quad Czc - (z + c) = 4, \quad Czd - (z + d) = 4$
See table 7.12.

There are no negative figures and therefore no further cost reductions (improvement to the solution) is possible.

Linear programming formulation for this example is

Let x_{ax} = amount of concrete sent from depot *a* to site *x*
Let x_{ay} = amount of concrete sent from depot *a* to site *y*
Let x_{az} = amount of concrete sent from depot *a* to site *z*
Let x_{bx} = amount of concrete sent from depot *b* to site *x*
. . . etc.

Minimise

$$6x_{ax} + 10x_{bx} + 6x_{cx} + 5x_{dx} + 4x_{ay} + 8x_{by} + 10x_{cy} + 4x_{dy} + 3x_{az} + 9x_{bz} + 8x_{cz} + 7x_{dz}$$

Subject to

$x_{ax} \geqslant 0$
$x_{bx} \geqslant 0$
$x_{cx} \geqslant 0$
.
.
.
$x_{dz} \geqslant 0$

and

$$x_{ax} + x_{ay} + x_{az} = 60$$
$$x_{bx} + x_{by} + x_{bz} = 50$$
$$x_{cx} + x_{cy} + x_{cz} = 50$$
$$x_{dx} + x_{dy} + x_{dz} = 45$$
$$x_{ax} + x_{bx} + x_{cx} + x_{dx} = 85$$
$$x_{ay} + x_{by} + x_{cy} + x_{dy} = 100$$
$$x_{az} + x_{bz} + x_{cz} + x_{dz} = 50$$

Example 4 – surplus availability

A premix company has to supply concrete to three different sites, *1*, *2* and *3*. The sites require 300, 400 and 550 m³ of concrete in a particular week. The firm has three plants *A*, *B* and *C* and these can provide 600, 250 and 450 m³ of concrete. Transportation costs (pence/m³) are shown in Table 7.13. Minimise costs of transporting concrete from each plant to each project.

Table 7.13

		PLANTS	
SITES	A	B	C
1	80	90	40
2	60	100	100
3	150	40	90

Solution

Since there is a surplus of 50 m³ available include a fictitious site to deal with this (Table 7.14).

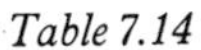
Table 7.14

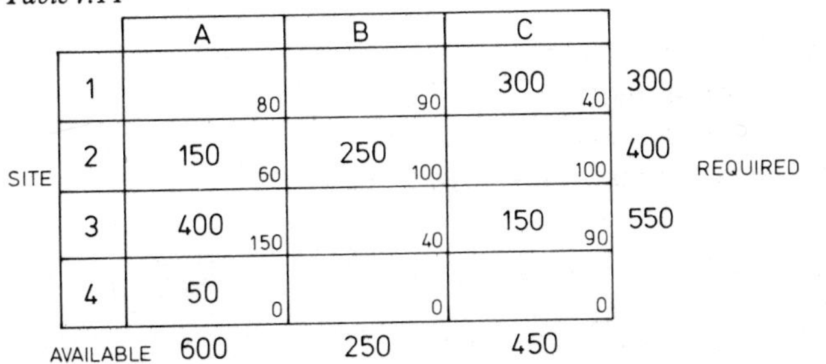

SITE	A	B	C	REQUIRED
1	(cost 80)	(cost 90)	300 (cost 40)	300
2	150 (cost 60)	250 (cost 100)	(cost 100)	400
3	400 (cost 150)	(cost 40)	150 (cost 90)	550
4	50 (cost 0)	(cost 0)	(cost 0)	
AVAILABLE	600	250	450	

Start by allocating maximum quantity to minimum cost, i.e. 300 m³ to *C1*.

Cost

300 ×	40p =	£120
150 ×	60p =	£90
250 ×	100p =	£250
400 ×	150p =	£600
150 ×	90p =	£135
		£1,195

Using shadow costs

$2 + A = 60$, $2 + B = 100$, $1 + C = 40$, $3 + A = 150$, $3 + C = 90$, $4 + A = 0$

Let $1 = 0$, therefore

$C = 40$, $3 = 50$, $4 = -100$, $A = 100$, $2 = -40$, $B = 140$

Table 7.15

			A	B	C	
		Shadow Costs	100	140	40	
	1	0	(80)	(90)	300 (40)	300
SITE	2	-40	+ 150 (60)	− 250 (100)	(100)	400
	3	50	− 400 (150)	+ (40)	150 (90)	550
	4	-100	50 (0)	(0)	(0)	
		AVAILABLE	600	250	450	

REQUIRED

Using $Cdx - (d + x)$ (see Table 7.15)

$$\begin{aligned}
\text{Route } A1 &= C_{A1} - (A + 1)\\
&= 80 - (100 + 0) = -20\\
B1 &= C_{B1} - (B + 1)\\
&= 90 - (140 + 0) = -50\\
C2 &= C_{C2} - (C + 2)\\
&= 100 - (40 + (-40)) = 100\\
B3 &= C_{B3} - (B + 3)\\
&= 40 - (140 + 50) = -150 \quad \text{(Use this route.)}\\
C4 &= C_{C4} - (C + 4)\\
&= 0 - (40 + (-100)) = +60\\
B4 &= C_{B4} - (B + 4)\\
&= 0 - (140 + (-100)) = -40
\end{aligned}$$

Table 7.16

		A	B	C	
	Shadow Costs	100	−10	40	
1	0	+ (cost 80)	(cost 90)	− 300 (cost 40)	300
2	−40	400 (cost 60)	(cost 100)	(cost 100)	400
3	50	− 150 (cost 150)	250 (cost 40)	+ 150 (cost 90)	550
4	−100	50 (cost 0)	(cost 0)	(cost 0)	
	AVAILABLE	600	250	450	

SITES (rows); REQUIRED (right-hand column)

Second feasible solution (Table 7.16)

New cost equals

$$
\begin{aligned}
400 \times 60\text{p} &= £240 \\
300 \times 40\text{p} &= £120 \\
150 \times 150\text{p} &= £225 \\
250 \times 40\text{p} &= £100 \\
150 \times 90\text{p} &= \underline{£135} \\
&\quad £820
\end{aligned}
$$

Third feasible solution

Let $1 = 0$, then

$1 + C = 40$, $2 + A = 60$, $3 + A = 150$, $3 + B = 40$, $3 + C = 90$,

$4 + A = 0$

$C = 40$, $3 = 50$, $B = -10$, $A = 100$, $2 = -40$, $4 = -100$

New cost equals

$$
\begin{aligned}
150 \times 80\text{p} &= £120 \\
150 \times 40\text{p} &= £60 \\
400 \times 60\text{p} &= £240 \\
300 \times 90\text{p} &= £270 \\
250 \times 40\text{p} &= \underline{£100} \\
&\quad £790
\end{aligned}
$$

Check if this is optimum solution (Table 7.17).

Table 7.17

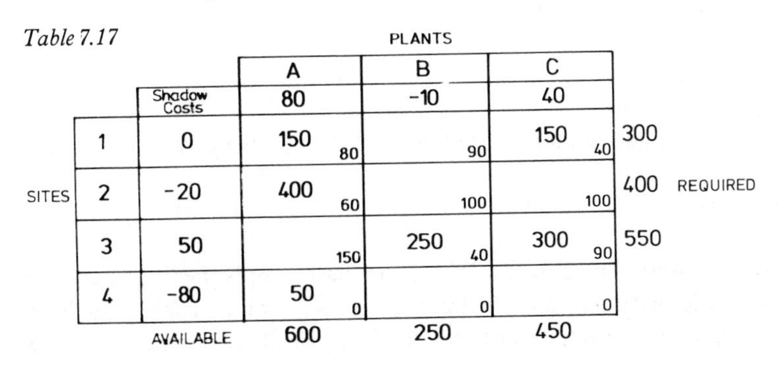

PLANTS

		A	B	C	
	Shadow Costs	80	−10	40	
1	0	150 (cost 80)	(cost 90)	150 (cost 40)	300
2	−20	400 (cost 60)	(cost 100)	(cost 100)	400
3	50	(cost 150)	250 (cost 40)	300 (cost 90)	550
4	−80	50 (cost 0)	(cost 0)	(cost 0)	
	AVAILABLE	600	250	450	

SITES (rows); REQUIRED (right-hand column)

Let $1 = 0$, then
$1 + A = 80$, $1 + C = 40$, $2 + A = 60$, $3 + B = 40$, $3 + C = 90$,
$4 + A = 0$
$A = 80$, $C = 40$, $2 = -20$, $3 = 50$, $B = -10$, $4 = -80$

Therefore: no minus, last iteration was optimum solution.

Cost (first feasible solution)	£1,195
Cost (final iteration)	£790
Saving	£405

7.5 The assignment problem

In the assignment problem a unique one-to-one pairing of resources and jobs is sought so as to minimise the sum of the measures of performance of each pairing that is made. Such problems arise, for example in assigning:

(a) machines or drivers to delivery routes;
(b) men to offices or projects;
(c) space to departments.

There is a set of jobs and a set of methods, a one-to-one correspondence. In an $n \times m$ matrix there are $n!$ ways of doing the matrix – permutation matrix.

Mathematical model – objective function

Maximise

$$Z = \sum_{i=1}^{m} \sum_{j=1}^{n} C_{ij} \times X_{ij}$$

subject to

$$\sum_{j=1}^{n} X_{ij} = 1, \quad j = 1, 2, \ldots n \text{ (columns)}$$

$$\sum_{i=1}^{m} X_{ij} = 1, \quad i = 1, 2, \ldots m \text{ (rows)}$$

$$X_{ij} \geqslant 0$$

The transportation technique cannot be used because the number of variables is not equal to $(n+m-1)$.

Example 5
Five men are available to do five different jobs. From past records the time that each man takes to do each job is known and the information is shown in Table 7.18. The C_{ij}'s are the times required for each man to perform the task and the x_{ij}'s are 1 or 0. If a

Table 7.18

Men	Jobs 1	2	3	4	5	Available
A	16	15	17	18	17	1
B	22	17	19	21	19	1
C	18	21	16	14	16	1
D	17	18	18	26	15	1
E	14	23	15	13	12	1

man is assigned to a job it is 1, otherwise 0. The aim is to achieve overall optimum assignments. This problem has 120 possible assignments (or feasible solutions), since there are five possible men to assign to job 1 and after man A is assigned a job there are four men left to assign to job 2, then three left to assign to job 3, two left to assign to job 4; and finally one remaining man to be assigned to job 5. Thus there are

$5 \times 4 \times 3 \times 2 \times 1 = 120$ feasible assignments

Procedure

1. Subtract from each column the minimum time element in the column (Tables 7.19 and 7.20).
2. Repeat the procedure for the rows (Table 7.21).
3. Obtain an allocation – find five zeros, so that there is one only in each row and one only in each column (Table 7.21).

Table 7.19

16	15	17	18	17
22	17	19	21	19
18	21	16	14	16
17	18	18	26	15
14	23	15	13	12

Table 7.20

Dealing with columns

2	0	2	5	5
8	2	4	8	7
4	6	1	1	4
3	3	3	13	3
0	8	0	0	0

Table 7.21

Dealing with rows →

2	0	2	5	5
6	0	2	6	5
3	5	0	0	3
0	0	0	10	0
0	8	0	0	0

4 lines, no allocation is possible

If number of lines is less than $n(5)$ then no allocation is possible. Find the smallest exposed number.

1. Subtract this number from all elements having no lines passing through.
2. Leave elements with lines passing through unchanged.
3. Add the number to all elements with two lines passing through (Table 7.22).

Table 7.22

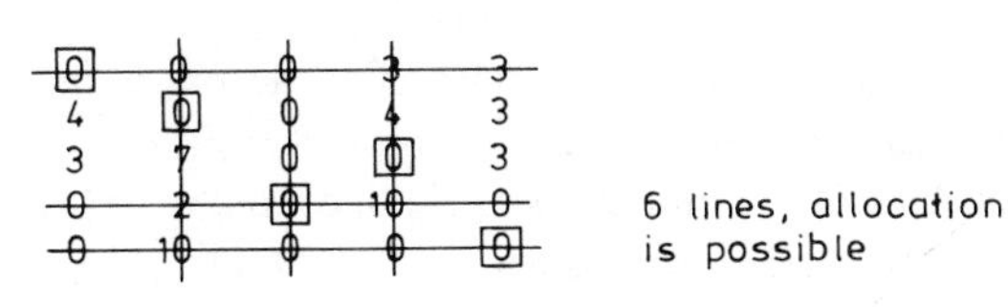

[0]	0	0	3	3
4	[0]	0	4	3
3	7	0	[0]	3
0	2	[0]	10	0
0	10	0	0	[0]

6 lines, allocation is possible

Man A to job 1
Man B to job 2
Man C to job 4 } Optimum assignment
Man D to job 3
Man E to job 5

The complete linear programming formulation for this is as follows. Let X_{A1} be 1 (if site agent A performs job 1) or 0 (if site agent A does not perform job 1), and similarly for $X_{B1}, X_{C1} \ldots X_{E5}$.

Minimise

$$\begin{aligned} &16X_{A1} + 15X_{A2} + 17X_{A3} + 18X_{A4} + 17X_{A5} \\ &+ 22X_{B1} + 17X_{B2} + 19X_{B3} + 21X_{B4} + 19X_{B5} \\ &+ 18X_{C1} + 21X_{C2} + 16X_{C3} + 14X_{C4} + 16X_{C5} \\ &+ 17X_{D1} + 18X_{D2} + 18X_{D3} + 26X_{D4} + 15X_{D5} \\ &+ 14X_{E1} + 23X_{E2} + 15X_{E3} + 13X_{E4} + 12X_{E5} \end{aligned}$$

subject to

$$\begin{aligned} X_{A1} + X_{A2} + X_{A3} + X_{A4} + X_{A5} &= 1 \\ X_{B1} + X_{B2} + X_{B3} + X_{B4} + X_{B5} &= 1 \\ X_{C1} + X_{C2} + X_{C3} + X_{C4} + X_{C5} &= 1 \\ X_{D1} + X_{D2} + X_{D3} + X_{D4} + X_{D5} &= 1 \\ X_{E1} + X_{E2} + X_{E3} + X_{E4} + X_{E5} &= 1 \\ X_{A1} + X_{B1} + C_{C1} + X_{D1} + X_{E1} &= 1 \\ X_{A2} + X_{B2} + X_{C2} + X_{D2} + X_{E2} &= 1 \\ X_{A3} + X_{B3} + X_{C3} + X_{D3} + X_{E3} &= 1 \\ X_{A4} + X_{B4} + X_{C4} + X_{D4} + X_{E4} &= 1 \\ X_{A5} + X_{B5} + X_{C5} + X_{D5} + X_{E5} &= 1 \end{aligned}$$

Example 6

A construction company has four crawler tractor cranes awaiting pick-up at four different sites. The plant depot has four low loaders, each capable of transporting any one of the cranes. The assignment of low loaders to cranes will affect the total distance travelled to pick up the cranes.

The plant manager would like an assignment of low loaders to cranes which minimises the total distance travelled to pick up the cranes. A low loader can only be assigned to pick up one crane. Table 7.23 gives the distance in kilometres between each crane – low loader combination.

The solutions (iterations) are given in Tables 7.23 to 7.28.

Example 7 (maximum allocation)

Five site agents are being considered for five projects. The suitability of a particular site agent for a specific project has been computed by the award of marks out of ten (Table 7.29). The higher the mark, the more suitable the site agent.

Procedure to date has required evaluation of the minimum total allocation, whereas this requires the maximum.

Table 7.23

Distance between cranes and low loader combinations

		Cranes			
		A	B	C	D
Low Loaders	1	20	14	30	15
	2	16	19	27	22
	3	33	23	19	29
	4	20	20	16	28

Table 7.24

Reducing the Columns

4	0	14	0
0	5	11	7
17	9	3	14
4	6	0	13

Table 7.25

Reducing the Rows — No allocation possible

4	0	14	0
0	5	11	7
14	6	0	11
4	6	0	13

Table 7.26

Deducting the smallest available number etc — No allocation possible

9	0	19	0
0	0	11	2
14	1	0	6
4	1	0	8

Table 7.27

Deducting the smallest available number etc — Optimum allocation

9	0	20	[0]
[0]	0	12	2
13	0	[0]	5
3	[0]	0	7

Table 7.28

Optimum Solution

Low Loader 1	to crane	D,	15 km
Low Loader 2	to crane	A,	16 km
Low Loader 3	to crane	C,	19 km
Low Loader 4	to crane	B,	20 km
	Total		70 km

Table 7.29

Site agent	Project A	B	C	D	E
A	10	8	9	5	8
B	7	4	6	8	6
C	9	5	4	3	5
D	6	6	7	6	8
E	5	3	2	4	2

The technique of solution is to invert this problem by making the maximum allocation equal to zero. All other elements then become the difference between the maximum and their value.

Table 7.30

Dealing with rows

0	2	1	5	2
3	6	4	2	4
1	5	6	7	5
4	4	3	4	2
5	7	8	6	8

Table 7.31

Dealing with columns

0	0	0	3	0
3	4	3	0	2
1	3	5	5	3
4	2	2	2	0
5	5	7	4	6

Table 7.32

Reducing the rows

0	0	0	3	0
3	4	3	0	2
0	2	4	4	2
4	2	2	2	0
1	1	3	0	2

Only four lines

No allocation

Table 7.33

Minimum exposed value is 1

Columns

1	0	[0]	4	1
3	3	2	[0]	2
[0]	1	3	4	2
4	1	1	2	[0]
1	[0]	2	0	2

Five lines
Allocation possible

Optimum allocation is
Agent A to project c
B to project d
C to project a
D to project e
E to project b

Example 8 – Inclusion of a dummy

A process can be carried out on any one of six machines by any one of six operatives. The average time taken by any operator on any specific machine is tabulated in Table 7.34. Consideration is being given to a new machine to replace one of the existing ones.

Is it advantageous at this stage to use the new machine? If so, which of the original machines should it replace and how should the operators be allocated?

Table 7.34

	Machine						
Operator	1	2	3	4	5	6	New
A	10	8	12	10	12	11	8
B	7	10	9	8	9	7	10
C	8	8	7	8	6	8	8
D	14	14	15	13	12	11	14
E	8	9	9	8	10	9	8
F	9	9	7	8	9	9	8

Since this is not a square matrix it cannot be solved as before. Make it a square matrix by introducing a dummy operator G and allocating times to him that will not make it worth while to give him preferential allocation. Then, in the end analysis, any machine that has G allocated to it is the least important to obtain the total minimum time.

Table 7.35

	10	8	12	10	12	11	8
	7	10	9	8	9	7	10
	8	8	7	8	6	8	8
	14	14	15	13	12	11	14
	8	9	9	8	10	9	8
	9	9	7	8	9	9	8
Say G	−30	30	30	30	30	30	30

Table 7.36

Reducing by columns

3	0	5	2	6	4	0
0	2	2	0	3	0	2
1	0	0	0	0	1	0
7	6	8	5	6	4	6
1	1	2	0	4	2	0
2	1	0	0	3	2	0
23	22	23	22	24	23	22

5 lines
No allocation

Table 7.37

Reducing by rows

3	0	5	2	6	4	[0]
[0]	2	2	0	3	[0]	2
1	0	0	0	[0]	1	0
3	2	4	1	2	0	2
1	1	2	[0]	4	2	0
2	1	[0]	0	3	2	0
1	[0]	1	0	2	1	[0]

7 lines
Allocation possible

The least important machine to obtain total minimum time is 2.

Allocations are:

Operator A to new machine
Operator B to machine 1
Operator C to machine 5
Operator D to machine 6
Operator E to machine 4
Operator F to machine 3
Operator G to machine 2

Chapter 8

Queueing theory

8.1 Introduction

Queueing theory is a valuable tool in management because many business problems can be characterised as arrival–departure problems. Queues are commonplace everyday experiences. For example:

1. In supermarkets.
2. At bus-stops.
3. Aircraft waiting to take off.

Less obvious examples are:

4. Replacement of stock at construction sites.
5. Waiting to speak to directory inquiries.
6. Machines waiting for repairs.

The term *customer* is used to refer to people, cars, aircraft, construction plant, etc.

Service facilities refer to, checkout counter, car parks, ready-mix concrete depots, material suppliers, construction sites, etc.

A queueing problem arises:

(a) If either the arrival rate of customers or the amount of service facilities available are subject to control.
(b) If there are costs associated with both the waiting time of customers and idle time of facilities.

Queueing problems usually require a 'compromise decision', i.e. either scheduling arrivals or providing facilities or both so as to minimise the sum of the cost of waiting customers and idle facilities. Since it is not possible to predict that a queue will or will not form it follows that to find a mathematical relationship we must consider probabilities. When some kind of 'event' occurs repeatedly but haphazardly (i.e. queues) we can use the Poisson distribution.

Let A_t = number of arrivals during the time interval t. If the following hold then A_t has a Poisson distribution:

1. Arrivals come independently – number of arrivals in a given time interval has no effect on the number of arrivals during any other time interval.

2. The probability of more than one arrival in a given time interval is negligible in comparison with the probability of one arrival in the same time interval.
3. The probability of an arrival in time interval t is approximately proportional to the time (for small t), i.e. the longer the time interval the more arrivals there tend to be.

If A_t has a Poisson distribution,

Values of $A_t = 0, 1, 2, \ldots$

Distribution of A_t : P (n arrivals in time t) is

$$\frac{e^{-\lambda t}(\lambda t)^n}{n!} \qquad n = 0, 1, 2, \ldots$$

Mean $\mu = \lambda t$ [expected number of events in time t] and variance $= \lambda t$

where parameter λ (lambda) is called the mean rate of arrival per unit time, and mean μ (mu) is called the mean rate of service per unit time.

When the variability of the number of customers arriving in a given time interval and the number of customers having completed service in that interval can be described by a Poisson distribution then the resulting system is called a *simple queue*. The most commonly used statistics in Poisson queueing theory are

$$\text{Traffic intensity } \rho \text{ (rho)} = \frac{\text{Mean rate of arrival}}{\text{Mean rate of service}} = \frac{\lambda}{\mu}$$

$$\text{Mean inter-arrival time} = \frac{1}{\lambda} = \text{mean service time} = \frac{1}{\mu}$$

so

$$\rho = \frac{1/\mu}{1/\lambda}$$

If $\lambda \geqslant \mu$ (mean rate of arrival greater or equal to the mean rate of service), then $\rho > 1$; system is overloaded more often than not – queues form. If $\lambda < \mu$ (mean rate of arrival less than mean rate of service), then $\rho < 1$; system is not overloaded – no queue.

Probability of n customers in system (average number of customers in system) $= \dfrac{\rho}{1-\rho}$ or $\dfrac{\lambda}{\mu-\lambda}$

Average number of customers in the queue including when queue length is zero $= \dfrac{\rho^2}{1-\rho}$ or $\dfrac{\lambda^2}{\mu(\mu-\lambda)}$

Average length of queue – excluding zero queues $= \dfrac{1}{1-\rho}$ or $\dfrac{\mu}{\mu-\lambda}$

Average queuing time – or average time a customer is in a queue $= \dfrac{\rho}{\mu(1-\rho)}$ or $\dfrac{\lambda}{\mu(\mu-\lambda)}$

Average time a customer is in the system $= \dfrac{1}{\mu(1-\rho)}$ or $\dfrac{1}{\mu-\lambda}$

Also for Poisson distribution with parameter $m = \lambda t$ (m = mean of distribution)

$$P(n \text{ events in time } t) = \frac{e^{-\lambda t}(\lambda t)^n}{n!} \qquad n = 0, 1, 2, \ldots$$

so λt = expected number of events in time t. When $t = 1$, λ is the expected number of events per unit time. Also if arrivals are Poisson the distribution of time between arrivals is the 'negative exponential'

$$f(t) = \lambda e^{-\lambda t} \qquad t \geqslant 0$$

where $f(t)$ is the probability density function.

$$\text{Mean} = \frac{1}{\lambda} \qquad \text{variance} = \frac{1}{\lambda^2}$$

8.2 Examples

Example 1
Suppose customers arrive at a service point (according to a Poisson distribution) and that the arrivals on average are twenty-four per hour. What is the probability of no arrivals in a given 5-minute interval?

$$\lambda t = 24 \times \frac{5}{60} = 2$$

P(number of arrivals in 5-minute period) $= e^{-\lambda t}$
$= e^{-2}$
$= 0.135$ (obtained from statistical tables: the exponential function)

Example 2
If the expected arrival rate is five per minute what is the expected time between arrivals of individual customers?

$\lambda = 5 \qquad \dfrac{1}{\lambda} = \dfrac{1}{5}$ minute or 12 seconds

Example 3
For a simple queue what is the average number of customers in the system for a traffic intensity of 0.8?

$$\rho = 0.8$$
$$1-\rho = 0.2$$

$$\frac{\rho}{1-\rho} = \frac{0.8}{0.2} = 4 \quad \text{(average number of customers)}$$

8.3 Single-server queueing model

Fig. 8.1 Single-server queueing model

In this system (Fig. 8.1) we assume a Poisson distribution and it therefore follows that the following assumptions must be made with respect to rates of arrivals and rate of service:

1. Random arrival and service.
2. First in, first served and no customer departs without service.
3. One service point.
4. Infinite number of customers.
5. One at a time at service.
6. Average rate of services greater than average rate of arrival.

Example 4
On average, cement lorries arrive at a factory every 15 minutes to be reloaded. A single hopper is capable of serving on average six lorries per hour. Service times and inter-arrival times follow a negative exponential distribution. Find

(a) Probability of a lorry having to wait for service.
(b) Probability of a lorry arriving and finding at least one lorry already at the factory.
(c) Average number of lorries at factory at any moment.
(d) What is the average time a lorry would expect to be at the factory?

Solution
If lorries arrive every 15 minutes, then on average four arrive per hour. Take 1 hour as unit of time:

$\lambda = 4$
$\mu = 6$

Since time and inter-arrival follow a negative exponential we can assume that a simple queue exists – therefore we can use Poisson distribution and its formula. Hence:

(a) Traffic intensity $\rho = \dfrac{\lambda}{\mu} = \dfrac{4}{6} = \dfrac{2}{3}$.

(b) This is also $\rho = \dfrac{2}{3}$.

(c) $\dfrac{\lambda}{\mu-\lambda} = \dfrac{4}{6-4} = 2.$

(d) $\dfrac{1}{\mu-\lambda} = \dfrac{1}{6-4} = \dfrac{1}{2}$ hr or 30 minutes.

Example 5

Packaged bricks are delivered to a site by contractor's lorries at a rate (average) of one every 12 minutes. A mobile crane plus labour can unload a lorry in 8 minutes and costs £4.20/hr whether working or waiting. The time that each lorry is on the site, either waiting or unloading, costs the contractor £5/hr. Alternative equipment and labour could be hired; this could unload 50 per cent faster but cost 50 per cent more per hour. Assuming random arrival of lorries, should we use the first or second unloading system?

(a) Small mobile crane (first system – random arrival – Poisson distribution for simple queues).

Rate of arrival $\lambda = \dfrac{60}{12} = 5$ lorries/hr

Rate of service $\mu = \dfrac{60}{8} = 7.5$ lorries/hr

$$\text{Average time a customer is in the system} = \frac{1}{\mu-\lambda}$$

$$= \frac{1}{7.5-5} = 0.4 \text{ hr}$$

Cost of mobile crane = £4.20/hr
Cost of 10 lorries on site = $10 \times 0.4 \times 5$ = £20/hr
Total cost = £24.20/hr

(b) Second system.
$\lambda = 5$
$\mu = 1.5 \times 7.5 = 11.25$ lorries/hr

$$\text{Average time a customer is in the system} = \frac{1}{11.25-5}$$

$$= 0.16 \text{ hr}$$

Cost of alternative equipment and labour = £5.30/hr
Cost of 10 lorries on site = $10 \times 0.16 \times 5$ = £8.00
Total cost = £13.30/hr

The time in the system is greatly reduced from $0.4 \rightarrow 0.16$. Hence the second system is cheaper.

8.4 Service points arranged in parallel

Suppose there are n identical service points arranged in parallel. The customer at the beginning of the queue is served at the service point which becomes available next (Fig. 8.2).

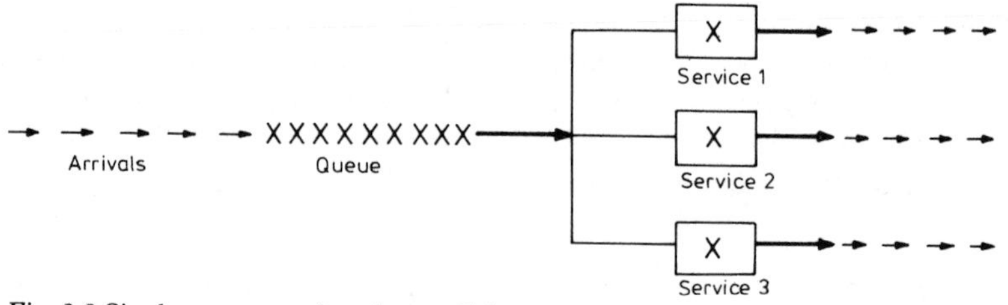

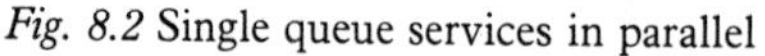
Fig. 8.2 Single queue services in parallel

Assume the system is identical with the simple queue the statistics for which are stated on pp. 168–9 then the statistics for this system must be as follows:

$$\text{Traffic intensity} = \frac{\text{Mean rate of arrival}}{n \times \text{mean rate of service}}$$

$$\rho = \frac{\lambda}{n\mu} \quad \text{or} \quad = \frac{1/\mu}{n/\lambda}$$

As before, if $\lambda \geqslant n\mu$, then $\rho > 1$ (system overloaded); and if $\lambda < n\mu$, $\rho < 1$ (usually no queue). Hence the statistics become:

P_0 = probability that there are no customers in the system

$$\text{Average number of customers in the system} = \frac{\rho(n\rho)^n}{n!(1-\rho)^2} P_0 + n\rho$$

$$\text{Average number of customers in the queue} = \frac{\rho(\rho n)^n}{n!(1-\rho)^2} P_0$$

Average time a customer is in the system $= \dfrac{(n\rho)^n}{n!(1-\rho)^2\, n\mu} P_0 + \dfrac{1}{\mu}$

Average time a customer is in the queue $= \dfrac{(n\rho)^n}{n!(1-\rho)^2\, n\mu} P_0$

Probability of a customer having to wait for service $= \dfrac{(n\rho)^n}{n!(1-\rho)} P_0$

The probability that there are no customers at any service point is given by the equation:

$$P_0 = \frac{n!\,(1-\rho)}{(n\rho)^n + n!(1-\rho) \sum\limits_{n=0}^{n-1} \dfrac{1}{n!} (n\rho)^n}$$

This is only valid when $n\mu > \lambda$.

Example 6

A ready-mix concrete plant has three identical loading bays each of which can fully load an average of six ready-mix lorries per hour. An average of twelve lorries arrive each hour. Find:

(a) Probability of a lorry having to wait to be loaded.
(b) Average number of lorries in the system.
(c) Average number of lorries in the queue.
(d) Average time a lorry is in the system.
(e) Average time a lorry is in the queue.

Solution

First we must calculate P_0 ($n = 3$, $\mu = 6$, $\lambda = 12$):

$$P_0 = \frac{3!(1-\frac{2}{3})}{(2)^3 + 3!(1-\frac{2}{3}) \sum\limits_{n=0}^{n=2} \frac{1}{3}! \,(2)^n}$$

$$= \frac{6 \cdot \frac{1}{3}}{8 + 6 \cdot \frac{1}{3}\,[\frac{1}{6} + \frac{2}{6} + \frac{4}{6}]}$$

$$= \frac{2}{8 + 2\frac{1}{3}} = \frac{6}{31} = 0.19 \text{ or } 19 \text{ per cent}$$

(a) Probability of a lorry having to wait is

$$\frac{(3 \cdot \frac{2}{3})^3}{3!(1-\frac{2}{3})} 0.19 = 0.76$$

(b) Average number of lorries in system is

$$\frac{\frac{2}{3}(2)^3}{3!(1-\frac{2}{3})^2}\,0.19 + 3\cdot\tfrac{2}{3} = 3.52$$

(c) Average number of lorries in the queue is

$$\frac{\frac{2}{3}(2)^3}{3!(1-\frac{2}{3})^2}\,0.19 = 1.52$$

(d) Average time a lorry is in the system is

$$\frac{(2)^3}{3!(\frac{1}{3})^2 3\cdot 6}\,0.19 + \frac{1}{6} = 0.29 \text{ hr}$$

(e) Average time a lorry is in the queue is

$$\frac{(2)^3}{3!(1-\frac{2}{3})^2 3\cdot 6}\,0.19 = 0.13 \text{ hr}$$

Chapter 9

Inventory models – deterministic demand

9.1 Introduction

Inventories provide the operating flexibility which ensure that an organisation's operations perform smoothly and efficiently. Proper inventory management can effect substantial savings to a company. The main decisions involved in inventory models are:

1. How many units to order.
2. When to order.

Although there are some similarities in all inventory systems each system is sufficiently unique to prevent the use of a general model for all situations.

9.2 Economic order quantity model (EOQ)

Certain advantages may be gained by buying goods in large quantities to save cost of ordering, handling and transportation, not to mention discounts. This, however, must be offset by increased holding costs and loss of interest on capital tied up in stock. A compromise is sought between too large or too small an order quantity. The quantity which minimises the total costs is called the economic order quantity. This is the best-known and most fundamental inventory decision model. Since it is too oversimplified to reflect most 'real life' situations it must be considered as completely theoretical. It does, however, give a good starting-point for further analysis of more realistic and complex models.

In this ideal model we assume the following are constant:

1. Demand for stock.
2. Rate of usage.
3. Delay time in delivery of stock.

Also we assume that stock is replenished at the exact moment when stock level becomes zero as shown in Fig. 9.1.

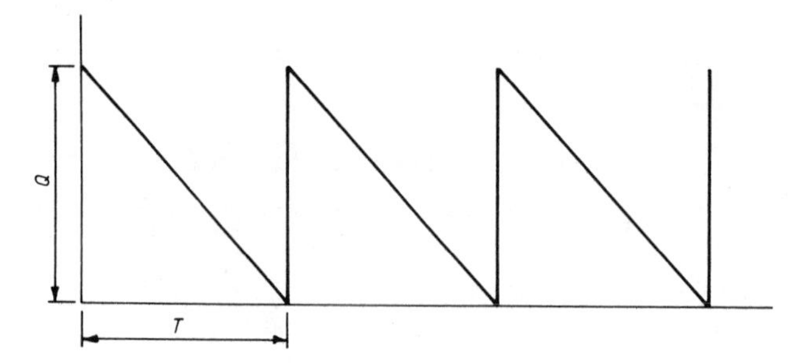

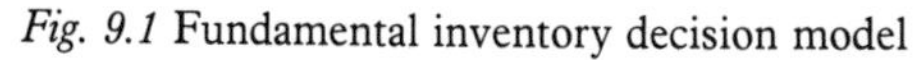

Fig. 9.1 Fundamental inventory decision model

Let Q = order quantity;
A = annual demand requirement (units/year);
T = length of time between orders;
N = number of orders per year;
C_h = annual cost of holding stock as a proportion of cost of stock;
C_q = purchase price per unit;
R = reorder cycle;
C_T = total annual cost of acquiring and holding stock;
P = marginal cost of placing and following through an order.

Now

C_T = holding cost per year + order cost per year

(a) Ordering cost.

Ordering cost = number of orders × marginal cost of placing and following through an order

$= N \times P$

$= NP$

But

$$N = \frac{\text{Annual demand}}{\text{Order quantity}} = \frac{A}{Q}$$

$$\Rightarrow \text{Ordering costs} = \frac{A}{Q}P \qquad \ldots [1]$$

(b) Holding costs.

Holding costs = average inventory × percentage holding cost

Average inventory $= \dfrac{Q}{2}$ (reorder cycle begins at Q and ends at 0

so average is $\dfrac{Q+0}{2} = \dfrac{Q}{2}$)

$$\text{Holding costs} = \frac{Q}{2} \cdot C_q \cdot C_h \qquad \ldots [2]$$

Combining equations [1] and [2] gives

$$C_T = \frac{A}{Q}P + \frac{Q}{2}C_q C_h$$

To find the optimal value of Q (that is, minimise C_T) we use simple maximum and minimum theory. Differentiate C_T with respect to Q and equate to zero

$$\frac{dC_T}{dQ} = -\frac{AP}{Q^2} + \frac{C_q C_h}{2} = 0$$

$$\Rightarrow Q^2 = \frac{2AP}{C_q C_h}$$

Hence the optimal order quantity is

$$Q^\star = \sqrt{\frac{2AP}{C_q C_h}}$$

This is sometimes called the square root formula or standard formula. Since $Q^\star$ is related to the square root of the constant, it is not very sensitive to any change in these constants and it is therefore seldom necessary to estimate them with great accuracy.

9.3 *The optimal number of orders N**

Now specify ordering costs and holding cost in terms of N. Since

$$Q = \frac{A}{N}$$

$$\text{Ordering costs} = NP$$

$$\text{Holding costs} = \frac{A}{2N} C_q \cdot C_h$$

Hence

$$C_T = NP + \frac{A}{2N} C_q \cdot C_h$$

Again use maximum and minimum theory. Differentiate C_T with respect to N and equate to zero.

$$\frac{dC_T}{dN} = P - \frac{A}{2N^2} C_q C_h = 0$$

$$P = \frac{A}{2N^2} C_q C_h$$

$$N^2 = \frac{A}{2P} C_q C_h$$

$$N^\star = \sqrt{\frac{A}{2P} C_q C_h}$$

9.4 To find the reorder cycle

$$N^\star = \frac{1}{R^\star}$$

Hence when $N^\star$ is known the reorder cycle $R^\star$ can easily be calculated. So $Q^\star$ – how much to order – and $R^\star$ – when to order – are both known.

Example 1

A large builder's supplier forecasts the demand for flush doors to be 15,000/year. The purchase cost of each door is £20 and the accounting department estimated that the cost of processing a purchase is £55. The holding costs are 20 per cent per annum. Assuming linear usage and deterministic conditions find the optimum order quantity.

Solution

$$Q^\star = \sqrt{\frac{2AP}{C_q C_h}}$$

$A = 15{,}000, \quad P = 55, \quad C_q = 20, \quad C_h = 0.20.$

$$Q^\star = \sqrt{\frac{2 \times 15{,}000 \times 55}{20 \times 0.20}} = 642.26 \text{ units/order}$$

Example 2

Builders' Suppliers Limited sell concrete paving stones. The sales forecast for the coming year is 50,000. The cost of processing each order is £50 and the purchase price is £3/stone. If the holding costs are 10 per cent per annum what is the optimal number of orders?

Solution

$A = 50{,}000, \quad C_q = 3, \quad C_h = 0.10, \quad P = 50.$

$$N^* = \sqrt{\frac{50{,}000 \times 3 \times 0.10}{2 \times 50}} = 12.25 \text{ orders/year}$$

9.5 Stocks-out

When using EOQ shortages were not allowed. However, it is often more economical for the inventory level to fall below zero than to incur large holding costs. Shortage cost must be considered but these are difficult to compute since they are usually related to 'goodwill'. When stocks-out occur the time period is considered in two parts (Fig. 9.2).

1. No stocks-out exist.
2. Stocks-out exist.

Let

t_1 = average time in stock (in years)
t_2 = average time out of stock (in years)

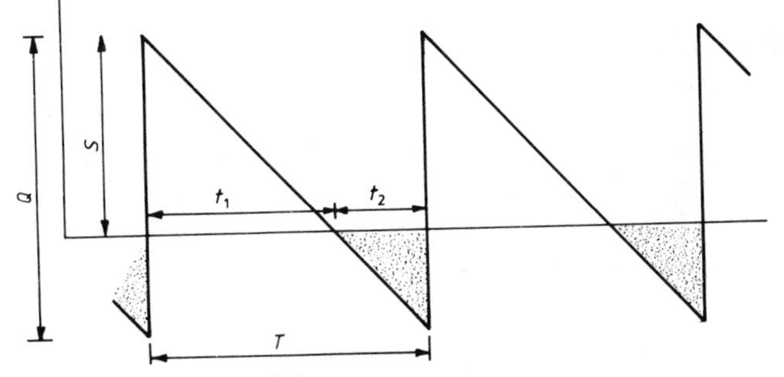

Fig. 9.2 Stocks-out model

By similar triangles

$$\frac{t_1}{T} = \frac{S}{Q}$$

$$t_1 = \frac{ST}{Q} \qquad \ldots [3]$$

and

$$\frac{t_2}{T} = \frac{Q-S}{Q}$$

$$t_2 = \frac{(Q-S)T}{Q} \qquad \ldots [4]$$

Now calculate costs:

(a) Holding costs per order = time × stock carried

$$\times \frac{\text{Holding cost/year} \times \text{unit cost of item}}{2}$$

= average inventory × per cent holding costs

$$= t_1 \frac{S}{2} \cdot C_h C_q$$

(b) Ordering cost = PA/Q.

(c) Cost of stock-out is

$$\frac{(Q-S)}{2} \cdot \frac{C_0 t_2 A}{Q}$$

where C_0 is the cost of a stock-out.
Hence:

C_T = ordering cost + holding cost + cost of stock-out

$$C_T = \frac{PA}{Q} + t_1 \frac{S}{2} C_h C_q + \frac{(Q-S)}{2} C_0 \frac{t_2}{Q} \cdot A$$

Substitute for t_1 and t_2 from equations [3] and [4] gives:

$$\text{Cost of stocks-out} = \frac{C_0(Q-S)}{2} \cdot \frac{A}{Q} \cdot \frac{(Q-S)}{Q} \cdot T$$

$$= \frac{C_0(Q-S)^2 \cdot AT}{2Q^2}$$

But $A = NQ$. Therefore:

$$\text{Cost of stocks-out} = \frac{C_0(Q-S)^2 \cdot NQT}{2Q^2}$$

$$= \frac{C_0(Q-S)^2 NT}{2Q}$$

But $NT = 1$ (year). Hence:

$$\text{Cost of stock-out} = \frac{C_0(Q-S)^2}{2Q}$$

Also,

$$\text{Holding costs} = \frac{C_q\, C_h S}{2}\, t_1$$

$$= \frac{C_q\, C_h S}{2} \cdot \frac{ST}{Q}$$

$$= \frac{C_q\, C_h\, S^2 T}{2Q}$$

Since there are N orders per year, the annual holding cost is N times. That is, holding cost per year is

$$\frac{C_q\, C_h\, S^2\, TN}{2Q}$$

But again $TN = 1$ (year). So holding cost per year is

$$\frac{C_q\, C_h\, S^2}{2Q}$$

Hence:

$$C_T = \frac{PA}{Q} + \frac{C_0(Q-S)^2}{2Q} + \frac{C_q\, C_h\, S^2}{2Q}$$

Since (P, A, C_0, C_h, C_q) are known we can now calculate the optimal order quantity $Q^\star$ using calculus.

Differentiate C_T partially with respect to Q:

$$\frac{\partial C_T}{\partial Q} = -\frac{PA}{Q^2} + \frac{C_0}{2}\left[\frac{2Q(Q-S)-(Q-S)^2}{Q^2}\right] - \frac{C_q\, C_h\, S^2}{2Q^2}$$

$$= -\frac{PA}{Q^2} + \frac{C_0}{2Q^2}\left[2Q^2 - 2QS - Q^2 - S^2 + 2QS\right] - \frac{C_q\, C_h\, S^2}{2Q^2}$$

$$= -\frac{PA}{Q^2} + \frac{C_0}{2Q^2}(Q^2 - S^2) - \frac{C_q\, C_h\, S^2}{2Q^2}$$

$$= -\frac{PA}{Q^2} + \frac{C_0}{2} - \frac{C_0 S^2}{2Q^2} - \frac{C_q\, C_h\, S^2}{2Q^2} \qquad \ldots [5]$$

Before proceeding we need to find S. Hence differentiate C_T partially with respect to S:

$$\frac{\partial C_T}{\partial S} = \frac{C_0}{2Q} \cdot 2(Q-S) - 1 + \frac{2C_q\, C_h\, S}{2Q} = 0 \quad \text{for optimal solution}$$

$$-2C_0Q + 2C_0S + 2C_qC_hS = 0$$

$$S(C_0 + C_qC_h) = QC_0$$

$$S^\star = \frac{Q \cdot C_0}{C_0 + C_hC_q}$$

Substitute this into [5] and equate to zero:

$$-\frac{PA}{Q^2} + \frac{C_0}{2} - \frac{C_0}{2Q^2} \cdot Q^2 \cdot \left[\left(\frac{C_0}{C_0 + C_hC_q}\right)\right]^2 - \frac{C_qC_h}{2Q^2} \cdot Q^2 \left[\frac{C_0}{C_0 + C_hC_q}\right]^2 = 0$$

$$-2PA + C_0Q^2 - C_0Q^2\left(\frac{C_0}{C_0 + C_hC_q}\right)^2 - C_qC_hQ^2\left(\frac{C_0}{C_0 + C_hC_q}\right)^2 = 0$$

$$Q^2\left[C_0 - C_0\left(\frac{C_0}{C_0 + C_hC_q}\right)^2 - C_qC_h\left(\frac{C_0}{C_0 + C_hC_q}\right)^2\right] = 2PA$$

$$Q^2\left[C_0 - \left(\frac{C_0}{C_0 + C_hC_q}\right)^2 (C_0 + C_hC_q)\right] = 2PA$$

$$Q^2\left[C_0 - \frac{C_0^{\,2}}{C_0 + C_hC_q}\right] = 2PA$$

$$Q^2\left[\frac{C_0 + C_hC_q - C_0}{C_0 + C_hC_q}\right] = \frac{2PA}{C_0}$$

$$Q^\star = \sqrt{\frac{2PA}{C_hC_q}} \cdot \sqrt{\frac{C_0 + C_hC_q}{C_0}}$$

Compare this result with that obtained or $Q^\star$ without shortages, i.e.:

$$Q^\star = \sqrt{\frac{2AP}{C_qC_h}}$$

They differ by the constant

$$\sqrt{\frac{C_0 + C_hC_q}{C_0}}$$

Hence when shortage costs are included $Q^\star$ is increased. Note also that as shortage costs C_0 become large relative to holding costs C_h the quantity $(C_0 + C_hC_q)/C_0$ approaches 1.

In other words, the back-order model and the regular EOQ model give similar results.

Also by substituting for $Q^\star$ we have

$$C_T^\star = \sqrt{C_h C_q PA}\sqrt{\frac{C_0}{C_h C_q + C_0}}$$

Example 3

A builders' supplier distributes roof trusses to a number of local contractors. It is estimated that the annual demand is 8,000 trusses per year and that the cost of processing an order is £5. The inventory holding cost is 15 per cent and the price per truss is £35. Shortage costs based on loss of goodwill, loss of potential profit and sales are estimated at five times the holding costs. Determine whether it is advisable to follow an inventory model that allows for some back orders (shortages) to occur.

Solution

From previously derived formula:

$$Q^\star = \sqrt{\frac{2AP}{C_q C_h}}\sqrt{\frac{C_0 + C_h C_q}{C_0}}$$

(with shortages permitted)

$A = 8{,}000$, $P = £5$, $C_q = 35$, $C_h = 0.15$, $C_0 = 5 \times 35 \times 0.15$.

So

$$Q^\star = \sqrt{\frac{2 \times 8{,}000 \times 5}{0.15 \times 35}}\sqrt{\frac{(5 \times 35 \times 0.15)+(0.15 \times 35)}{5 \times 35 \times 0.15}}$$

$= 135$ units

Also

$$C_T^\star = \sqrt{2 C_h C_q PA}\sqrt{\frac{C_0}{C_h C_q + C_0}}$$

$$= \sqrt{2 \times 0.15 \times 35 \times 5 \times 8{,}000}\sqrt{\frac{5 \times 35 \times 0.15}{(0.15 \times 35)+(5 \times 0.15 \times 35)}}$$

$= £592$

Compare these results with those for $Q^\star$ and $C_T^\star$ with no shortages permitted:

$Q^\star = 123$ units

$C_T^\star = £648$

Hence allowing shortages gives a saving of £56 or approximately 8.6 per cent in cost from no shortages. This comparison is, however, based on an accurate assessment of stock-out cost which in reality is extremely difficult to estimate as deterioration in goodwill and lost sales must also be considered.

Part 2

Construction economy studies

Chapter 10

Primary economic comparisons

Economy studies are concerned with the difference in economic results from alternative courses of action. In general terms, there are two broad classes of construction economy problems:

1. Primary economic comparisons.
2. Time-based studies.

Time-based studies, i.e. cash flow forecasting and investment appraisal, will be considered later in the text. Primary economic comparison implies that all the factors influencing the decision are already present. The effects of time are usually irrelevant or at most only short-range cost relationships are considered.

Primary comparisons can be subdivided into:

(a) present economy studies;
(b) break-even analysis.

10.1 Present economy studies

Typical applications of present economy studies in construction are:

1. Selection of plant and equipment.
2. Selection of materials and construction methods.

In the following examples, to illustrate the principles of present economy studies, it is unnecessary to consider the time value of money but rather the manner in which the various resources may be combined to give alternative cost solutions. To obtain a true comparison between one alternative and another it is important to ensure that each alternative under consideration is equivalent in every detail.

Examples of present economy studies

1. *Plant/equipment selection*

(a) Material excavation – haul length at 350 m

	Cost/hr	Output/hr	Cost/m^3
Towed scrapers	£40	250 m^3	16p★
Motor scrapers	£130	600 m^3	21p
Dump trucks	£30	150 m^3	20p

★Therefore use towed scrapers.

(b) Material excavation – haul length 550 m

	Cost/hr	*Output/hr*	*Cost/m^3*
Towed scrapers	£40	100 m^3	40p
Motor scrapers	£130	450 m^3	29p★
Dump trucks	£30	80 m^3	38p

★Therefore use motor scrapers.

2. *Material selection*

	Material A	*Material B*
Strength	25,000 units	35,000 units
Cost	£75/tonne	£85/tonne
Ratio	£0.0030 unit/tonne	£0.0024 unit/tonne

Therefore choose material B – lower cost/strength ratio.

Effect of changing conditions

It is important to recognise that the correct solution today may not apply tomorrow if the conditions on which the data were compiled have changed. Thus, concrete pipes for a sewage scheme may be economic today, but at some future date a cheaper alternative (e.g. plastic pipes) could become available.

10.2 Break-even analysis

An important method of analysis which can be employed in the selection among alternatives is that of break-even analysis. The break-even chart shows how the costs of using different processes or plant, for accomplishing the same objective, are affected by a common variable, and for a certain value of the variable (e.g. time, cubic metres, etc.) the costs will be equal. This value of the variable is known as the break-even point (Fig. 10.1(a) and (b)). Break-even analysis is also applicable to a manufacturing situation where revenue/volume relationships are to be considered with respect to fixed and variable costs of production (Fig. 10.1(c)). Another way of presenting the same information is the profit–volume chart (Fig. 10.2).

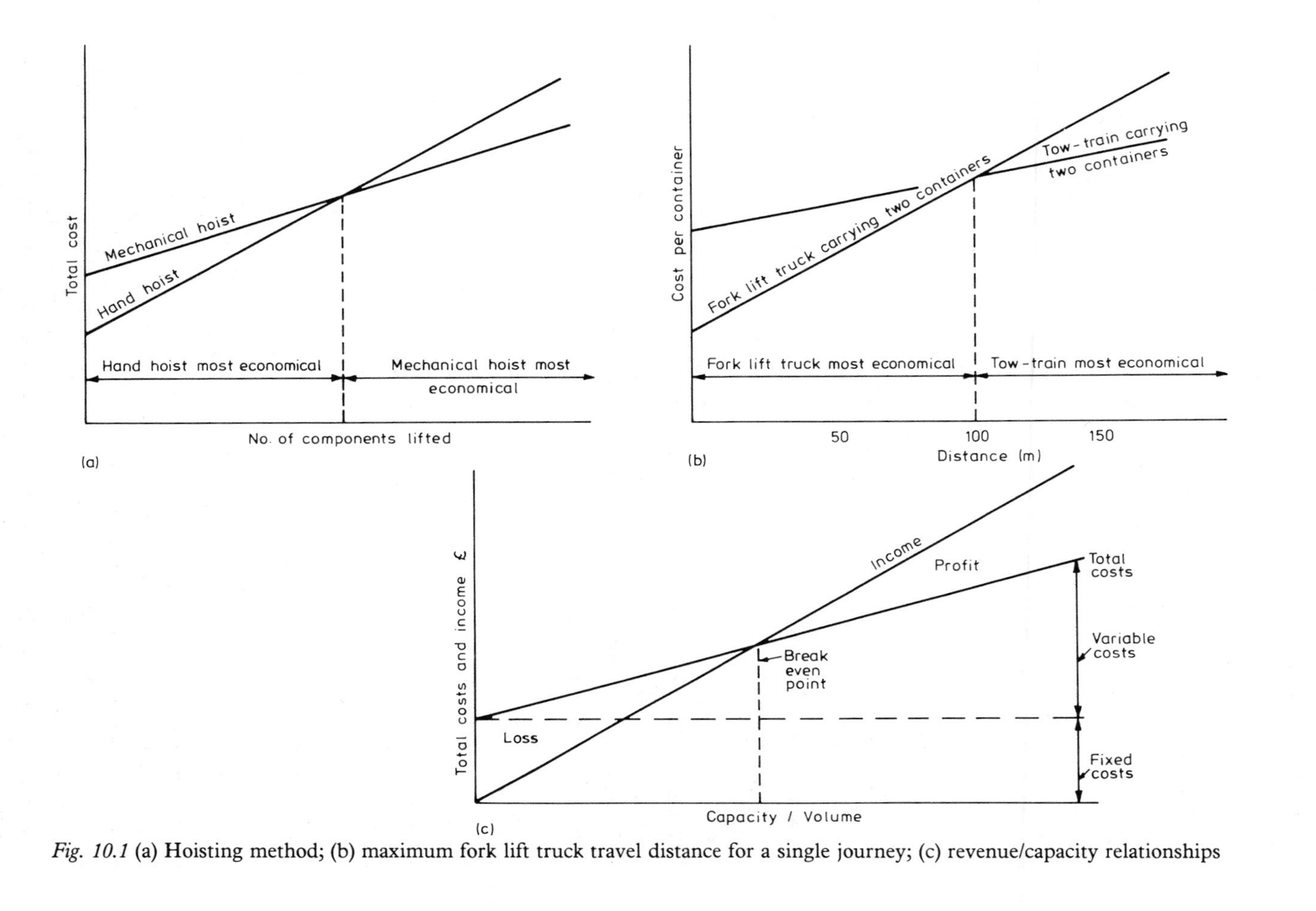

Fig. 10.1 (a) Hoisting method; (b) maximum fork lift truck travel distance for a single journey; (c) revenue/capacity relationships

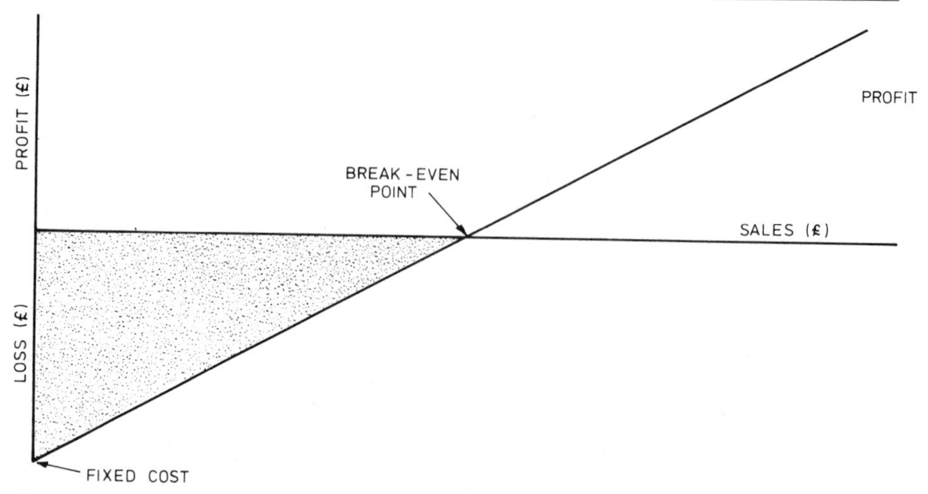

Fig. 10.2 Profit–volume chart

The profit–volume chart is used in place of or in addition to the break-even chart. Profits and losses are shown in the vertical scales, units or percentages on the horizontal scales.

The profits and losses at various sales levels are then plotted on the chart and where this line crosses the horizontal zero line, it gives the break-even point.

In this form, the break-even chart and profit–volume chart provide a useful way of studying the profit factors of a manufacturing business. There are four ways in which profit can be increased (or loss decreased).

1. Increase selling price per unit.
2. Decrease variable costs per unit.
3. Decrease fixed costs.
4. Increase volume.

The effect of the change in any of the four variables can be depicted on the break-even graph (Fig. 10.3).

Example 1

A haulage contractor operates a rough terrain tipper which he hires out for a fixed charge of £60.00 per week, plus a mileage charge of 20p per mile run. The weekly cost of operating the vehicle based on a mileage of 300 weekly is made up of:

Fixed charges £80.00
Running charges £20.00

1. Prepare a break-even chart showing the mileage at which the revenue earned in a week equals the total operating costs for a week.
2. Indicate on the chart, and state the profit or loss which would have arisen had the mileage run been: (1) 400/week, (2) 700/week. (Assume fixed charges remain constant.)

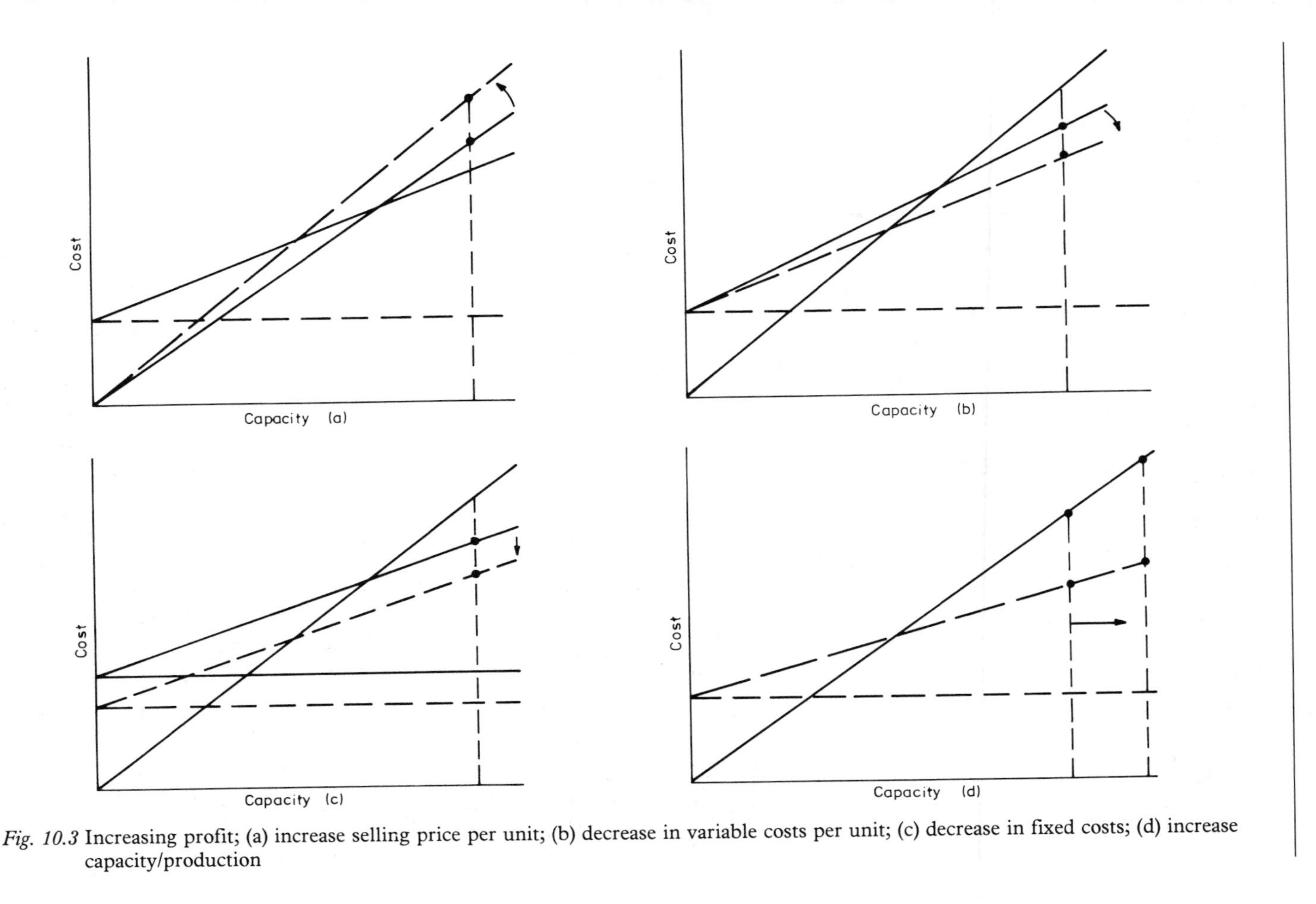

Fig. 10.3 Increasing profit; (a) increase selling price per unit; (b) decrease in variable costs per unit; (c) decrease in fixed costs; (d) increase capacity/production

3. Ascertain from the chart, and state, the amounts of the fixed charge and mileage charge for which the contractor would have to hire out the vehicle in order to make a profit (whatever mileage is run) of: (1) £30/week, (2) £40/week. The solution is given in Fig. 10.4.

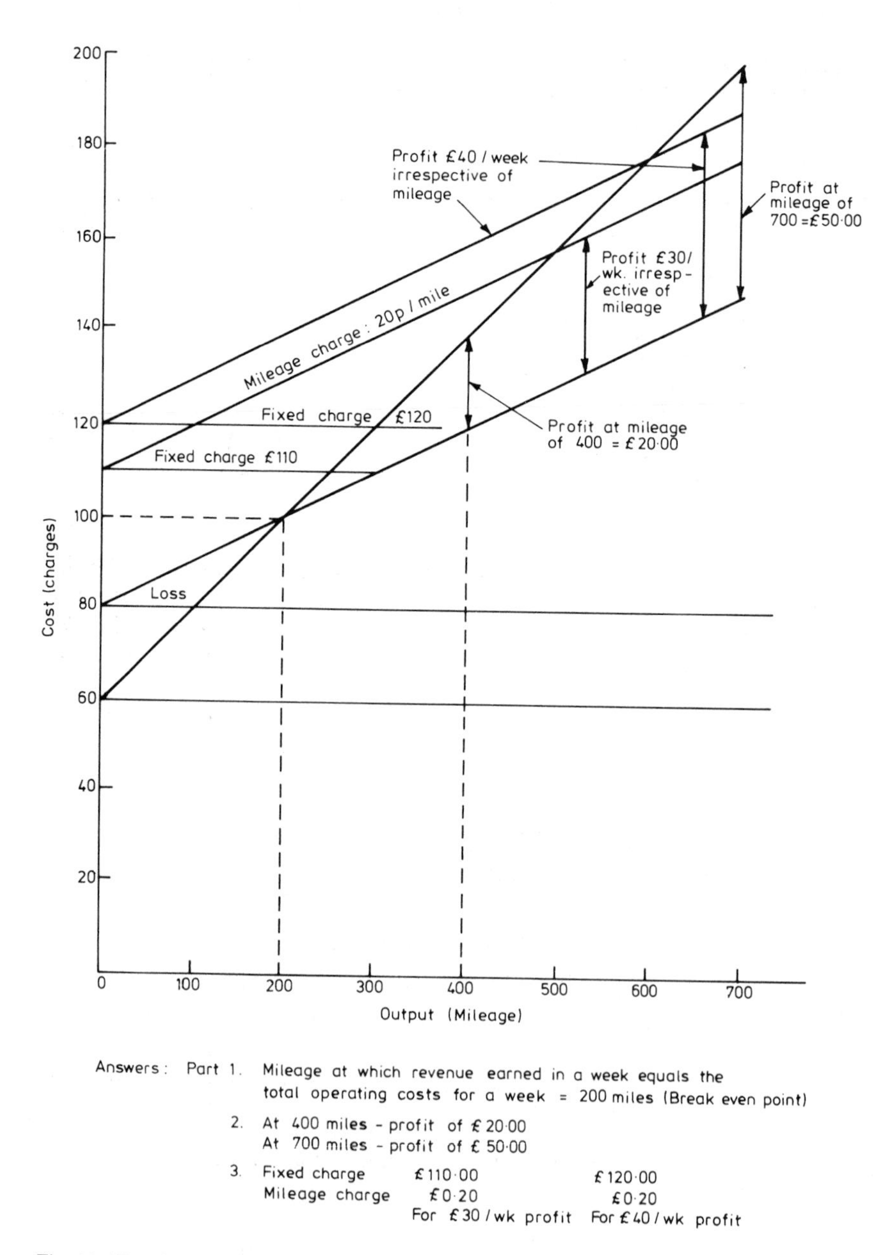

Fig. 10.4 Break-even chart. Solution example 1

Example 2
A joinery manufacturer's current costs and sales at 100 per cent normal capacity are:

Annual sales at 100% normal capacity	£250,000
Fixed costs	£75,000
Variable costs	£150,000

The manager proposes to increase productive capacity by acquiring 25 per cent additional workshop space and plant. He estimates that fixed costs are likely to rise by £15,000 per annum.

Determine from a break-even chart:

(a) The break-even points for the existing factory and after the proposed extension.
(b) At what capacity utilisation the profit will be the same as at 100 per cent capacity utilisation before the extension?

The solution is given in Fig. 10.5.

Use of ratios

The *break-even volume; profit–volume ratio* and *margin of safety* can all be calculated from data used to plot break-even charts.

Example 3
Break-even volume. Wood Craft Company Limited, operating at 100 per cent capacity, estimate their sales turnover to be £1,200,000. Fifteen per cent of turnover is expected to be profit at full capacity. Variable costs at 100 per cent capacity are £700,000. Calculate the company's expected break-even volume, profit–volume ratio and margin of safety based on these data.

The break-even volume is calculated using

$$\text{Break-even volume} = \frac{\text{Total fixed cost}}{1 - \left[\dfrac{\text{Total variable cost}}{\text{Total sales}}\right]}$$

$$= \frac{320{,}000}{\left[1 - \left(\dfrac{700{,}000}{1{,}200{,}000}\right)\right]} = £767{,}938$$

Profit–volume ratio. This is the rate at which profit changes with changes in volume/output.

$$\text{Profit–volume ratio} = 1 - \frac{\text{variable costs}}{\text{sales}} \quad \text{or} \quad \frac{\text{sales} - \text{variable costs}}{\text{sales}}$$

$$= 1 - \frac{700{,}000}{1{,}200{,}000} = 0.416, \text{ say 42 per cent}$$

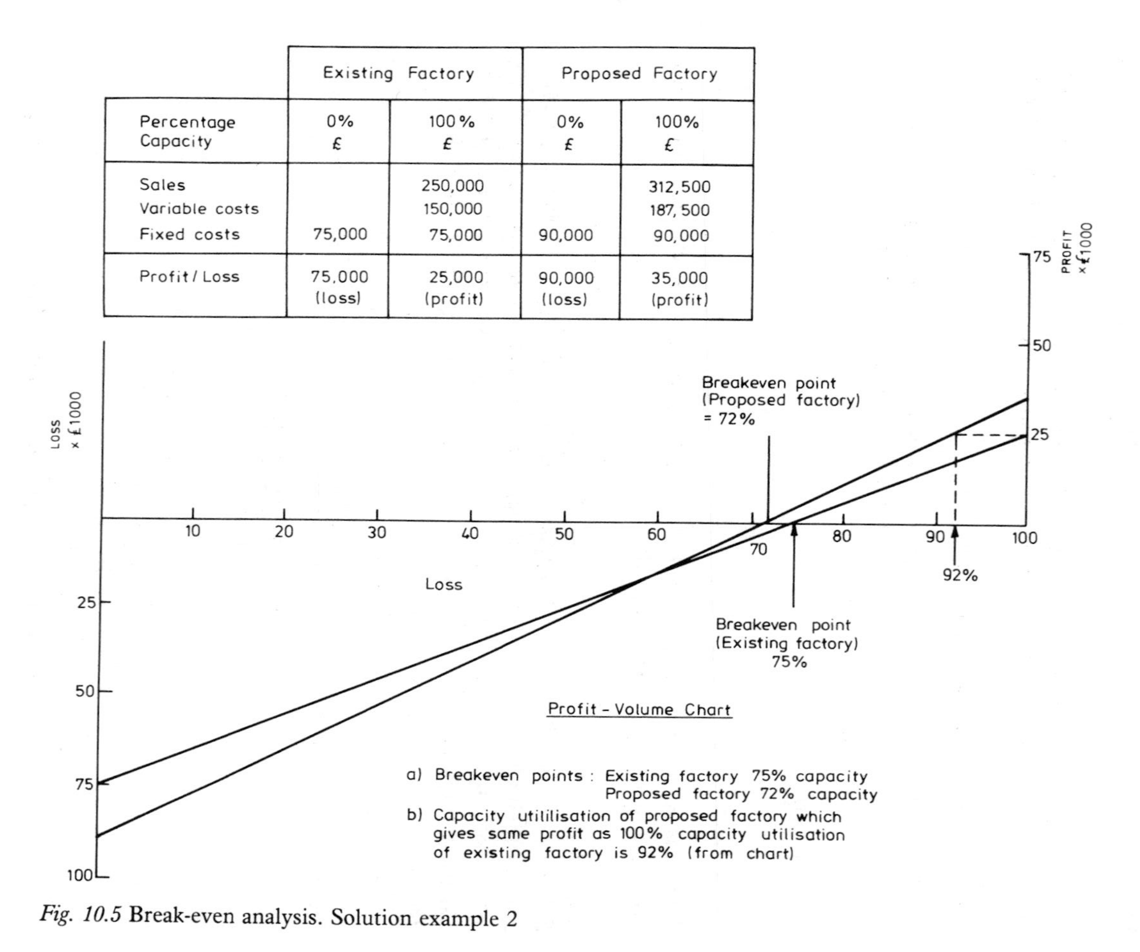

	Existing Factory		Proposed Factory	
Percentage Capacity	0% £	100 % £	0% £	100% £
Sales		250,000		312,500
Variable costs		150,000		187, 500
Fixed costs	75,000	75,000	90,000	90,000
Profit / Loss	75,000 (loss)	25,000 (profit)	90,000 (loss)	35,000 (profit)

Fig. 10.5 Break-even analysis. Solution example 2

The profit–volume ratio indicates the likely change in profit resulting from a change in volume of production. It represents the slope of the profit line on the break-even chart.

Margin of safety. The margin of safety is the sales in excess of the break-even volume.

$$\text{Margin of safety} = \frac{\text{Actual sales} - \text{sales at break-even point}}{\text{Actual sales}}$$

$$= \frac{1{,}200{,}000 - 768{,}012}{1{,}200{,}000} = 0.36 \text{ or } 36 \text{ per cent}$$

This is shown graphically in Fig. 10.6.

Example 4
Data for two pre-cast concrete manufacturing companies are given below.

Company	*Sales turnover at 100% capacity*	*Variable cost at 100% capacity*	*Expected profit*
A	£1,900,000	£ 900,000	14.74% turnover
B	£1,900,000	£1,110,000	15.00% turnover

Using break-even analysis assess the likely effects of increased or decreased business for each company in the future.

Solution
Step 1: Draw break-even chart (this determines fixed cost lines).
Step 2: Using ratios or chart determine break-even sales for both companies.
Step 3: Determine the profit–volume ratio.
Step 4: Determine the margin of safety (see graph, Fig. 10.7).

Break-even sales

$$\text{Company A} = \frac{720{,}000}{\left[1 - \left(\frac{900{,}000}{1{,}900{,}000}\right)\right]} = £1{,}368{,}041$$

$$\text{Company B} = \frac{505{,}000}{\left[1 - \left(\frac{1{,}110{,}000}{1{,}900{,}000}\right)\right]} = £1{,}214{,}526$$

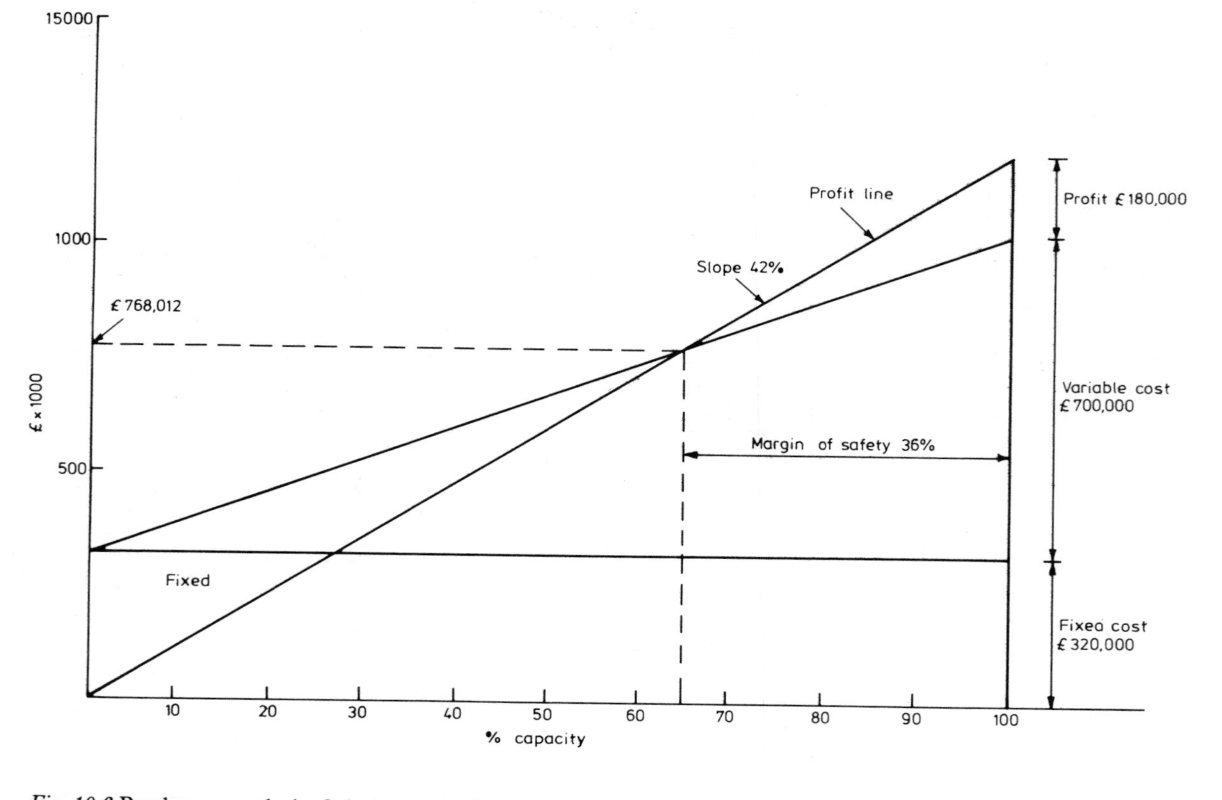

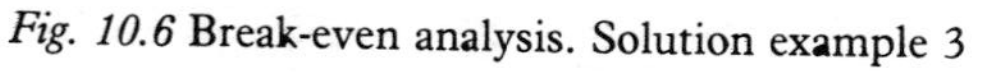
Fig. 10.6 Break-even analysis. Solution example 3

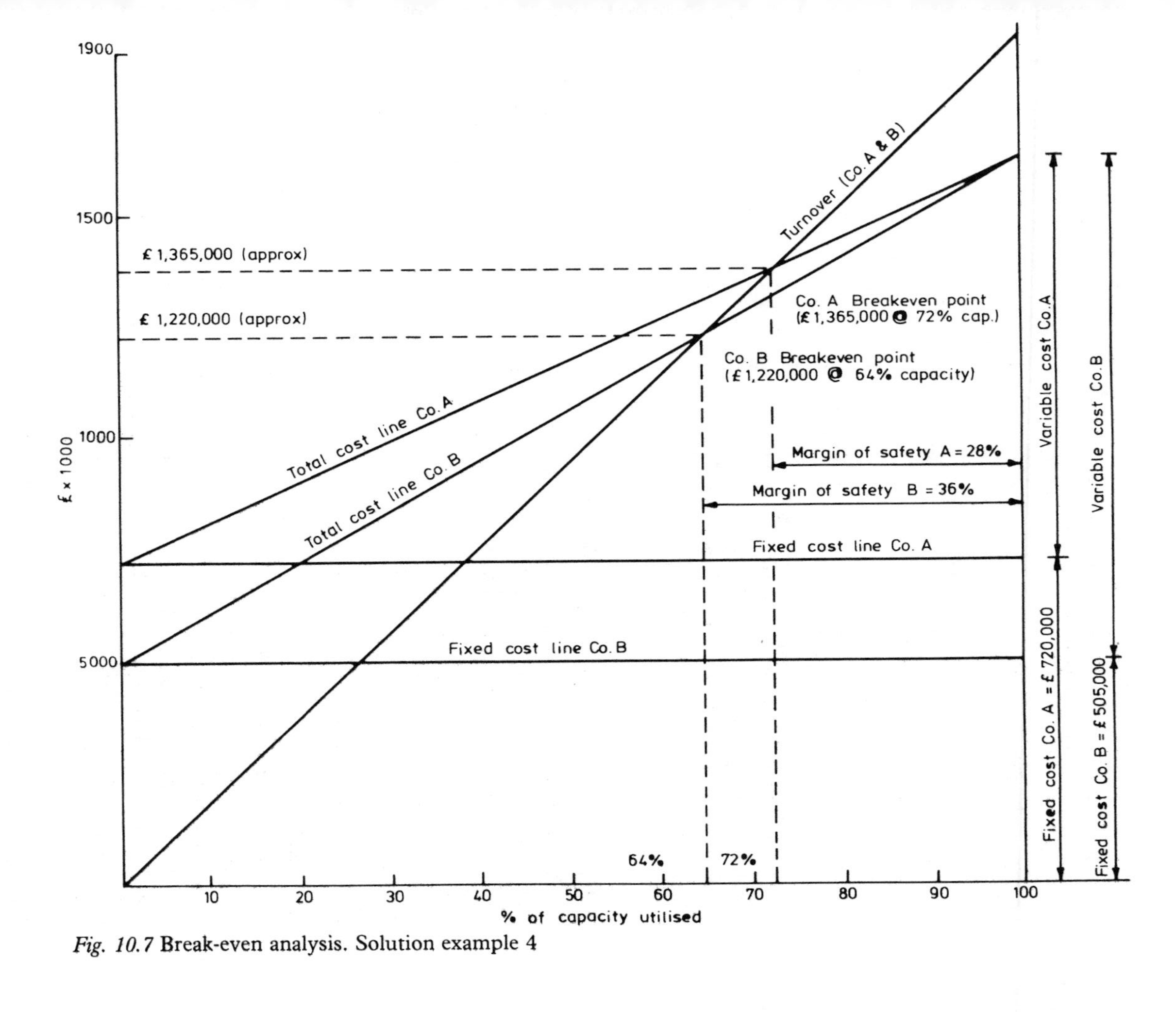

Fig. 10.7 Break-even analysis. Solution example 4

Profit–volume ratio

$$\text{Company A} = 1 - \left[\frac{900{,}000}{1{,}900{,}000}\right] = 52 \text{ per cent}$$

$$\text{Company B} = 1 - \left[\frac{1{,}110{,}000}{1{,}900{,}000}\right] = 42 \text{ per cent}$$

Margin of safety

$$\text{Company A} = \frac{1{,}900{,}000 - 1{,}368{,}041}{1{,}900{,}000} = 28 \text{ per cent}$$

$$\text{Company B} = \frac{1{,}900{,}000 - 1{,}214{,}526}{1{,}900{,}000} = 36 \text{ per cent}$$

The profit–volume ratio is the most informative regarding effects of increased or decreased business.

With an increase or decrease in sales for Company A the profit would increase or decrease by 52 per cent. For Company B the figure is 42 per cent.

Example 5

A joinery manufacturer decides to produce high-class exterior doors. Using existing workshop space and machinery it is possible to produce 2,500 doors per annum working at 100 per cent capacity.

Basic costs are estimated as follows:

Fixed costs	£40,000/year
Variable costs	£40/door
Selling price	£72/door

Calculate the following:

(a) Profit or loss and manufacturing cost per door at:
  (i) 25 per cent capacity;
  (ii) 80 per cent capacity;
  (iii) 100 per cent capacity.

(b) The number of doors to be manufactured in order to break even.

(c) The profit–volume ratio (assume 100 per cent sales).

(d) Margin of safety (assume 100 per cent sales).

Solution

See break-even chart (Fig. 10.8).

(a) Manufacturing cost per door

At 25 per cent capacity:

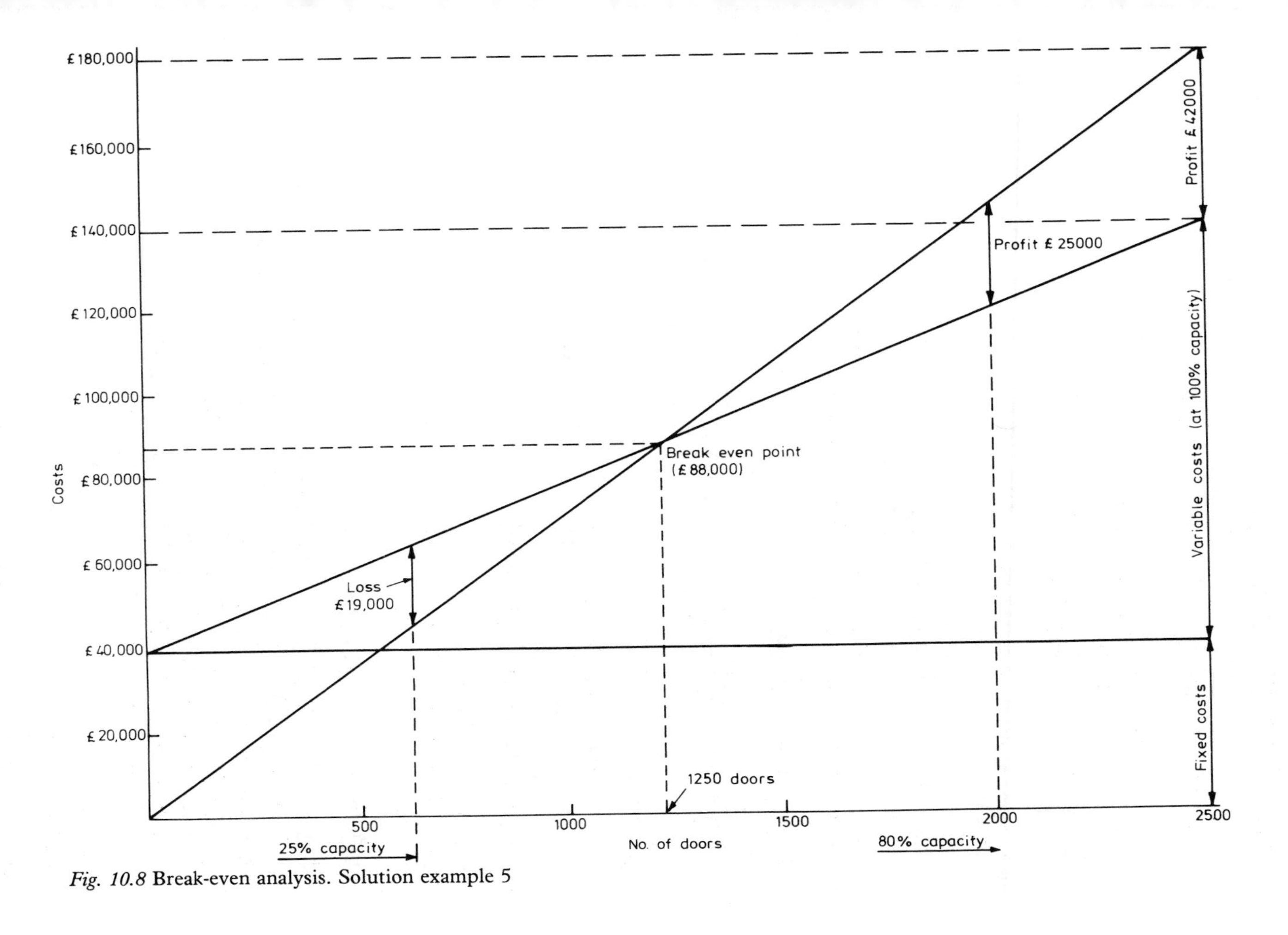

Fig. 10.8 Break-even analysis. Solution example 5

$$\text{Cost} = \frac{40{,}000 + (625 \times 40)}{625} = £104$$

At 80 per cent capacity:

$$\text{Cost} = \frac{40{,}000 + (2{,}000 \times 40)}{2{,}000} = £60$$

At 100 per cent capacity:

$$\text{Cost} = \frac{40{,}000 + (2{,}500 \times 40)}{2{,}500} = £56$$

Profit/loss situation (from break-even chart)
At 25 per cent: loss of £19,000 (£20,000 by calculation).
At 80 per cent: profit of £25,000 (£24,000 by calculation).
At 100 per cent: profit of £42,000 (£40,000 by calculation).

(b) See break-even chart

$$\text{Check } \frac{40{,}000 \times 2{,}500}{180{,}000 - 100{,}000} = 1{,}250 \text{ doors}$$

(c) Profit–volume ratio

$$\frac{180{,}000 - 100{,}000}{2{,}500} = 32 \text{ per cent (100 per cent sales)}$$

(d) Safety margin (100 per cent sales)

$$\frac{180{,}000 - 88{,}000}{180{,}000} = 51 \text{ per cent}$$

Validation of break-even charts

1. Break-even charts are only true within the actual limits on which they are based. The break-even point does not provide a standard of performance.
2. Fixed costs may change at different levels of output.
3. Variable costs may not give a straight-line chart.
4. The income curve may bend due to extra discounts, etc. in order to increase sales.
5. Managerial decisions may alter the fixed and variable costs.

Chapter 11

Cash flow forecasting*

11.1 Importance of timing

A cash flow is the movement or transfer of money into or out of a company. Contractors should be alert to the timing of both payments and receipts for work done. In construction work there is a disparity between timing of payments that may distort the true financial position.

Typical timing of payment in construction is shown below.

	Labour	*Weekly*
	Materials	Usually 3 months' credit facility
Payments (cash out)	Subcontractors	Some credit facility usually available, i.e. 1 or 2 months
	Plant	Some credit facility usually available

Receipts (cash in)

The conditions of contract usually indicate when payments will be made to the contractor for work done and also the amount of retention both during and after the contract. On most projects the contractor is entitled to payment based on an interim valuation usually on a monthly basis. From this valuation would be deducted a mutually agreed percentage retention value.

To illustrate the basic principles one construction operation will be considered (Fig. 11.1).

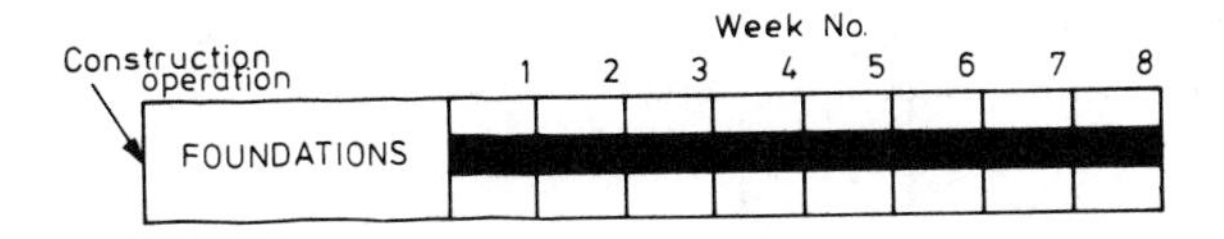

Fig. 11.1 Single activity

* Adapted from Pigott (1971).

Operation cost £8,800

This operation cost is broken down into elements of:

Labour	£1,600	paid weekly	
Plant	£4,000	2 months' credit facility	All different timings
Materials	£800	3 month's credit facility	
Subcontractors	£2,400	1 months' credit facility	

Procedure

1. Using the estimate and project plan (bar chart), develop a time-scaled plan for payments (cash out) for labour, materials, plant and subcontractors. Include direct costs only in this calculation.
2. Add to this plan the expected expenditure on preliminaries and general overheads for each project.
3. Establish expected receipts in the form of payments from clients and other sources for each project. (Assume 50% of cost received week 5, remainder week 9.)
4. Prepare a cumulative statement indicating the next requirements for cash.
5. Compare actual expenditure and receipts on an ongoing basis, and revise each plan if necessary.

A chart is prepared showing when each payment by the contractor is due (Table 11.1).

Table 11.1

Contractor's payment chart (Cash out)

OPERATION	1	2	3	4	5	6	7	8	9	10	11	12	13	14	15	16
FOUNDATIONS																
LABOUR		200	200	200	200	200	200	200	200							
PLANT		2 month payment delay							500	500	500	500	500	500	500	500
MATERIALS		3 month payment delay											100	100	100	100
SUB-CONTRACTOR		1 month			300	300	300	300	300	300	300	300				
Total Payments	-	200	200	200	500	500	500	500	1000	800	800	800	600	600	600	600

OPERATION	17	18	19	20	21	22	23	24
FOUNDATIONS								
LABOUR								
PLANT								
MATERIALS	100	100	100	100				
SUB-CONTRACTOR								
Total Payments	100	100	100	100				

Receipts (cash in)

Assume there is a monthly interim valuation and that work will progress according to programme. The timing is shown in Table 11.2. Compare payments and receipts on an ongoing basis (Table 11.3).

Table 11.2

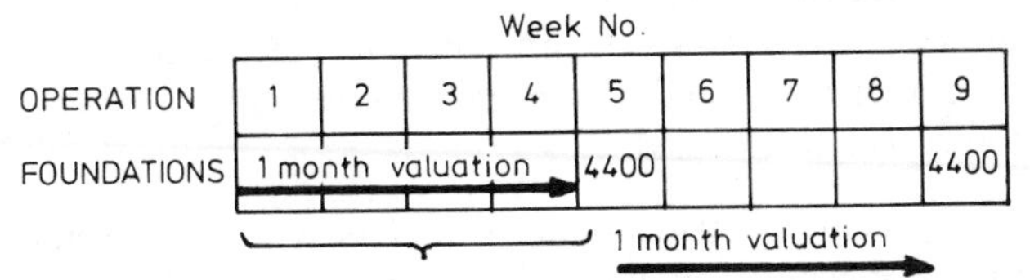

Table 11.3

Cash Flow Statement		2	3	4	5	6	7	8	9	10	11	12	13
PAYMENTS	-	200	200	200	500	500	500	500	1000	800	800	800	600
RECEIPTS					4400				4400				
NET CASH FLOW		-200	-400	-600	3300	2800	2300	1800	5200	4400	3600	2800	2200

14	15	16	17	18	19	20
600	600	600	100	100	100	100
1800	1000	400	300	200	100	-

This cash flow reflects the amount of cash the contractor will require to finance this operation and when the cash is required. It indicates that £600 is required by the fourth week.

The shortfall that may occur between the supply of funds and the need for cash is often met by short-term bank loans or overdraft facilities. In recent years however, the credit facilities extended by financial institutions have been subject to more strict controls and this has often resulted in cash shortages in firms who may not suspect a threat from this source. The resulting shortage of cash may often force liquidation of assets and foreclosure by the company's creditors. A contractor may be forced to avail himself of short-term borrowing at very high interest rates.

The main purpose of cash flow planning may be summarised as follows:

1. It ensures that sufficient cash is available to meet demands.
2. It provides a reliable indicator to lending institutions that advances made can be repaid according to an agreed programme.
3. It ensures that cash resources are fully utilised to the benefit of the owner and investor in the company.

The three main ingredients in the determination of cash flows are:

1. *Payments (cash outflow)*: This is the aggregate of the payments which a contractor will make over a period of time for the resources he uses in a project, e.g. labour, plant, materials, subcontractors.
2. *Receipts (cash inflow)*: This is the receipts which a contractor will receive over a period for the work he has completed.
3. *Timing of payments*: In cash flow analysis we are interested in the timing of payments so that the size of deficiency can be estimated and the time period over which this deficiency applies can be assessed.

The timing of payments and receipts is related to the work done by the contractor. In contract work the conditions of contract indicate when interim and final receipts may be expected. The size of these receipts reflects progress to date and the project plan is the principal source of these data. The project plan which may be prepared in a network form and then converted to bar-chart format indicates the starting and finishing date of each operation. An examination of these operations indicates what payments will be made and received over a period of each operation. While the project plan indicates, in physical terms, the programme for construction, the estimate is used to expand this to give the data for each element.

11.2 Case study 1

Example: house construction

Assume that the contractor has estimated the direct costs as shown in Table 11.4. In addition, the contractor estimates that his general overheads will be £12,500 and requires a profit of 7 per cent of direct costs (£2,560). The total cost to the client is therefore £51,610.

The main conditions of contract are that the client agrees to release:

- 50 per cent of cost when the roof is completed;
- 30 per cent of cost at completion of internal finishes;
- 15 per cent of cost at practical completion;
- 5 per cent retention payable 6 months after completion.

Table 11.4

Site Preparations, Foundations, Footings, & Ground Floor	£ 6000·00
External Walls	£ 9000·00
Internal Walls and Partitions	£ 3750·00
Roof Trusses and Coverings	£ 3000·00
Finishes	£ 7250·00
Plumbing and Electrical	£ 5100·00
Drainage and Siteworks	£ 2450·00
Total Direct Cost	£ 36550·00

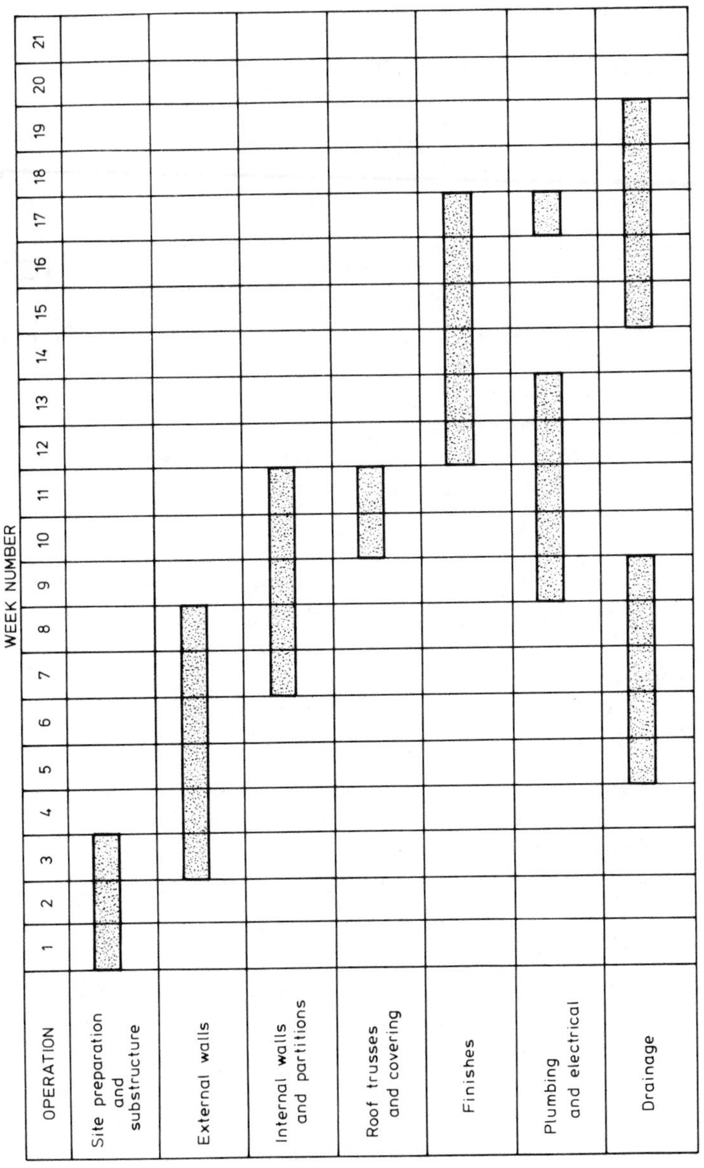

WEEK NUMBER

OPERATION	1	2	3	4	5	6	7	8	9	10	11	12	13	14	15	16	17	18	19	20	21
Site preparation and substructure																					
External walls																					
Internal walls and partitions																					
Roof trusses and covering																					
Finishes																					
Plumbing and electrical																					
Drainage																					

Fig. 11.2 Bar chart for house construction

Table 11.5 Breakdown of estimate

	Labour (£)	*Materials* (£)	*Plant* (£)	*Total* (£)
Site preparation and substructure	2,000	2,500	1,500	6,000
External walls	4,000	5,000		9,000
Internal walls and partitions	1,750	2,000		3,750
Roof trusses and coverings	1,000	2,000		3,000
Finishes	4,500	2,750		7,250
Plumbing and electrical	2,200	2,900		5,100
Drainage and siteworks	1,200	750	500	2,450

A simple bar chart for the construction of this house is shown in Fig. 11.2. In order to plan his cash flows the contractor must break down the original estimate into the components of labour, materials and plant (Table 11.5). The reason for this is that each has a different time scale.

Assumptions
Labour: paid 1 week in arrears (cost is spread evenly over the contract).
Materials: 3 months' credit allowed.
Plant: 2 months' credit allowed.
From this information a statement of project costs is prepared (Table 11.6).

With a planned construction time of 19 weeks it is reasonable to assume that project overheads will be incurred evenly over the period at a rate of £657/week, and this must be added to the direct costs to give the following statement for net cash outflow (or payments) (Table 11.7).

The cash inflow or receipts, based on expected payments from client (as stated in conditions of contract), is as follows:

- 50 per cent of cost when roof is completed: £25,805 on week 12;
- 30 per cent of cost at completion of internal finishes: £15,483 on week 18;
- 15 per cent of cost at practical completion; £7,741 on week 20;
- 5 per cent retention payable 24 weeks after practical completion: £2,580 on week 44.

The weeks in which the cash inflow or receipts may be expected are found by inspection of the bar chart. The cash flow statement is then prepared on a cumulative basis, taking into account the timing of expected receipts (Table 11.8). The cash flow statement indicates the *amount of cash* the contractor will require to finance the project and *when* this cash will be required. A maximum sum of £18,437 is required by the eleventh week.

It is possible that this amount of finance can be provided from the contractor's own resources, but normally a bank is the usual source. This type of statement of expected cash requirements is invaluable in explaining proposals for loan overdraft requirements to a bank. It immediately shows the bank manager how it is intended to repay the loan and interest and he can be reasonably assured that the risk involved is minimal.

Table 11.6 Statement of project costs

Week No.

OPERATION	1	2	3	4	5	6	7	8	9	10	11	12	13	14	15	16	17	18	19	20	21	22	23	24	25	26	27	28	29	30	31	32
SITE PREPARATION & SUBSTRUCTURE £																																
LABOUR 2000		666	666	666																												
MATERIALS 2500													833	833	833																	
PLANT 1500									500	500	500																					
EXTERNAL WALLS																																
LABOUR 4000			666	666	666	666	666	666																								
MATERIALS 5000														833	833	833	833	833	833													
INTERNAL WALLS AND PARTITIONS																																
LABOUR 1750								350	350	350	350	350																				
MATERIALS 2000																			400	400	400	400	400									
ROOF TRUSSES AND COVERINGS																																
LABOUR 1000										500	500																					
MATERIALS 2000																					1000	1000										
FINISHES																																
LABOUR 4500													750	750	750	750	750	750														
MATERIALS 2750																								458	458	458	458	458	458			
PLUMBING AND ELECTRICAL																																
LABOUR 2200										366	366	366	366	366				366														
MATERIALS 2900																					483	483	483	483	483				483			
DRAINAGE AND SITEWORKS																																
LABOUR 1200						120	120	120	120	120						120	120	120	120	120												
MATERIALS 750																	75	75	75	75	75						75	75	75	75	75	
PLANT 500													50	50	50	50	50						50	50	50	50	50					
	—	666	1332	1332	666	786	786	1136	970	1836	1716	716	1199	2832	2466	1753	1828	2144	1428	595	1958	1883	933	991	991	508	583	533	1016	75	75	

Table 11.7 Net cash outflow (payments)

WEEK No.	1	2	3	4	5	6	7	8	9	10	11	12	13	14	15	16	17	18	19	20	21	22
DIRECT COSTS		666	1330	1330	666	786	786	1136	970	1836	1716	716	1990	2832	2466	1753	1828	2144	1428	595	1958	1883
OVERHEADS	657	657	657	657	657	657	657	657	657	657	657	657	657	657	657	657	657	657	657			
TOTAL	657	1323	1987	1987	1323	1443	1443	1793	1627	2493	2373	1373	2647	3489	3123	2410	2485	2801	2085	595	1958	1883

23	24	25	26	27	28	29	30	31	32
933	991	991	508	583	533	1016	75	75	
933	991	991	508	583	533	1016	75	75	

Table 11.8 Cash flow statement

WEEK No.	1	2	3	4	5	6	7	8	9	10	11
PAYMENTS	657	1323	1987	1987	1323	1440	1440	1793	1627	2493	2373
RECEIPTS											
CASH FLOW	-657	-1980	-3961	-5948	-7271	-8711	-10151	-11944	-13511	-16064	-18437

WEEK No.	12	13	14	15	16	17	18	19	20	21
PAYMENTS	1373	2647	3489	3123	2410	2485	2801	2085	595	1958
RECEIPTS	25805						15483		7741	
CASH FLOW	+5995	3348	-141	-3264	-5674	-8159	4523	2438	9584	7626

WEEK No.	22	23	24	25	26	27	28	29	30	31
PAYMENTS	1883	933	991	981	508	583	533	1016	75	75
RECEIPTS										
CASH FLOW	5743	4810	3819	2828	2320	1737	1204	188	113	38

WEEK No.	44
PAYMENTS	
RECEIPTS	2560
CASH FLOW	2598

The preparation of the estimate and project plan is essential to the calculation of cash flows. In addition, the method of preparing the estimate is important so that the costs are established in a form that facilitates the preparation of both the project plan and the cash plan. The key to this is operational estimating which isolates the particular elements of cost and provides a realistic basis for developing the project plan either in network or bar-chart form.

11.3 The use of S curve presentation

One of the simplest aids in cash flow planning is the use of the S curve. It is used to forecast the cash requirements of a project. The cumulative expenditure for a project normally takes the shape of a flattened letter S. By plotting cumulative direct costs (cash out) against cumulative receipts (cash in) it is possible to calculate the project cash requirements. The receipt curve is stepped. Each step represents the receipt of a certificated value of latest application for payment.

Research indicates that there is an appropriate form of S curve geometry for certain types of work. A typical form is shown in Fig. 11.3.

This characteristic shape arises because of the fact that during the early contract stage there is a lead in to construction activity and the expenditure curve is relatively flat. As other activities commence expenditure increases and the curve develops a steeper middle portion. Towards the end of the project many activities will be completed and there will be a run-down of construction work resulting in a flattened curve.

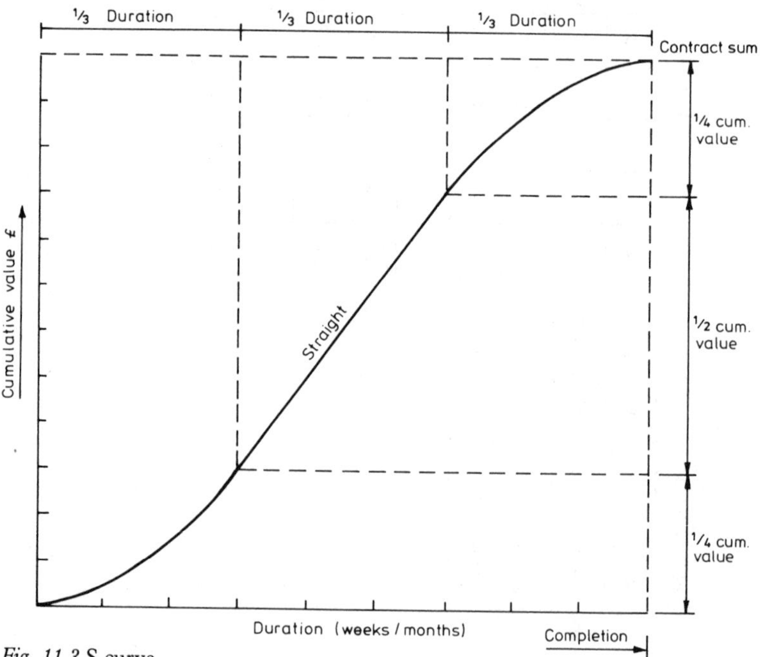

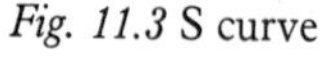
Fig. 11.3 S curve

In Fig. 11.3 one-quarter of the expenditure represents the activity build-up period of one-third the contract duration and a further quarter occupies the run-down period of one-third. Half of the accumulated value is gained over a centre third of the maintained progress.

A curve based on the cash flows from the previous house construction example is shown in Fig. 11.4.

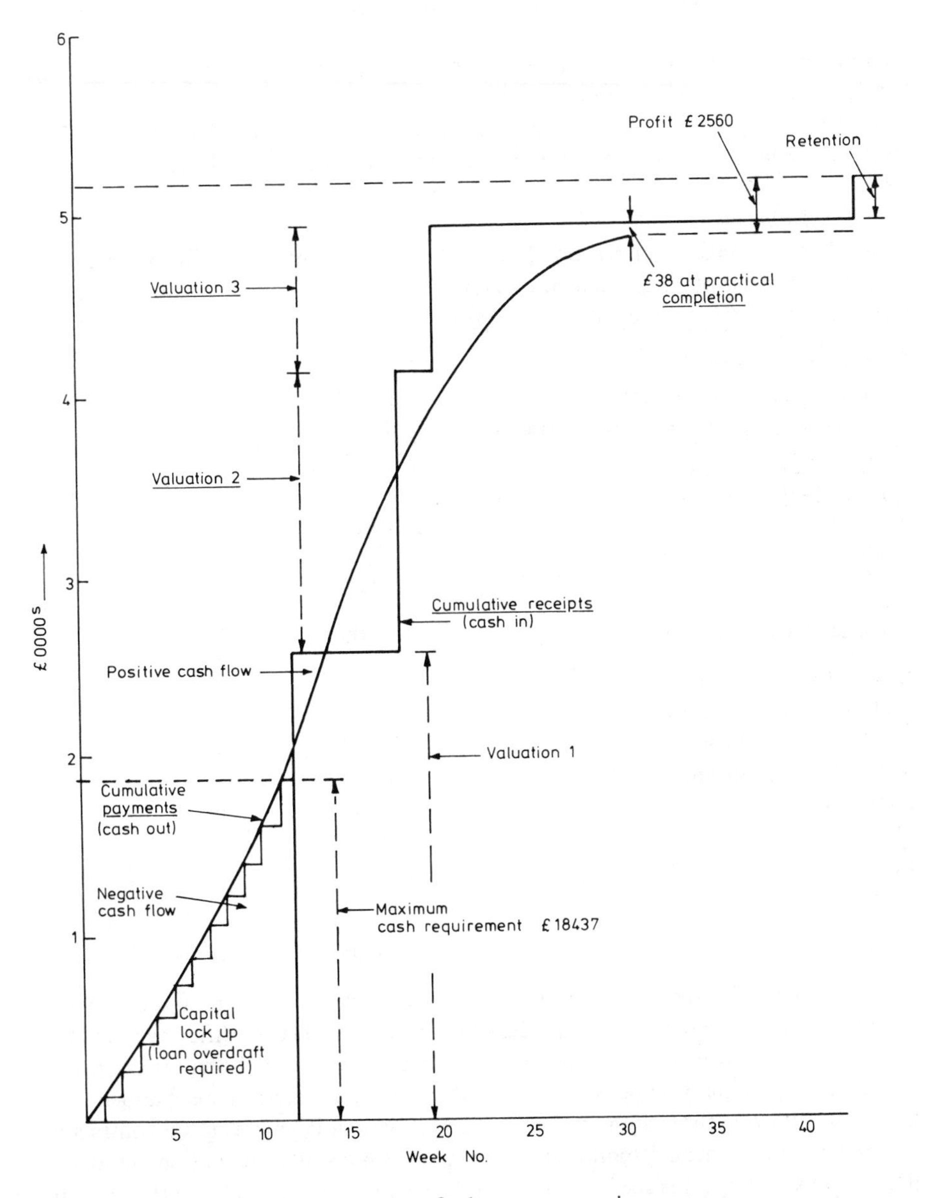

Fig. 11.4 Cash inflow and outflow curves for house construction contract

11.4 Case study 2

Composite cash flow statement

A contracting company has two projects on hand and the details are as follows:

Contract no. 1. A small school for which the tender figure is £693,000. The contract time is 14 months and it has been running for 4 months.

Contract no. 2. This contract is about to commence and is for a factory building. The contract time is 12 months and the tender figure is £1,535,000.

The contract programmes in bar-chart form for each project are shown in Figs. 11.5 and 11.6, together with the estimate breakdowns (Tables 11.9 and 11.10).

Step 1 – payments
Translate the estimate and project plan for each contract into a plan for payments for labour, materials, plant and subcontractors. Initially the components of the tender figure are required and they are as follows.

Contract no. 1 (school)	
Preliminaries (job overheads)	£23,000
Prime cost (PC) sums and subcontractors	£142,000
General overheads	£25,000
Direct labour	£187,500
Materials	£273,200
Plant	£7,300
Profit	£35,000
Total tender figure	£693,000

Contract no. 2 (factory)	
Preliminaries and profit	£215,000
PC sums and subcontractors	£271,200
General overheads and profit	£85,000
Direct labour	£371,800
Materials	£495,500
Plant	£21,500
Profit	£75,000
Total tender figure	£1535,000

The estimate figures are then transferred to bar-chart form to indicate when payments will be made. Certain considerations influence when specific payments will be made. For instance, it can be assumed that labour payments will occur in the period a particular operation is expected to occur. Payments for materials are assumed to be made after allowing the customary 3-month credit period. Payments to subcontractors are assumed to be made 1 month after the completion of the subcontractors' operations. In this example it is also assumed that the payments are spread over the duration of an operation.

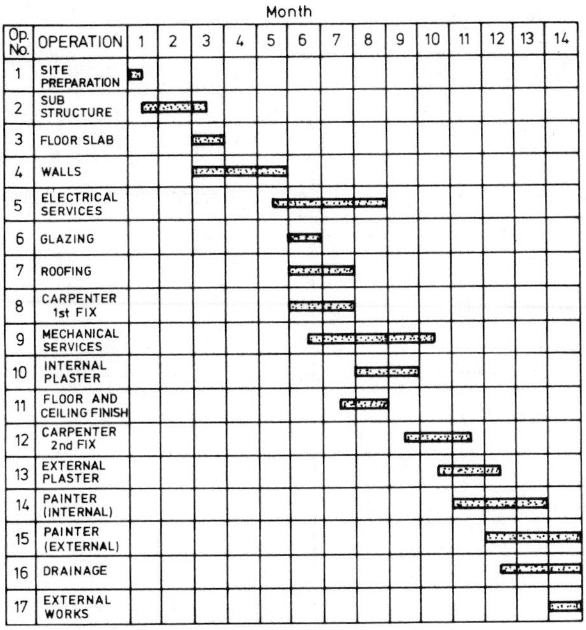

Month

Op. No.	OPERATION	1	2	3	4	5	6	7	8	9	10	11	12	13	14
1	SITE PREPARATION														
2	SUB STRUCTURE														
3	FLOOR SLAB														
4	WALLS														
5	ELECTRICAL SERVICES														
6	GLAZING														
7	ROOFING														
8	CARPENTER 1st FIX														
9	MECHANICAL SERVICES														
10	INTERNAL PLASTER														
11	FLOOR AND CEILING FINISH														
12	CARPENTER 2nd FIX														
13	EXTERNAL PLASTER														
14	PAINTER (INTERNAL)														
15	PAINTER (EXTERNAL)														
16	DRAINAGE														
17	EXTERNAL WORKS														

Fig. 11.5 Bar chart for Contract No. 1 – school

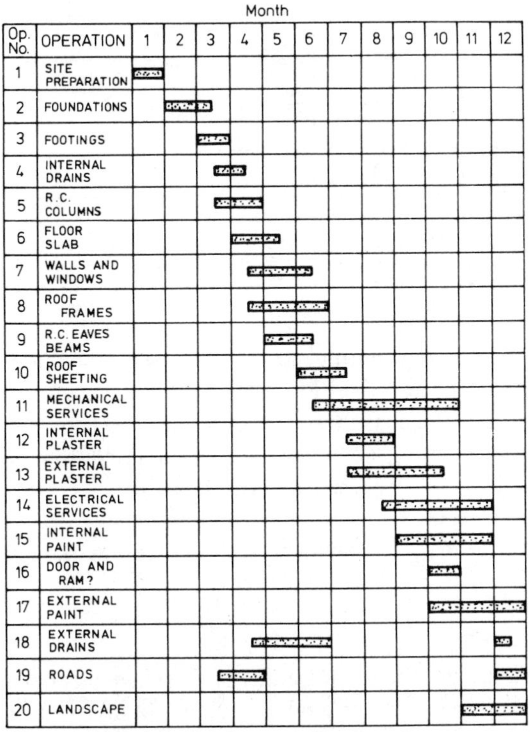

Month

Op. No.	OPERATION	1	2	3	4	5	6	7	8	9	10	11	12
1	SITE PREPARATION												
2	FOUNDATIONS												
3	FOOTINGS												
4	INTERNAL DRAINS												
5	R.C. COLUMNS												
6	FLOOR SLAB												
7	WALLS AND WINDOWS												
8	ROOF FRAMES												
9	R.C. EAVES BEAMS												
10	ROOF SHEETING												
11	MECHANICAL SERVICES												
12	INTERNAL PLASTER												
13	EXTERNAL PLASTER												
14	ELECTRICAL SERVICES												
15	INTERNAL PAINT												
16	DOOR AND RAM?												
17	EXTERNAL PAINT												
18	EXTERNAL DRAINS												
19	ROADS												
20	LANDSCAPE												

Fig. 11.6 Bar chart for Contract No. 2 – factory

Table 11.9 Estimate breakdown – contract no. 1

OP NO	OPERATION	LABOUR	MATERIALS	PLANT	SUB CONTRACT
1	Site preparation	3000	700	2500	
2	Sub structure	11000	17000	300	
3	Floor slab	14000	24000	400	
4	Rising walls	35000	57000		
5	Elect. services	2000			42000
6	Glazing				7000
7	Roofing	24000	43000		
8	Carpenter 1st fix	9000	28000		
9	Mechanical serv.	2500			60000
10	Plasterer (Int)	17000	15000		
11	Floor & clg. finish				21000
12	Carpenter 2nd fix	12000	30000		12000
13	External plaster	15000	12500		
14	Painter (Internal)	11500	4500		
15	Painter (External)	12000	5500		
16	Drainage	10500	24000	2000	
17	External works	9000	12000	2100	
		187500	273200	7300	142000

The treatment of PC items, provisional and contingency sums may present a problem in terms of payments and receipts. In the case of PC items, the timing should be established as soon as details are available in consultation with the architect and consultants for the project. Provisional and contingency sums should be omitted from the calculations until details are available on when these sums will be used.

The payment charts for both contracts are shown in Tables 11.11 and 11.12.

To these payments must be added the expected payments for job preliminaries and general overheads under step 2.

Step 2 – preliminaries and general overheads

The payment plan just prepared should now be implemented by the payments likely to arise from preliminaries (job overheads) and general overheads. Many of these payments can be expected to be uniform over the duration of the contract while others

Table 11.10 Estimate breakdown – contract no. 2

OP NO	OPERATION	LABOUR	MATERIALS	PLANT	SUB CONTRACT
1	Site preparation	9000	3500	7000	20100
2	Foundations	28000	45000	9500	
3	Footing	25000	34000		
4	Internal drains	10000	21000		
5	R.C. columns	9000	16000		
6	Floor slab	46000	75000	5000	
7	Walls & windows	65000	70000		
8	Roof frames				46000
9	R.C. eaves beam	15000	28000		
10	Roof sheeting				65000
11	Mech. services	3300			15000
12	Internal plaster	40000	35000		
13	External plaster	42000	30500		
14	Elect. services	1800			101100
15	Internal paint	19000	16000		
16	Door & ramp	4200	15000		
17	External paint	12500	5500		
18	External drains	25000	42000		
19	Roads	12500	50000		24000
20	Landscape	4500	9000		
		371800	495500	21500	271200

will be made for specific items of overhead. For example, the expenditure on foremen and site clerks will be reasonably uniform, while items such as temporary works, loan and insurance premiums will be paid at specific times. This assessment can best be done with a checklist of overhead items and with the preliminaries section of the BOQ.

The combination of direct and overhead payments for both contracts is shown in Table 11.13. It will be noticed that certain payments are deferred outside the 15-month period under consideration (i.e. retentions) and the last 10 months of contract no. 1 is being considered. The payments deferred should be noted and included in the cash flow for the following period.

Table 11.11 Payments (cash out) chart for contract no. 1

OP. No.	OPERATION		MONTH 1	2	3	4	5	6	7	8	9	10	11	12	13	14	15	16	17
1	SITE Preparation	Labour	3000																
		Material				700													
		Plant	2500																
		Subcontract																	
2	SUB STRUCTURE	L	2750	5500	2750														
		M				4250	8500	4250											
		P	75	150	75														
		SC																	
3	FLOOR SLAB	L			14000														
		M						24000											
		P			400														
		SC																	
4	RISING WALLS	L			11667	11667	11667												
		M						19000	19000	19000									
		P																	
		SC																	
5	ELECTRICAL SERVICES	L					286	571	571	571									
		M																	
		P																	
		SC						6000	12000	12000	12000								
6	GLAZING	L																	
		M																	
		P																	
		SC							7000										
7	ROOFING	L						12000	12000										
		M									21500	21500							
		P																	
		SC																	
8	CARPENTER 1ST FIX	L						4500	4500										
		M									14000	14000							
		P																	
		SC																	
9	MECHANICAL SERVICES	L						312	624	624	624	312							
		M																	
		P																	
		SC							7500	15000	15000	15000	7500						
10	PLASTERER (INTERNAL)	L								8500	8500								
		M											7500	7500					
		P																	
		SC																	
11	FLOOR & CEILING FINISH	L																	
		M																	
		P																	
		SC								7000	14000								
12	CARPENTER 2ND FIX	L									3000	6000	3000						
		M												7500	15000	7500			
		P																	
		SC										3000	6000	3000					
13	EXTERNAL PLASTER	L										5000	10000						
		M													4166	8333			
		P																	
		SC																	
14	PAINTER (INTERNAL)	L											3833	3833	3833				
		M														1500	1500	1500	
		P																	
		SC																	
15	PAINTER (EXTERNAL)	L												4000	4000	4000			
		M															1833	1833	1833
		P																	
		SC																	
16	DRAINAGE	L												2100	4200	4200			
		M															4800	9600	9600
		P												333	666	666			
		SC																	
17	EXTERNAL WORKS	L														9000			
		M																	12000
		P														2100			
		SC																	
			8325	5650	28892	16617	20453	70633	63195	62695	88624	64812	37833	28266	31865	37299	8133	12933	23433

Table 11.12 Payments (cash out) chart for contract no. 2

OP No.	OPERATION		MONTH 1	2	3	4	5	6	7	8	9	10	11	12	13	14	15
1	SITE PREPARATION	L	9000														
		M			3500												
		P	7000														
		SC		20100													
2	FOUNDATIONS	L		18667	9333												
		M				30000	15000										
		P		6334	3166												
		SC															
3	FOOTINGS	L			25000												
		M						34000									
		P															
		SC															
4	INTERNAL DRAINS	L			5000	5000											
		M						10500	10500								
		P															
		SC															
5	R.C. COLUMNS	L			3000	6000											
		M						5333	10666								
		P															
		SC															
6	FLOOR SLAB	L				30666	15333										
		M							50000	25000							
		P				3334	1666										
		SC															
7	WALLS & WINDOWS	L				16250	32500	16250									
		M							17500	35000	17500						
		P															
		SC															
8	ROOF FRAMES	L															
		M															
		P															
		SC					9800	18400	18400								
9	R.C. EAVES BEAMS	L					10000	5000									
		M								18666	9333						
		P															
		SC															
10	ROOF SHEETING	L															
		M															
		P															
		SC							43333	21666							
11	MECHANICAL SERVICES	L						376	733	733	733	733					
		M															
		P															
		SC							1666	3333	3333	3333	3333				
12	INTERNAL PLASTER	L							13333	26666							
		M										16666	23333				
		P															
		SC															
13	EXTERNAL PLASTER	L							7000	14000	14000	7000					
		M										5083	10166	10166	5083		
		P															
		SC															
14	ELECTRICAL SERVICES	L								257	514	514	514				
		M															
		P															
		SC										14442	28885	28885	28885		
15	INTERNAL PAINT	L									6333	6333	6333				
		M												5333	5333	5333	
		P															
		SC															
16	DOOR & RAMP	L										4200					
		M													15000		
		P															
		SC															
17	EXTERNAL PAINT	L										4166	4166	4166			
		M													1833	1833	1833
		P															
		SC															
18	EXTERNAL DRAINS	L				4166	8334	8334						4166			
		M							7000	14000	14000						7000
		P															
		SC															
19	ROADS	L			2500	5000								5000			
		M						10000	20000								20000
		P															
		SC				4800	9600								9600		
20	LANDSCAPE	L											2250	2250			
		M														4500	4500
		P															
			16000	45101	51499	105216	101633	108184	200131	159321	65746	62470	78980	59960	65734	11666	33333

Table 11.13

MONTH

		1	2	3	4	5	6	7	8	9	10	11	12	13	14	15
CONTRACT 1	DIRECT	20453	70633	63195	62695	88624	64812	37833	28266	31865	37299	8466	12933	23433		
	PRELIMINARIES	1643	1643	1643	1643	1643	1643	1643	1643							
	OVERHEADS	1786	1786	1786	1786	1786	1786	1786	1786							
CONTRACT 2	DIRECT	16000	45101	51499	105216	101633	108184	200131	159381	65746	62470	78980	59960	65734	11666	33333
	PRELIMINARIES	17917	17917	17917	17917	17917	17917	17917	17917	17917	17917	17917	17917			
	OVERHEADS	7083	7083	7083	7083	7083	7083	7083	7083	7083	7083	7083	7083			
	Total Payments	64882	144163	143123	196340	218686	201425	266393	216016	122611	124769	112146	97893	89167	11666	33333

Step 3 – receipts
The requisite information on expected receipts can be developed from the estimate and project plan in conjunction with the conditions of contract applicable to the projects on hand. The conditions of contract should indicate how often receipts may be expected and the amount and period of retention.

In both contracts the following conditions apply:

1. Period for honouring certificates – 14 days.
2. Period for interim certificates – 1 month.
3. Percentage of certified value retained – 5 per cent.
4. Defects liability period – 6 months.

It is appreciated that payment for work done is not always made in accordance with the conditions of contract, for a variety of reasons. It is necessary, therefore, to make an assessment, at this stage, of the likely delays and include these in the receipts statement. In this example it is assumed that payments are made when due. Progress statements are prepared by the contractor on the first of each month and payment is received in 14 days.

Payments are made for work done on the basis of unit rates expressed in the BOQ. Payments are also made for materials delivered to the site and intended for the job being done. In order to estimate receipts the Bill rates must be used and an estimate made of the unfixed materials on site. In this example it is assumed that the preliminaries include job and general overheads. The receipts statements for contracts nos. 1 and 2 derived from the BOQ and the project plan are shown in Tables 11.14 and 11.15.

The combined receipts for the 15-month period are given in Table 11.16.

There is one element of the cash flow which has been omitted in this example because it may not always be sizeable in contractors' organisations. This is the cash inflow which arises from the depreciation of assets. If a contractor is in the habit of depreciating assets and including this depreciation in the estimate for work, then this depreciation is, in effect, a receipt for which there is no balancing payment in cash terms. It should be added to the general overheads figure and spread over the duration of the contract.

Step 4 – cash flow
This is the core of the exercise and the cash requirements and surplus are calculated on a cumulative basis. The results are shown in Table 11.17. The negative sign indicates a cash requirement and the positive sign is a cash surplus. The maximum cash requirement is £66,135 in the second month, and there is a maximum surplus of £110,080 in the sixth month.

From a contractor's point of view this cash flow would be less than satisfactory. Banks would not be too pleased with it because it shows that the loan could be difficult to repay within a few months. It should be emphasised that in this example receipts were shown in the month in which they were due. As most contractors are aware, this is not always the case and delays may occur for a variety of reasons. If an assumption is made that receipts will be 1 month overdue the cash flow position alters radically.

Table 11.14 Receipt statement – contract no. 1

OP No.	OPERATION	1	2	3	4	5	6	7	8	9	10	11	12	13	14	15	
1	SITE PREPARATION		6200														
2	SUB STRUCTURE		7075	14105	7075												
3	FLOOR SLABS				38400												
4	RISING WALLS				30666	30666	30666										
5	ELECTRICAL SERVICES						6286	12571	12571	12571							
6	GLAZING							7000									
7	ROOFING							33500	33500								
8	CARPENTER 1st FIX							18500	18500								
9	MECHANICAL SERVICES							7813	15625	15625	15625	7813					
10	PLASTERER (INTERNAL)									16000	16000						
11	FLOOR & CLG FINISHES								7000	14000							
12	CARPENTER 2nd FIX										13500	27000	13500				
13	PLASTERER (EXTERNAL)											9167	18333				
14	PAINTER (INTERNAL)												5333	5333	5333		
15	PAINTER (EXTERNAL)													5833	5833	5833	
16	DRAINAGE													7300	14600	14600	
17	EXTERNAL WORKS															23100	
	TOTAL		13275	14105	76141	30666	36952	79384	87196	58196	45125	43980	37166	18466	25766	43533	
	Add Preliminaries and Claimed Overheads		5930	5930	5930	5930	5930	5930	5950	5930	5930	5930	5930	5930	5930	5930	
			19187	20035	82071	36596	42882	85314	93126	64126	51055	49910	43096	24396	31696	49463	
	Less Retentions at 5%		560	1000	4100	1830	2140	4265	4656	3206	2553	2495	2155	1220	1585	2473	
	Add Retentions Paid																34250
	Nett Receipts		18627	19035	77971	34766	40742	81049	88476	60920	48502	41415	40941	23176	30111	46990	34250

Table 11.15 Receipt statement – contract no. 2

OP No.	OPERATION	MONTH 1	2	3	4	5	6	7	8	9	10	11	12	13	14
1	SITE PREPARATION		39600												
2	FOUNDATIONS			55000	27500										
3	FOOTINGS				59000										
4	INTERNAL DRAINS				15500	15500									
5	R.C. COLUMNS				8333	16666									
6	FLOOR					84000	42000								
7	WALLS & WINDOWS					33750	67500	33750							
8	ROOF FRAMES					9200	18400	18400							
9	R.C. EAVES BEAMS						28666	14333							
10	ROOF SHEETING							4333	21666						
11	MECHANICAL SERVICES							2033	4066	4066	4066	4066			
12	INTERNAL PLASTER								25000	50000					
13	EXTERNAL PLASTER								12083	24166	24166	12083			
14	ELECTRICAL SERVICES									14700	29400	29400	29400		
15	INTERNAL PAINTING										11666	11666	11666		
16	DOOR & RAMP											19200			
17	EXTERNAL PAINTING											6000	6000	6000	
18	EXTERNAL DRAINS					11166	22333	22333						11166	
19	ROADS				17300	34600								34600	
20	LANDSCAPE												6750	6750	
	TOTAL	–	39600	55000	127633	204882	178899	134182	62815	92932	69298	82415	53816	58516	
	Add Preliminaries and Claimed Overheads		31250 70950	31250 86250	31250 158383	31250 236132	31250 210149	31250 165432	31250 94065	31250 124182	31250 100548	31250 113665	31250 85066	31250 89766	
	Less Retentions at 5%		3548	4313	7444	11806	10507	8272	4703	6209	5027	5653	4253	4488	
	Add Retentions Paid														76223
	Net Receipts		64702	81937	150939	224326	199642	157160	89362	117973	95521	107982	80813	85278	76223

Table 11.16 Combined receipts

CONTRACT			3	4	5	6	7	8	9	10	11	12	13	14
CONTRACT No. 1	34766	40742	81049	88416	60920	48502	47415	40941	23176	30111	46990	34250		
CONTRACT No. 2	—	67402	81937	150937	224326	199642	157160	89362	117973	95521	107982	80813	85278	76223
TOTAL NET RECEIPTS	34766	108144	162986	239413	285246	248144	204575	130303	141149	125632	154972	115063	85278	76223

Table 11.17

	MONTH														
	1	2	3	4	5	6	7	8	9	10	11	12	13	14	15
Payments	64882	144163	143123	196340	218686	201425	266393	216016	122611	124769	112146	97893	89167	11666	33333
Receipts	34766	108144	162986	239413	285246	248144	204575	130303	141149	125632	154972	115063	85278	76223	—
Net cash flow	-30116	-66135	-46272	-3199	+63361	+110080	+48262	-37451	-18913	-18050	+24776	+41946	+38057	+102614	+69281

Table 11.18

	MONTH														
	1	2	3	4	5	6	7	8	9	10	11	12	13	14	15
Payments	64882	144163	143123	196340	218686	201425	266393	216016	122611	124769	112146	97893	89667	11666	33332
Receipts	—	34766	108144	162986	239413	285246	248144	204575	130303	141149	125632	154972	115063	85278	76223
Net cash flow	-64882	-174279	-209258	-242612	-221885	-138064	-156813	-167754	-160062	-143682	-130196	-73117	-47721	+25891	+68782

Table 11.18 shows the effect of delaying payment for work alone for 1 month. Thus the payments are shown as before and receipts 1 month later.

The maximum cash requirement is £242,612 (month 4) and cash is required for most of the 15-month period. A contractor must be sure that this cash resource is available otherwise his creditors might become uneasy.

Step 5
It is of course necessary to ensure that both payments and receipts are running according to plan. Unless this is done the initial exercise of planning the cash flow may be useless. In contract work it is necessary to update the cash flow at regular intervals and some updating may require estimates of both payments and receipts for contracts not yet signed. It is suggested that cash flows be updated at 3-monthly intervals and apply to the 12 or 15 months ahead. Thus contracts which are finalised will be accounted for without delay.

11.5 Cost of borrowing

Cash requirements (negative cash flows) during a project result in a contractor either having to borrow money to meet his obligations or removing funds from the company's reserves, which may have been more profitably employed elsewhere.

Inevitably there must be a charge against the project for the use of these funds. A means of determining the amount of interest to be charged during a contract is to calculate the area between the receipts and payments curves (Fig. 11.7). The area is in units of £/month and is known as *captim (capital × time).*

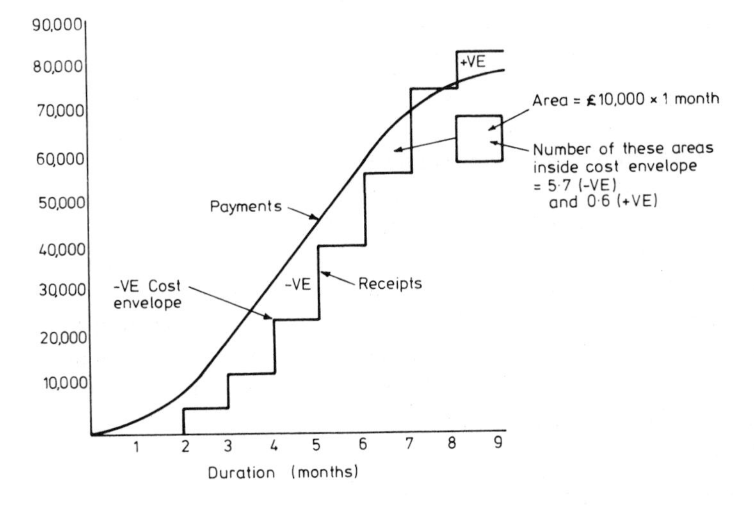

Fig. 11.7 Cash flow diagram

In Fig. 11.7 both positive and negative captims are shown. When calculating the cost of borrowing, interest earned by the positive captim is deducted from the interest paid on the negative captim. Unfortunately contractors often pay higher rates on negative captims than they gain on positive captims.

Example 1
The cash flow diagram in Fig. 11.8 represents the receipts and payments curves for a construction project. During construction, money will be borrowed from the bank as required at an interest rate of 15 per cent per year. Income from the project earns interest at a rate of 12 per cent per year. Calculate the net interest to be charged to the project.

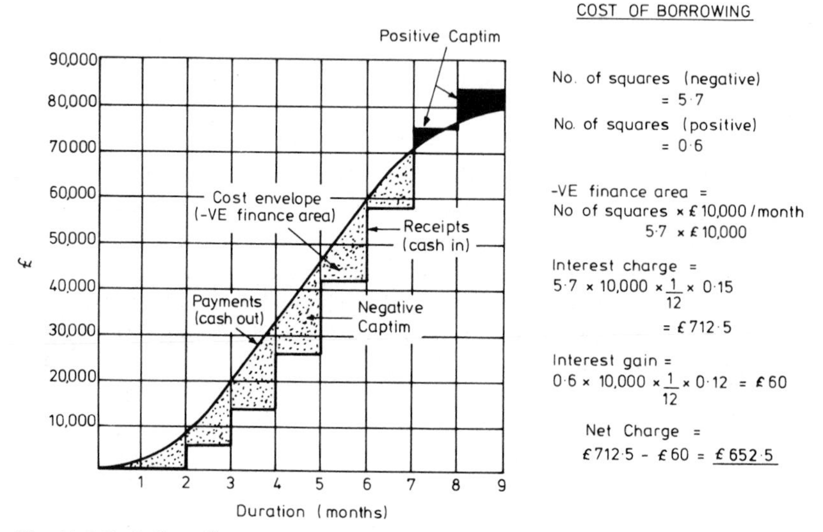

Fig. 11.8 Cash flow diagram

Example 2
Considering the cash flows for the combined contracts 1 and 2, dealt with earlier, it can be seen from Tables 11.17 and 11.18 that if payments are not delayed the maximum cash requirement is £66,135 and maximum cash surplus is £110,080. Delaying payments by one month alters the situation considerably, i.e. maximum cash requirement is £242,612 and maximum cash surplus is reduced to £31,558.

Figure 11.9 represents the cumulative cash flows for these contracts. Calculations of cost of borrowings have been carried out for both normal payments and one month delayed payments. Table 11.19 highlights the influences of delays and interest charges on a contractor's profit figure.

Table 11.19

	Payments to contractor (receipts) made on time (£)	*Payments to contractor (receipts) delayed 1 month (£)*
Maximum cash requirement	66,135	242,612
Maximum cash surplus (before interest charges)	110,080	39,948
Net surplus (after interest charges)	60,531	39,948

Finance area between payments and receipts curves = No. of squares × £ 100,000 month

Assume 16% interest rate

Interest charge = *9 × 100,000 × $\frac{1}{12}$ × 0·16 = 12 000 (Negative captim)

Positive captim (at 13% interest) = 3 × 100000 × $\frac{1}{12}$ × 0·13 = £ 3250

Net Profit = Gross profit − Interest charges = £69280 − (12000 − 3250)
= £60530

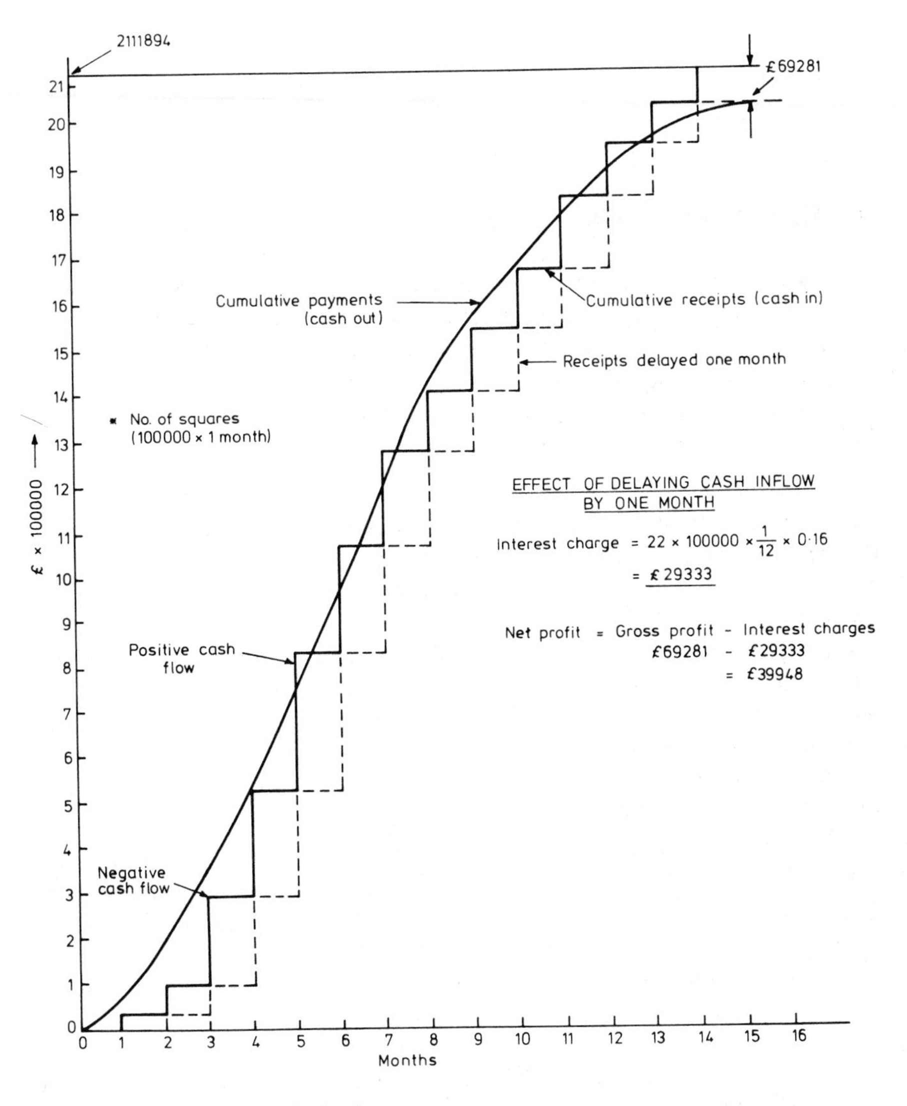

Fig. 11.9 Cash flow curves for both contracts

Chapter 12

Capital expenditure evaluation

12.1 Capital investment situation

A capital expenditure situation is one in which the company's funds are committed to projects which will return the invested funds and profits during future periods. The objective is to evaluate from alternatives the project (investment) which will promote the profitability and long-range growth of the enterprise.

Managerial problems requiring time-based economic appraisal may be broadly classified as:

1. New construction.
   (a) whether to build a new factory or expand existing facilities;
   (b) whether to build a new bridge, embankment, etc.
2. Replacement and modernisation, e.g. replacement of old plant and machinery by modern and more efficient models.
3. Economic choice. In (1) and (2) there may be several ways of achieving the desired result, e.g. building work: steel-framed building v. reinforced concrete, each offering different life spans, initial costs, running costs, etc. Civil engineering work: soil embankment v. concrete bridge v. steel bridge.
4. Financing problems: whether to finance projects from loans, retained profits, trade credit on a short-term basis, etc.

Before considering the various investment appraisal methods it must be appreciated that these do not give a *definite* decision. They simply act as a guide and help communicate useful information to the decision-maker. The actual decision is based on many diverse factors which cannot be incorporated into an overall formula or technique. In making economic comparisions it is important to ensure that the various alternatives are substantially *equivalent*, particularly with regard to their technical specification and performance characteristics.

Factors other than income or expenditure may override a calculated optimum solution. Personal and social factors, for example, may dictate a 'prestige' construction project which would be difficult to evaluate in purely monetary terms. Notwithstanding the importance of such factors, there are three basic approaches to capital expenditure evaluation which will be considered:

Payback.
Return on capital employed. } Traditional methods of appraisal.
Discounted cash flow method.

12.2 *Traditional methods of appraisal*

Payback

The aim of this method is to determine the number of years it takes to pay back the original investments from profits arising from the investment.

Projects can be considered on:

(a) an accept–reject basis depending on the payback period; or
(b) project ranking, where the fastest paying-back project is accepted from a number of mutually exclusive projects.

Example 1 – accept or reject situation
Considering the expected cash flow from project 1 (Table 12.1) and assuming that the basis for acceptance is a 3-year maximum payback period, then this project is economically acceptable.

It pays back the initial outlay of £90,000 in the 3-year period.

Table 12.1

Project 1		
Year	*Cash flow*	
0	−90,000	
1	20,000	
2	30,000	
3	50,000	Payback period ↑
4	40,000	
5	20,000	

Example 2 – project ranking
If projects 2, 3 and 4 (Table 12.2) are mutually exclusive, project 4 has the fastest speed of payback and is therefore economically worthwhile.

This is a simple appraisal method concentrating on speed of return. There are, however, serious disadvantages with this method. Negative cash flows create a problem in that the payback period may be ill defined, e.g. Table 12.3. Is the initial cash outflow £100,000 or £150,000? (These are both capital outflows). Project 6 (Table 12.4) highlights similar ambiguities in that it contains negative cash flows.

In the examples given in Tables 12.5 and 12.6, each successive project is more profitable than the preceding one although the payback period does not show this. It is clear from these results that the payback method does not make any allowance for the time value of money.

Table 12.2

Project	*2*	*3*	*4*
Year	*Cash flow*	*Cash flow*	*Cash flow*
0	−100,000	−20,000	−80,000
1	40,000	8,000	40,000
2	50,000	7,000	45,000
3	60,000	7,000	20,000
4	60,000	8,000	30,000

Table 12.3

Project 5	
Year	*Cash flow*
0	−100,000
1	−50,000
2	+50,000
3	+60,000
4	+60,000
5	+50,000

Payback period?

Table 12.4

Project 6	
Year	*Cash flow*
0	−100,000
1	+60,000
2	−20,000
3	+30,000
4	+50,000
5	+80,000

Payback period?

Table 12.5

Project	*A*	*B*	*C*
Year	*Cash flow*	*Cash flow*	*Cash flow*
1	50,000	50,000	50,000
2	50,000	50,000	50,000
3		50,000	50,000
4			50,000
Payback period	2 years	2 years	2 years

Table 12.6

Project	*A*	*B*	*C*
Year	*Cash flow*	*Cash flow*	*Cash flow*
1	10,000	25,000	40,000
2	20,000	25,000	30,000
3	30,000	25,000	20,000
4	40,000	25,000	10,000
Payback period	4 years	4 years	4 years

This technique should not be considered as an investment appraisal technique as such but rather a means of providing information on the speed of return of the initial outlay.

Return on capital employed (rate of return)

With this method the annual profit is expressed as a percentage of the capital required to produce that profit. There are various methods of computation according to the definition of capital and profit. Generally, however, investment is taken as the initial outlay on the project, while profit is calculated as an average annual figure, after deduction of depreciation, operating costs and expenses, over the life of the project. For example, consider project A with an initial investment of £100,000 and estimated profits as shown in Table 12.7.

Table 12.7

Year	*Project A Profit*
1	40,000
2	40,000
3	30,000
4	30,000

$$\text{Initial investment} = £100{,}000$$

$$\text{Average annual profit} = \left(\frac{40{,}000 + 40{,}000 + 30{,}000 + 30{,}000}{4}\right) = £35{,}000$$

$$\text{Return on capital employed} = \frac{35{,}000}{100{,}000} \times 100 = 35 \text{ per cent}$$

The return on capital employed method can assist management in two ways:

1. If a minimum target is set then those projects exceeding it will be accepted and those below rejected.
2. Projects may be ranked in order of investment preference. The higher a project's percentage rate of return the more it is preferred.

Disadvantages

1. The readily observed weakness of this method (Table 12.8) is that it cannot take into account the fact that earnings from the initial investment may vary from year to year, neglecting the higher returns in the earlier years.

Table 12.8

Project	*A*	B	C
Year	*Profit*	*Profit*	*Profit*
1	10,000	25,000	40,000
2	20,000	25,000	30,000
3	30,000	25,000	20,000
4	40,000	25,000	10,000
Rate of return (%)	25	25	25

2. Another disadvantage is the vague nature of the method. There is no firm agreement on how capital employed should be calculated or how profit is defined.
3. Since the measure of a potential investment is an absolute rather than a relative measure it is unable to take into account the financial size of a project when alternatives are compared.
4. It ignores the possibility of differing lengths of project lives. Most of the disadvantages of such methods as payback and return on capital employed are avoided by the use of discounted cash flow (DCF) techniques. The two principal advantages of DCF methods for determining whether a given project is worth while, or for comparing alternative projects are:
   (a) They take account of the *time value* of money where incomes and expenditures may vary over the anticipated life of the project.
   (b) The significant effects of investment incentives and taxation can be allowed for.

12.3 Discounted cash flow method

(See p. 266 – section 12.13, App. 1).

The basic principle of this method is that a sum of money received today is worth more than the same sum received at some future date. This is not allowing for future changes in the real value of money, but is simply an acknowledgement that a sum of money received now can be used to earn more money, while nothing can be earned on money which has not yet been received. Conversely, a sum of money spent today costs more than the same sum spent in, say, 10 years' time.

Interest is a reward for forgoing the use of money or a payment for its use. Discounting cash flow methods recognise this and show that future earnings from an investment must be sufficient to repay the original investment with compound interest added.

If £1,000 is invested at 10 per cent it will 'grow' as follows:

Start of year 1 – invested	£1,000
End of year 1 – interest	100
	£1,100
End of year 2 – interest	110
	£1,210
End of year 3 – interest	121
	£1,331

Assuming 10 per cent compound interest, the sum required now to provide £1,000 in 3 years is therefore:

$$£1{,}000 \times \frac{1{,}000}{1{,}331} = £751.3$$

Following this principle, tables can be constructed for various rates of interest of *present value factors* which give the present value of £1 receivable or payable at future points in time. Thus at 10 per cent the factors would be as shown in Table 12.9.

Table 12.9

Timing	*Calculation*	*Present value factor*
End of year 1	1,000 ÷ 1,100	0.909
End of year 2	1,000 ÷ 1,210	0.826
End of year 3	1,000 ÷ 1,331	0.751

Table 12.10

Time	*Cash flow (£)*	*Factor*	*PV (£)*
End of year 1	10,000	0.909	9,090
End of year 2	20,000	0.826	16,520
End of year 3	25,000	0.751	18,775
End of year 4	35,000	0.683	23,905
	£90,000		£68,290

Using these factors, it is possible to calculate the total *present value* of a series of future receipts if the going rate of interest is 10 per cent (Table 12.10). Put in everyday terms, the key figures could be repayments to a building society where a person borrows say £1,000 and agrees to pay £440 at the end of the first, second and third years, and nothing thereafter. The interest charged by the building society in this example is

15 per cent. The meaning of this 15 per cent is demonstrated below, where the annual repayments are divided between the interest charge and the repayment of the capital sum:

Capital £1,000
Year 1 repayment = £440
= £150 (15 per cent interest on £1,000)

Leaving £290 (capital repayment)
Capital (£1,000 − 290) = £710
Year 2 repayment = £440
= £107 (15 per cent interest on £710)

Leaving £333
Capital (£710 − 333.50) = £376.50
Year 3 repayment = £440
= £57 (15 per cent interest on £376.50)

Leaving £383 (which will repay the capital outstanding)
Interest recovery is high in early years − 150→107→57
Capital recovery is high in later years − 290→333→383

440 440 440

From these principles, three basic methods of using DCF have been developed:

1. Net present value (present worth).
2. Equivalent annual cost (annual cost)
3. Internal rate of return (yield, DCF rate of return).

12.4 Net present value (NPV)

Given the choice between £1,000 now or £1,000 in 10 years' time most people would take the £1,000 now because the cash could be invested or simply gain compound interest in a savings account. If the savings account interest rate is 10 per cent the £1,000 would be worth £2,593 in 10 years' time.

The compound interest formula is $A(1+r)^n$, where A is the initial amount invested or deposited, r the annual rate of interest and n the number of years for which the amount is invested or deposited.

NB: $(1+r)^n$ is the single-payment compound amount factor (SPCAF) (see section 12.16)

In the previous example:

$$A = £1{,}000$$
$$r = 0.10$$
$$n = 10$$
$$£1{,}000 \times (1+0.10)^{10} = 2{,}593$$

By arranging the compound interest formula it is possible to calculate the *present value* of £2,593 received in 10 years' time:

$$£2{,}593 \times \frac{1}{(1+0.10)^{10}} = £1{,}000$$

NB: $1/(1+r)^n$ is the single-payment present worth factor (SPPWF) (see section 12.16).

The formula $A(1+r)^n$ is the terminal value in n years' time of an amount A which is now invested at an annual compound interest rate of r; $A \cdot [1/(1+r)^n]$ is the present value of an amount A received in n years' time.

Cash flows are discounted at a predetermined rate of interest. If the present value of inflows is greater than the present value of outflows the project is regarded as being economically worth while (Table 12.11). 'Economically worth while' merits further consideration.

Data given in Table 12.12 indicate that the project is economically *not* worth while, i.e. merits consideration only if there are overriding non-financial considerations.

Table 12.11

Time	*Cash flows (£)*	*Factor (from tables or formula 2)*	*PV (£)*
Year 0	−10,000	1.000	−10,000
Year 1	+4,000	0.909	+3,636
Year 2	+5,000	0.826	+4,130
Year 3	+4,000	0.751	+3,004
	£13,000		£770

Table 12.12

Time	*Cash flows (£)*	*Factor*	*PV (£)*
Year 0	−10,000	1.00	−10,000
Year 1	—	0.909	
Year 2	2,000	0.826	1,652
Year 3	11,000	0.751	8,261
	Total £13,000		−£87

The general mathematical expression of NPV of an investment project as the sum of its DCF is given as:

$$\sum_{t=0}^{n} \frac{A_t}{(1+r)^t}$$

where A_t is the project's cash flow, t the time period (value 0 to n) and r the annual rate of discount. Investing £10,000 for 3 years in a deposit account at an interest rate of 10 per cent produces £13,309, whereas investing this sum in the previous project will

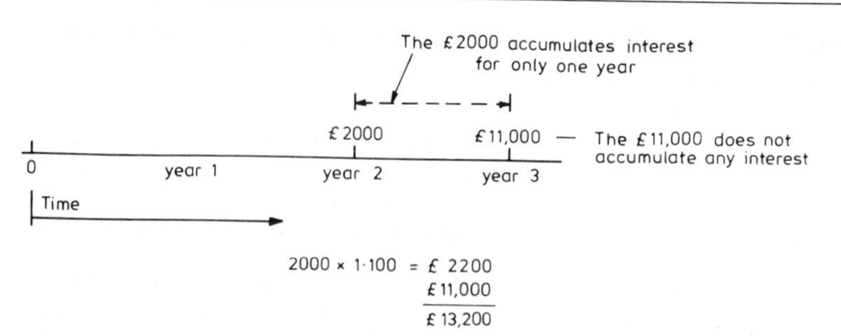

Fig. 12.1 Present value model

produce only £13,200 (Fig. 12.1). Therefore the *project* investment is not an economically worth while alternative.

The present value of each of the sums is as given in Table 12.13. The difference between the two sums (£10,000 − £9,913) is £87.00 which is the amount of the project's negative NPV.

Table 12.13

Type of investment	*Sum (£)*	*Discount formula or from tables*		*PV (£)*
Project	13,200	$\frac{1}{(1+0.10)^3}$	Factor 0.75131	9,913
Deposit	13,309	$\frac{1}{(1+0.10)^3}$	0.75131	10,000
			Deposit	£87

If the result is zero or positive then the project will be economically worth while. The negative value in the previous example indicates that this project is not economically worth while. Why does this advise against investment? The project requires a capital outlay of £10,000 and produces cash flows in years 2 and 3. If, however, the money was deposited in a savings account or invested elsewhere at an interest rate of 10 per cent, at the end of 3 years this would produce:

$$£10,000 \times (1+0.10)^3 = £13,310$$

If the cash inflows from this project (2,000 at year 2, £11,000 at year 3) were placed on deposit as these arose, the value that would accumulate by the end of year 3 would be as shown in Table 12.14.

Table 12.14

Year	*Cash inflow (£)*	*Compound interest factor*	*PV (£)*
2	2,000	1.1000	2,200
3	11,000	1.0000	11,000
		Total value	£13,200

12.5 Equivalent annual cost (annuity method)

Future cash flows are first discounted at a predetermined rate of interest to find their total or NPV. This present value is then divided by the sum of the discount factors (known as the cumulative discount factor) for the life of the project to find the equivalent annual cost. Consider the data in Table 12.15.

Table 12.15

Time	*Amount (£)*	*Factor*	*PV (£)*
Year 1 end	200	0.909	182
Year 2 end	900	0.826	743
Year 3 end	300	0.751	225
		2.486	1,150

Predetermined rate of interest – say, 15 per cent

$$\text{Equivalent uniform annual cost} = \frac{1{,}150}{2.486} = £462.6 \text{ per year}$$

When using this method in choosing between alternatives, the treatment of any salvage values warrants careful attention. For example, a particular alternative may require a new machine costing £250,000 with an anticipated salvage value of £15,000 after a period of 5 years (Fig. 12.2). Assume 12 per cent is the going interest rate.

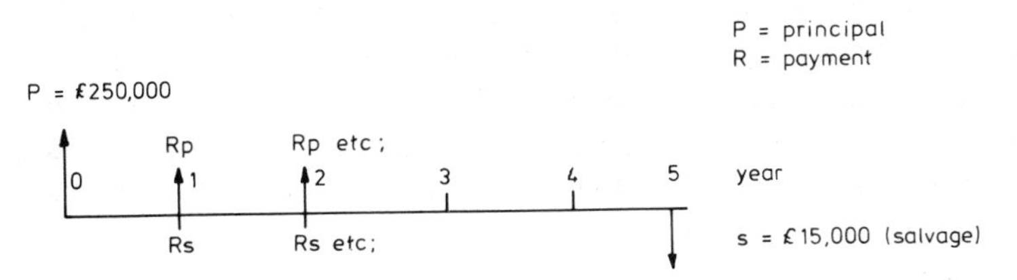

Fig. 12.2 Equivalent annual cost model

The annual capital recovery cost of the machine would be:

$$(\text{Initial cost} - \text{salvage value}) \times \left[\frac{i(1+i)^n}{(1+i)^n - 1}\right] + (\text{salvage value} \times \text{interest rate})$$

$$= (A - S) \times \left[\frac{i(1+i)^n}{(1+i)^n - 1}\right] + Si$$

$$= 235{,}000\ (0.27740) + 1{,}800$$
$$= £66{,}989$$

Alternatively:

$R_A - R_S$
$A(\text{USCRF}) - S(\text{USSFDF})$

$$\frac{A(i(1+i)^n}{(1+i)^n - 1} - \frac{S(i)}{(1+i)^n - 1}$$

£250,000 (0.27740) − 15,000 (0.15740)
£69,350 − £2,361 = £66,989

As the salvage value becomes available at the end of the fifth year it is only necessary to 'charge' to each annual cost the interest on that amount. (For explanation of abbreviations, see section 12.14).

12.6 Internal rate of return (IRR)

In the NPV method when the result is positive, it means that the return is greater than the interest rate used. The internal rate of return of a project is the rate of discount that, when applied to a project cash flows, produces a zero NPV.

The IRR is the value for v which satisfies the expression:

$$\sum_{t=0}^{n} \frac{A_t}{(1+r)^t} = 0$$

This formula can be used for projects, the cash flows of which extend up to three periods. If they extend beyond this, complex polynomial equations are required for solution. Fortunately a good approximation of a project's IRR can be found through linear interpolation. In this method, cash flows are discounted at a series of 'trial' rates of interest to determine the rate of interest at which the value of incomes is exactly equal to the present value of expenditures.

Example – two-period cash flow

Year	*Cash flow*
1	−£1,000
2	+£1,250

using

$$\sum_{t=0}^{n} \frac{A_t}{(1+r)^t} = 0$$

$$-1{,}000 + \frac{1{,}250}{1+r} = 0$$

$$1{,}250 = 1{,}000 + 1{,}000r$$
$$250 = 1{,}000r$$

$$r = \frac{250}{1{,}000} = 0.25 = 25 \text{ per cent}$$

If this project is now discounted by 25 per cent the NPV will be zero.

Year	*Cash flow*	*Present value factor*	*Present value*
1	−1,000	1.000	−£1,000
2	+1,250	0.80	+£1,000
			£0

Example 2 – linear interpolation

From Table 12.16 it is clear that the rate of return lies between 5 and 10 per cent. A graph is drawn (Fig. 12.3) and the approximate rate of return is read off. The IRR can

Table 12.16

Year	*Net cash flow (£)*	*PWF 5%*	*PW of new cash flow (£)*	*PWF 10%*	*PW of new cash flow (£)*
0(*A*)	−10,000	1.000	−10,000	1.000	
1	+4,000	0.952	3,808	0.909	3,636
2	+4,000	0.907	3,628	0.826	3,304
3	+4,000	0.864	3,456	0.751	3,004
(B)	£12,000		£10,892		£9,944
Ratio *A/B*	0.833		0.918		1.01

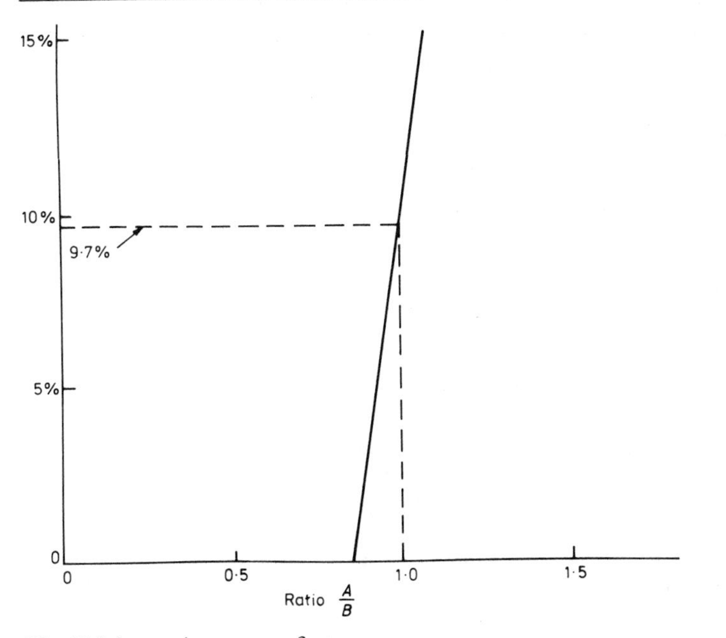

Fig. 12.3 Approximate rate of return

also be calculated as follows:

$$\text{IRR} = 5 + \frac{(10{,}892 - 10{,}000)}{(10{,}892 - 9{,}944)}(5)$$

$$= 5 + 4.7 = 9.7 \text{ per cent}$$

This confirms our graphical result.

A third alternative is to use the discounted NPV obtained from Table 12.16. For example, discounting at 5 per cent produces an NPV of £892, and discounting by 10 per cent produces an NPV of −£56. The NPV profile is shown graphically in Fig. 12.4.

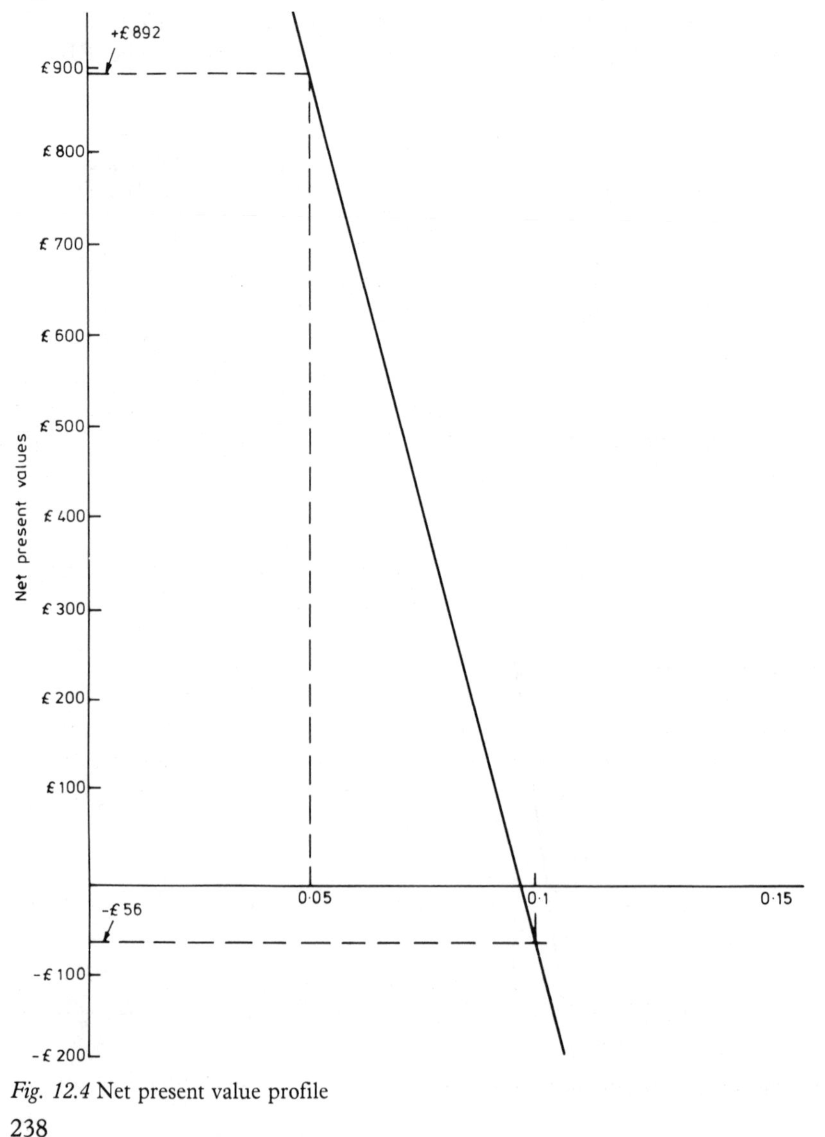

Fig. 12.4 Net present value profile

The gradient of the NPV profile is

$$\frac{-56-892}{0.1-0.05} = -18{,}960$$

Using the equation of a straight line,

$$y = mx + C \quad \text{hence} \quad y = -18{,}960x + C \qquad \ldots [1]$$

At the point (0.05, 892),

$$892 = -18{,}960 \times 0.05 + C$$
$$892 = -948 + C$$
$$1{,}840 = C$$

Substituting into equation [1] where $y = 0$, so that

$$x = \frac{1{,}840}{18{,}960} = 0.097$$

IRR = 9.7 per cent

12.7 Dealing with multiple alternatives

Most construction and business projects can be accomplished by more than one method. Invariably, the alternative methods require differing levels of capital investment, give rise to different incomes and may, in turn, have varying economic lives. The investment of additional amounts of capital in various alternatives usually results in increased capacity, increased revenue, decreased operating expenses, or increased life and thus enhances the manager's chances of obtaining a greater profit.

If mutually exclusive projects are to be compared, then it is not possible to select the correct project solely on the basis of an internal rate of return analysis.

Mutually exclusive projects

Suppose a company requires one new factory and three possible sites are under consideration. The projects are mutually exclusive since only one new factory is required, so if it is built on site 1, sites 2 and 3 are therefore excluded and vice versa.

Mutually exclusive projects can be defined as those which compete for acceptance in that it is not possible or desirable to accept more than one, e.g.:

(a) two or more schemes for the use of one plot of land;
(b) two or more financially acceptable contracts when a contractor has sufficient management staff or labour for only one.

It is not possible to compare such projects by IRR analysis because the IRR does not give any indication of either the amount of capital involved in an investment or the duration of the investment.

Discrimination can easily be made, however, on the basis of incremental cash flows.

Incremental analysis

Incremental analysis consists essentially of two stages:

1. The subtraction of one cash flow from another.
2. The application of a DCF method to the differences between them (Table 12.17).

Table 12.17

Year	*Project A (£)*	*Project B (£)*	*Increment A–B (£)*
0	−2,000,000	−3,050,000	−1,050,000
1	+1,000,000	1,500,000	500,000
2	1,000,000	1,500,000	500,000
3	1,000,000	1,200,000	200,000
	23.4%	20%	8%

Example
The project with the smaller capital investment is subtracted from the project with the larger capital investment. Project A seems initially to be preferable to project B. Project B, however, uses more capital than project A, so it is necessary to know the profitability of the extra investment. This is 8 per cent, which, depending on the attractive rate of return, may or may not be acceptable. Incremental analysis in this way is useful for comparison of projects of different sizes and lives because it directly relates the extra profitability to the extra investment of life.

This incremental approach should also be used for comparing three or more alternatives in order to obtain a true relationship between the merits of each alternative.

Example
Scheme A initial investment = £5,000, annual income £1,500
Scheme B initial investment = £25,000, annual income £6,000
Scheme C initial investment = £35,000, annual income £9,000
Scheme D initial investment = £48,000, annual income £12,000
Life of each scheme is five years; attractive rate of return 8 per cent.

	Scheme			
	A	B	C	D
Capital (£)	−5,000	−25,000	−35,000	−48,000
Additional capital (£)		20,000	10,000	13,000
Annual income (£)	1,500	6,000	9,000	12,000
IRR (%)	15	6	9	8
Increase in income (£)		4,500	3,000	3,000
IRR on increased outlay (%)		4	15	5

Choose scheme C.

Postponing an investment

The incremental rate of return method of choosing between alternatives can be extended to determine whether or not an investment in a project should be postponed. This problem can only arise where the project's inflow will be higher (or the capital outlays lower) if the project is postponed.

For example, capital outlays on new machinery or plant will often need to be postponed until an increasing level of repairs or operating diseconomies on the old plant make for profitable replacement.

Example

The choice may be between investing in a new building of £100,000 now or in 2 years' time. By delaying the investment for two years maintenance costs on the existing building of £15,000 per annum are anticipated. For the new building, maintenance costs are estimated as follows: first year £2,500, second year £5,000 and £10,000 for each succeeding year to the end of an 8-year study period (Table 12.18).

Table 12.18

Year	*Proposal*	*Postponed proposal*	*Incremental new cash flow*	*14% PWF*	*PW of new cash flow (£)*
0	−100,000	−15,000	−85,000	1.000	−85,000
1	−2,500	−15,000	+12,500	0.877191	+10,964.8
2	−5,000	−100,000	+95,000	0.76946	+73,098.7
3	−10,000	−2,500	−7,500	0.67497	−5,062.2
4	−10,000	−5,000	−5,000	0.59208	−2,960.4
5	−10,000	−10,000			
6	−10,000	−10,000			
7	−10,000	−10,000			
8	−10,000	−10,000			
					−£8,959

If the cost of capital is taken at 14 per cent, the series of cash flows in comparing the two proposals offers a NPV of −£8,959, thus indicating that postponing and starting in two years is the better alternative. (N.B. Present worth of 'proposal' is greater than present worth of 'postponed proposal' and therefore the discounted investment required in year 2 is less than in year 0.

Consider the following. The choice of plant is between two alternatives x and y (Tables 12.19 and 12.20). Plant y would be preferred because it has the larger positive NPV. However, there is a replacement problem to contend with. Plant x has a two-year life, whereas plant y has a 4-year life. In order to compare these alternatives properly a common life span should be taken – say, 4 years (lowest common denominator of the life spans). A replacement of plant x would be required at the end of year 2 (Table 12.21).

Table 12.19 Plant *x*

Year	*Cash flow (£)*	*PWF 14%*	*PV (£)*
0	−45,000	1.000	−45,000
1	+24,000	0.87719	21,052
2	+38,000	0.76946	+29,240
			NPV £5,293

Table 12.20 Plant *y*

Year	*Cash flow (£)*	*PWF 14%*	*PV (£)*
0	−53,000	1.000	−53,000
1	15,000	0.87719	+13,158
2	25,000	0.76946	+19,237
3	30,000	0.67497	+20,249
4	10,000	0.59208	+ 5,921
			NPV £5,565

Table 12.21

	Cash flow			
Year	*Plant x_1*	*Plant x_2*	*PWF 14%*	*PV (£)*
0	−45,000		1.0000	−45,000
1	+24,000		0.87719	+21,052
2	+38,000	−45,000	0.76946	−5,386
3		+24,000	0.67497	+16,200
4		+38,000	0.59208	+22,500
				NPV £9,366 (preferred)

When comparing the projects over a common life span of four years, plant x with a replacement in year 2 is preferred on the basis of having a larger positive NPV.

Projects having differing lives can be compared more easily using the *equivalent annual cost* method. Considering the above two alternatives:

Plant x – at a discount rate of 14 per cent = £5,293

Applying the equivalent annual cost factor

£5,293 × $^{2}\text{USCRF}^{14\%}$ = 5,293 × 0.60728 = £3,214 (preferred)

Plant y – at a discount rate of 14 per cent = £5,565
£5,565 × 0.34320 = £1,910

x is the better alternative.

12.8 Inflation

This is a term used to describe a situation where prices rise with the passage of time, resulting in the reduced purchasing power of money.

Management is faced with the additional problem of trying to estimate likely future rates of inflation and allowing for this in investment decisions. If the prevailing inflation rate is, say, 12 per cent then to replace £1,000 in 1 year's time will require £1,000 $(1+0.12) = £1{,}120$. To this must be added the cost of capital (annual interest rate), say, 10 per cent. The total amount required at the end of 1 year is $£1{,}120 \times (1+0.10) = £1{,}232$. The total annual money interest rate is therefore $(1+0.12) \times (1+0.10)-1 = 0.23$ or 23 per cent. An alternative to this method is to work from the total estimated annual money interest rate and calculate the 'real' discount rate.

For example, let the total annual money interest rate (m) = 23 per cent; let the general rate of inflation (C) = 12 per cent; let the real rate of interest = r. Then:

$$(1+r)(1+C)-1 = m$$
$$(1+r)(1+12)-1 = 23$$

$$r = \frac{(1+m)}{(1+C)} - 1$$

$$= \frac{1+0.23}{1+0.12} - 1 = 0.10$$

Real interest rate = 10 per cent.

When considering the effects of inflation in project appraisal the main essential is that all estimates should be made on a consistent basis. All cash flows should be expressed either in terms of purchasing power of money at the time when the estimate is made, or in terms of purchasing power, allowing for inflation, when the individual flows are estimated to occur.

If a constant rate of inflation C is included, an amount A will produce $A/(1+C)^n$ in terms of real purchasing power after n years, with no rate of interest. At a total money rate of interest m, this amount A will produce

$$F = \frac{A}{(1+C)^n}(1+m)^n \text{ after } n \text{ years}$$

The effective rate r, or rate of return in real terms, will be:

$$F = A(1+r)^n \qquad (1+r)^n = \frac{F}{A} = \frac{(1+m)^n}{(1+C)^n}$$

or

$$r = \frac{1+m}{1+C} - 1 \qquad r = \frac{1+m-1-C}{1+C} \cong m - C \text{ when both rates are low}$$

So the rate of interest in real terms is approximately equal to the money interest rate less the rate of inflation.

Example
A project requires an initial outlay of capital of £200,000. Cash receipts less expenses are estimated to be £120,000, £80,000 and £70,000 after 1, 2 and 3 years respectively. The company's (money) cost of capital which is appropriate to this project is 10 per cent per year. If the rate of inflation is 7 per cent per year what is the present worth of the project?

Solution

$$m = 0.10 \qquad C = 0.07 \qquad r = \frac{1+m}{1+C} - 1 = 0.028$$

or $r \simeq m - C = 0.03$

$$\text{Present worth} = -200{,}000 + \frac{120{,}000}{1.03} + \frac{80{,}000}{1.03^2} + \frac{70{,}000}{1.03^3}$$

$= £55,973$

12.9 Taxation

The influences of taxes should be taken into consideration when assessing the feasibility of a project in practice. Payment of taxes can be considered as an outward cash flow like any other payment of expenses. A developer may be liable to pay tax under the following.

1. Local taxation (rates) levied by local authority on the value of land and premises.
2. Taxation levied by the central government on net accounting profits.
3. Levy on the gain in the value of land as a result of development.
4. Capital gains tax.
5. Capital transfer tax.

There are important differences between the positions of private firms and public authorities in regard to tax liability. For example, all private concerns have a general responsibility for all forms of taxation, while local authorities are liable to pay rates on certain classes of their assets including housing, but not highways, sewers and public parks. Nationalised industries are at present subject to both local taxation and central government taxation as if they were ordinary businesses.

United Kingdom taxation affects the cost and manner of raising finance for an enterprise. Interest paid to the holders of fixed interest securities is allowed as a deductible expense in computing the amount of profit on which taxation is assessed. Suppose that a loan stock is issued and repayable at par, and has an interest cost of 8 per cent per year, and that the enterprise is a company taxed at 30 per cent on profits. The loan then has an effective net cost of only 6.2 per cent since 1.8 per cent is recovered in the form of taxation relief. Dividends paid to shareholders, however, are not allowable as a deduction in computing the taxation liability. In computing the weighted average external cost of capital of a company it is therefore necessary to combine the net of tax interest cost of fixed interest securities with the gross cost of ordinary share capital. The matter is further complicated because the present UK system of taxation makes it more worth while to finance an investment from retained earnings than from a new issue of ordinary shares.

Taxation forms an important negative element in the estimation of future cash flows. Because tax rates depend upon the action of government, this estimation is subject to a very high degree of uncertainty.

12.10 Sensitivity analysis

If accurate values of variables are not known, sensitivity analysis can be used. Calculations are performed using the best approximate values of all variables. Then the calculations are repeated, but with one variable altered by a known amount. The sensitivity of the final solution to the changing value of each variable in turn is thus found. Those variables to which it is most sensitive can be examined more thoroughly in an attempt to find more accurate values of them.

Accurate estimates are more difficult to make for the very long term than for the near future. It is fortunate that decisions can often be insensitive to distant estimates if the discount rate is not too low. For example, a large warehouse is required for a period of 50 years. Two schemes are proposed:

	Scheme A	*Scheme B*
Initial cost	£180,000	£400,000
Life of building	25 years	50 years
Salvage value	£20,000	£60,000
Cost of a second warehouse after 25 years	£350,000	
Annual maintenance and operating costs	£25,000	£15,000
Rate of interest	12%	12%
Present worth for 50-year period	£378,902	£524,358
	difference of £145,456	

1. *Influence of salvage value of scheme B.* If the salvage value of scheme B warehouse were increased twentyfold to £1,200,000 this would only reduce the present value of scheme B by

$$£1{,}200{,}000 \times {}^{12\%}\underset{(0.00346)}{\mathrm{SPPWF}^{50}} = 4{,}152$$

Therefore, new NPV of B = £520,414 and scheme A is still preferable.

2. *Influence of estimated replacement cost of scheme A.* The estimated cost of a replace-

ment warehouse after 25 years is £350,000. If this is increased fivefold to £1,750,000 the present worth of scheme A would be increased by

$$1{,}750{,}000 \times \underset{(0.05882)}{{}^{12\%}\text{SPPWF}^{25}} = £102{,}935$$

New present worth = £461,250 and scheme A is still preferable.

3. Influence of discounting rate. The sensitivity of the decision to changes in the discounting rate can be tested. The lower the rate, the greater will be the influence of long-term assessments.

In general, if prices and replacement costs are expected to rise considerably, alternatives having longer lives are preferable. On the other hand, anticipated technological improvements, changes in requirements and price reductions favour shorter-life alternatives.

Discounting methods compared

Correctly used, all three methods of discounting are numerically correct and would lead to the same investment decision. The choice between the methods is principally one of ease of application, each separate use to be considered on its merits as outlined below.

Net present worth (NPW)

1. It is necessary to establish an interest rate before calculations can proceed.
2. Useful for analysing all the possible ways of undertaking an investment in the early stages of the appraisal process.

Annual cost method

1. Convenient where cash flows follow a regular and defined pattern.
2. It is necessary to establish interest rate before calculations can proceed.

Internal rate of return (IRR)

1. Easy to understand.
2. Does not readily rank projects (incremental analysis is needed).
3. A firm interest rate is calculated from the established cash flows.

12.11 Depreciation and valuations

Depreciation is the decrease in value of an asset (plant, buildings, etc.) through any cause, e.g.:

1. Passage of time.
2. Replacement by new machines, building having improved efficiency.
3. Diminished performance due to wear, corrosion, decay.

4. Changes in amount and type of service required.
5. Changes in public legislation causing the assets no longer to conform with requirements.
6. Destruction of the asset due to accident.
7. Difficulty in obtaining spares.

Different types of assets have different economic lives. The economic life is the number of years which it is estimated it will be economic to keep the asset. At the end of this life, it may only have a salvage value and may be purchased by another organisation, or it may remain in the same organisation and be used for a less demanding job.

Typical lives are as follows:

	Range in years
Dams, tunnels, docks, harbours, breakwaters	50–100
Permanent buildings, roads, bridges	40–60
Outdoor steel structures such as cranes, lock gates	35–50
Railway tracks	20–35
Diesel-electric stations	15–20
Portable tools and office furniture	12–15
Construction plant	5–10

In construction economy studies it is often necessary to predict the salvage value of an asset at the end of, or during, its economic life.

Since depreciation represents a loss or expense of carrying on a business, it must be recorded in the books of account in order that the time profit or loss position may be ascertained, and that the reduced value of the asset may be shown in the balance sheet. Depreciation, therefore, is that part of an asset which is not recoverable when the asset is taken out of service. Provision against this loss of capital must be made. Funds must be retained in the business to finance replacements at the end of the asset's economic life.

Depreciation and valuation methods

In considering depreciation it is helpful to picture a charge for depreciation as being a series of payments made to a specific fund for the eventual replacement of the asset under consideration.

The four most common methods available are:

1. The straight-line method.
2. The sum of the digits method.
3. The declining balance method.
4. The sinking fund method.

All these methods are based on time, i.e. a piece of plant used every day has the same depreciation charge as one used only once every year. Many managers and accountants advocate that depreciation should be based on the amount of use as well as the

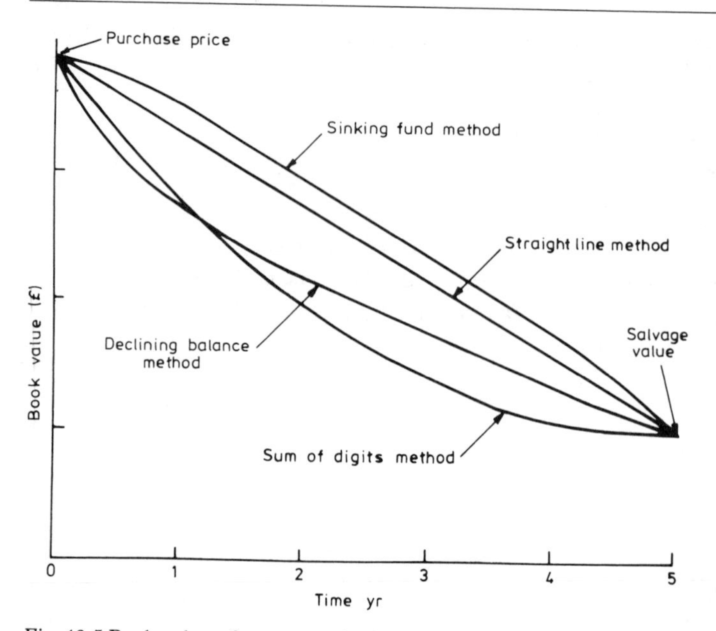

Fig. 12.5 Book value of an asset calculated by different depreciation methods.

estimated economic life. Each of the above methods have certain unique features. Some managers prefer a method which recovers most of the money invested early in the life of the asset (Fig. 12.5), e.g. declining balance or sum of the digits method, while others prefer the simpler accounting procedure of constant annual charges throughout the asset's life. All depreciation methods should:

1. Recover the capital investment of an asset.
2. Maintain a book value closely corresponding to the actual value of the asset throughout its life.
3. Be easy to apply.
4. Be acceptable for tax relief purposes.

Valuations of plant/assets

It is sometimes necessary to calculate the value of an asset to determine what amount would be reasonable to pay to purchase it. In order to do this the following information is required.

Cash flow {
1. Estimate amounts and dates of expected receipts as a result of owning the asset.
2. Estimate the amount and dates of expenses as a result of owning it.
}
3. Estimate a reasonable economic life for the asset.

The valuation of estimate of a reasonable purchase price can depend on the present worth of all these factors. The same set of calculations is also useful in judging how much an asset is worth to its present owner.

The main depreciation and valuation methods will be illustrated through the following example.

Example 1
A contractor has purchased excavating plant costing £25,000. The estimated economic life is 5 years. Salvage value after 5 years' use is £5,000 and the current interest rate is 8 per cent. Determine:

(a) The depreciation charges during years 1 and 2.
(b) The depreciation reserve accumulated by the end of year 3.
(c) The book value at the end of year 3.

Symbols used in formulae:

P = purchase price (present worth) of asset;
S = salvage value;
n = economic life (years);
N = number of years of depreciation or use from date of purchase;
i = interest rate.

Straight-line depreciation
This is the simplest method of calculating the depreciation as a function of time and is one of the most widely used methods. The initial cost less the prospective salvage value is divided by the asset's estimated economic life to obtain the annual depreciation charge. This annual charge is constant. The book value is the difference between the purchase price and the product of the number of years in service times the annual depreciation charge.

$$\text{Annual depreciation charge} = \frac{P-S}{n}$$

$$\text{Book value at end of year } N = P - \frac{N}{n}(P-S)$$

Since the annual depreciation charge is constant, the charges for years 1 and 2 are:

(a) Depreciation charge

$$\frac{P-S}{n} = \frac{25{,}000-5{,}000}{5} = £4{,}000/\text{year}$$

(b) The depreciation reserve at the end of year 3 is 4,000 × 3 = £12,000.
(c) The book value at the end of 3 years is

$$25{,}000 - \frac{3}{5} \times (25{,}000-5{,}000) = £13{,}000$$

or

$$25{,}000-12{,}000 = £13{,}000$$

Sum of the digits
This method provides for depreciation charges which are higher in the early years and lower in the later years. The name is taken from the calculation procedure.

(a) Annual depreciation charge

$$= \frac{n-N+1}{1+2+3\ \ldots\ +n}(P-S)$$

$$= \frac{2(n-N+1)}{n(n+1)}(P-S)$$

Book value at end of year N

$$= 2\,\frac{[1+2+3+\ \ldots\ +(n-N)]}{n(n+1)}(P-S)+S$$

The sum of the digits for the five-year life is

$$1+2+3+4+5 = 15$$

$$\frac{n(n+1)}{2} = \frac{5(6)}{2} = 15$$

Depreciation charge during year 1

$$= \frac{n-N+1}{15}(P-S) = \frac{5-1+1}{15}(25{,}000-5{,}000)$$

$= £6,666.6$ (year 1)

After the first year only 4 years remain, therefore depreciation charge during year 2 is

$$\frac{4}{15}\times 20{,}000 = £5{,}333.3$$

(b) Depreciation reserve at end of year 3

$$= \frac{5+4+3}{15}\times 20{,}000$$

$= £16,000$

(c) Book value at end of year 3

$P-$depreciation reserve $= £25,000-£16,000 = £9,000$
or
$0.2 \times 25,000 = £4,000 + £5,000 + £9,000$

Declining balance method

This provides higher charges in the early years with a corresponding lower annual charge near the end of the asset's economical life. The annual depreciation charge is calculated by taking a constant percentage of the declining undepreciated balance. The salvage value with this method must always be greater than zero.

$$\text{Depreciation rate} = 1 - \left(\frac{S}{P}\right)^{1/n}$$

$$\text{Book value at end of year } N = P(1 - \text{depreciation rate})^N$$

$$= P\left[1-(1-\left(\frac{S}{P}\right)^{-n}\right]^N$$

$$= P\left(\frac{S}{P}\right)^{N/n}$$

$$\text{Annual depreciation charge} = \left[P(1 - \sqrt[n]{\left(\frac{S}{P}\right)}\right]$$

(a) $$\text{Depreciation charge year 1} = £25{,}000 \times \left[1-\left(\frac{5{,}000}{25{,}000}\right)^{1/5}\right]$$

$$= £25{,}000 \times (1 - \sqrt[5]{0.20})$$
$$= £25{,}000 \times 0.276$$
$$= £6{,}900$$

Year 2 depreciation $= (£25{,}000 - £6{,}900) \times 0.276 = £4{,}996$

(b) Depreciation reserve accumulated by end of year 3
$= £6{,}900 + £4{,}996 + [(25{,}000 - 11{,}896) \times 0.276]$
$= £15{,}513$

$$\text{Accumulated depreciation} = P \times \left[1-\left(\frac{S}{P}\right)^{N/n}\right] = £25{,}000 \times \left[1-\left(\frac{5{,}000}{25{,}000}\right)^{3/5}\right]$$

$$= £25{,}000 \times 0.6206$$
$$= £15{,}515$$

(c) Book value at end of year 3

$$P\left(1-(1-\sqrt[n]{\frac{S}{P}})\right)^N = £25{,}000 \times (1-0.276)^3$$

$$= £25{,}000 \times (0.724)^3$$
$$= £9{,}485$$

or

$$P(S/P)^{N/n} = 25{,}000 \times \left(\frac{5{,}000}{25{,}000}\right)^{3/5}$$

$$= £9{,}485$$

Sinking fund method

The annual depreciation charges calculated by this method are lowest in the early years and largest in the later years. It is based on the concept that the depreciation recoveries each year are placed in a sinking fund. This fund earns interest and the actual amount to be deposited in the fund so that the depreciable cost of the asset is accumulated is dependent on the interest rate. The book value is the difference between the purchase price and the depreciation reserve. Book values using this method are greater than the straight-line method. Increasing the interest rate increases the difference.

$$\text{Annual depreciation charge} = (P-S)\underbrace{\frac{2}{(1+i)^n-1}}_{\text{(USSFDF)}}$$

$$\text{Book value at end of year } N = P-(P-S) \times \underbrace{\frac{i}{(1+i)^n-1}}_{\text{(USSFDF)}} \times \underbrace{\frac{(1+i)^N-1}{i}}_{\text{(CAF)}}$$

(a) Constant annual charge for depreciation is (25,000 − 5,000) 0.17054 = £3,411. Year 2 – capital recovery includes the depreciation charge (£3,411) plus the interest earned by the first year's depreciation charge (£3,411 × 0.08 = £273) i.e. £3,684.

(b) Depreciation reserve at end of year 3 is
$(P-S) \times {}^{8\%}\text{USSFDF}^5 \times {}^{8\%}\text{USCAF}^3$
20,000 × 0.17054 × 3.246 = £11,071
or
£3,411 + £3,684 + (£3,684 × 0.08 + £3,684) = £11,073

(c) Book value at end of year 3 is
$P-$accumulated depreciation = £25,000 − £11,071
= £13,929
See Table 12.22 for comparisons of methods.

(For explanation of abbreviations, see section 12.14).

Table 12.22 Summary of results highlighting features of the depreciating methods

	Depreciation method			
	Straight-line method	*Sum of the digits*	*Declining balance*	*Sinking fund*
Depreciation charge				
Year 1	4,000	6,667	6,900	3,411
Year 2	4,000	5,333	4,996	3,684
Accumulated depreciation reserve year 3	12,000	16,000	15,515	11,071
Book value at end of year 3	13,000	9,000	9,485	13,929

Example 2

A hydraulic excavator can be purchased now for £18,000. It is estimated that its economic life will be 5 years and that a similar replacement will then cost £22,000. Alternatively, a more expensive vehicle costing £30,000 can be bought and it will last 8 years. Considering an 8 year period from now, calculate salvage values using the declining balance method with a 20 per cent rate of depreciation. Use present worth (PW) calculations and a 9 per cent interest rate to decide which scheme is best. Assume operating and maintenance costs for both vehicles are similar.

Solution

	Cost
Scheme 1	
Initial cost = £18,000	£18,000
Cost of replacement = 22,000 × $^{9\%}\text{SPPWF}^5$ = £14,299	£14,299
	£32,299
Salvage value after 5 years:	
18,000 $(1-0.2)^5$ = £5,898	
P = 5,898 × $^{9\%}\text{SPPWF}^5$ = £3,833	£ 3,833
Salvage value of second machine:	
22,000 $(1-0.20)^3$ = £11,264	
P = £11,264 × $^{9\%}\text{SPPWF}^8$ = £5,653	£ 5,653
	£ 9,486
NPW = £32,299 − £9,486 = £22,813	
Scheme 2	
Purchase of £30,000 now	
Initial cost £30,000	£30,000
Salvage value:	
30,000 $(1-0.2)^8$ £5,033	
P = 5,033 × $^{9\%}\text{SPPWF}^8$ = £2,526	−£ 2,526
	£27,474

Therefore it is better to purchase the £18,000 machine now, as in Scheme 1.

Example 3

A property speculator has recently purchased an office block. Basing your calculations on the data given below, determine the annual rent to be charged per m^2 of floor area, to justify a net return of 12 per cent after deduction of taxes and insurances.

Floor area	2,500 m^2
Cost of office block	£1,000,000
Annual maintenance	£7,000
Annual taxes	2 per cent of initial cost
Annual insurance costs	£4 per £5,000 of initial cost
Estimated life	50 years
Salvage value	£850,000

Solution

Costs

Construction	£1,000,000
Maintenance £7,000 × $^{12\%}\text{USPWF}^{50}$ (8.3044)	£58,131
Taxes	£20,000
Insurance costs $\dfrac{£1,000,000}{5,000} \times 4$	£800
	£1,078,931

Income

Salvage value £850,000 × $^{12\%}\text{SPPWF}^{50}$ (0.00346)	£2,941
Rent 2,500R(8.3044)	£20,761R

To break even, £2941 + £20,761R − £1,078,931 = 0.

R = £52/m^2

Therefore rent = £52/m^2. (For explanation of abbreviations, see section 12.14.)

12.12 Construction engineering applications – worked examples

Case study 1

Two designs are under construction for a four-storey office building. Design A provides additional strength in columns and foundations to support a further two storeys to be added later. The initial cost is £900,000 and it will cost £450,000 to add two storeys later. Design B does not make provision for additional storeys and its initial cost will be £850,000. If a further two storeys are added later this additional work will cost £600,000, including the cost of strengthening the existing building. Building maintenance costs will be the same for each scheme. Assume both schemes have a 50-year life. Use an 8 per cent interest and PW calculations.

Calculate the year up to which the scheme for providing initially to support two additional storeys is still more economic. Ignore taxation and inflation.

Solution

	Scheme A	*Scheme B*
Initial cost	£900,000	£850,000
Costs for extensions	£450,000	£600,000

Express costs of extensions at year x discounted to present-day values.

	Scheme A	*Scheme B*
Initial cost	£900,000	£850,000
Extensions at year x	$\dfrac{£450,000}{1.08^x}$	$\dfrac{£600,000}{1.08^x}$

At break-even point

$$\overset{A}{£900,000 + \frac{450,000}{1.08^x}} = \overset{B}{£850,000 + \frac{600,000}{1.08^x}}$$

$$£50,000 = \frac{600,000}{1.08^x} - \frac{450,000}{1.08^x}$$

$$£50,000 = \frac{150,000}{1.08^x}, \quad 1.08^x = 3$$

Take logs to both sides

$$x \log_e 1.08 = \log_e 3$$

$$x = \frac{\log_e 3}{\log_e 1.08} = \frac{1.09861}{0.076961}$$

$$= 14.275, \text{ say, } 14 \text{ years}$$

Case study 2

A contractor is considering the purchase of a rough terrain fork-lift truck. Its cost and operational data are estimated to be as follows:

Initial cost		£65,000
Maintenance cost during first year of service		£2,500
Annual increase in cost of maintenance during the life of the truck		£700 per year
Expected useful life		5 years
Salvage value		£12,000
Site utilisation (hours) –	year 1	2,500 hr/year
	year 2	2,100 hr/year
	year 3	1,900 hr/year
	year 4	1,500 hr/year
	year 5	1,300 hr/year

Using the present worth method, decide whether a hire charge of £12/hr, not including charges for fuel, operators and overheads, will be profitable. Assume that costs and benefits can be calculated at the end of each year and an interest rate of 12 per cent.

Solution

Cost		*Present worth*
Initial cost		£65,000
Maintenance, end of year:		
(1) 2,500 × $^{12\%}SPPWF^{1}$ (0.89285)	= £2,232	
(2) 3,200 × $^{12\%}SPPWF^{2}$ (0.79719)	= £2,551	
(3) 3,900 × $^{12\%}SPPWF^{3}$ (0.71178)	= £2,776	
(4) 4,600 × $^{12\%}SPPWF^{4}$ (0.63551)	= £2,923	
(5) 5,300 × $^{12\%}SPPWF^{5}$ (0.56742)	= £3,007	
	£13,489	£13,489
	Total cost	£78,489

Receipts		
End of year:		
(1) 12 × 2,500 × 0.89285	= £26,786	
(2) 12 × 2,100 × 0.79719	= £20,090	
(3) 12 × 1,900 × 0.71178	= £16,229	
(4) 12 × 1,500 × 0.63551	= £11,440	
(5) 12 × 1,300 × 0.56742	= £ 8,852	
	£83,397	
	Total cost	£83,397

Salvage value	
12,000 × $^{12\%}SPPWF^{5}$ = £6,809 (0.56742)	£ 6,809
	£90,206

NPV = £90,206 − £78,489 = £11,717

Therefore, the hire charge will be profitable.

To find hire rate h for no profit:

Benefit
End of year:

(1) $2{,}500h \times 0.89285 = 2{,}232h$
(2) $2{,}100h \times 0.79719 = 1{,}674h$
(3) $1{,}900h \times 0.71178 = 1{,}352h$
(4) $1{,}500h \times 0.63551 = 953h$
(5) $1{,}300h \times 0.56742 = 738h$

£6,949h

For benefit to equal cost
$6{,}949h + 6{,}809 = 78{,}489$

$$h = \frac{71{,}680}{6{,}949} = £10.32/\text{hr}$$

Therefore the hire charge of £12/hr will be profitable.

Case study 3

An embankment is to be built to reduce flooding of agricultural land. Determine its optimum height using the following information:

Life of embankment	70 years
Loss from each flood which flows over embankment	£800,000
Interest rate	10 per cent

Height of top above datum (m)	Present worth on project including maintenance costs (£)	Estimated frequency of flooding: f (years)
8	850,000	5
10	1,000,000	12
12	1,300,000	30
14	1,700,000	75
16	2,300,000	100

The optimum height is when cost plus losses is least, all values being expressed in terms of PW.

The PW of losses can be calculated approximately as follows: a loss L will in practice occur at irregular intervals and it is not possible to know exactly when. Statistically, however, it will occur every f years. Assume then that a loss occurs annually which equals the average annual loss $A = L/f$. The value A can be considered equivalent to a series of equal payments, made at the end of equal periods, i.e. an annuity. Its PW,

$$P = A\frac{(1+i)^n - 1}{i(1+i)^n} \quad \text{or} \quad A\ {}^i\text{USPWF}^n$$

When n is large,

$$P = \frac{A}{i} = \frac{L}{fi} \qquad \ldots [1]$$

or

$$P = \frac{800{,}000}{f \times 0.1} = \frac{8{,}000{,}000}{f} \qquad \ldots [2]$$

The PW of losses from equation [2] are calculated and added to the PW of costs for different proposed embankment heights as follows:

Height of embankment above datum (m)	Losses (£)	Costs (£)	Losses and costs (£)
8	1,600,000	850,000	2,450,000
10	666,667	1,000,000	1,666,667
12	266,667	1,300,000	1,566,667
14	106,667	1,700,000	1,806,667
16	80,000	2,300,000	2,380,000

This is shown graphically in Fig. 12.6.

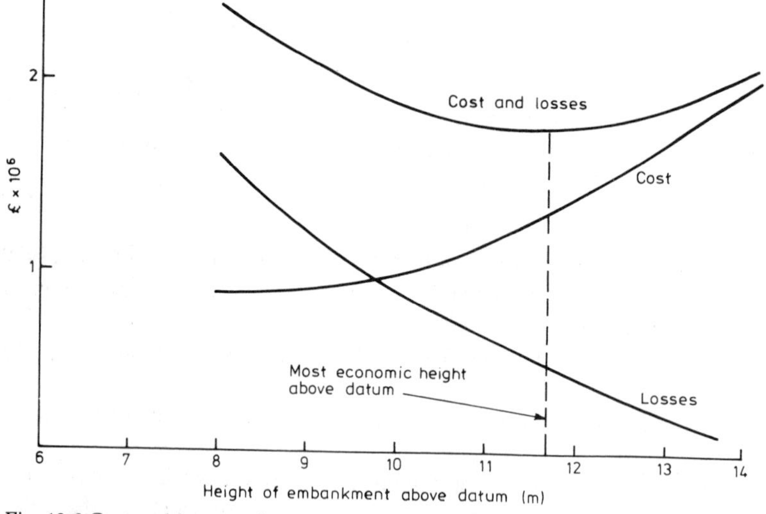

Fig. 12.6 Cost and loss graph

Case study 4

A tunnel is necessary as part of an aqueduct for a city water supply. It is estimated that a tunnel built to half the capacity of the aqueduct will be adequate for 20 years.

Because part of the cost of construction is due to initial expenses of purchasing plant, it is estimated that a full-capacity tunnel can be built now for £690,000 and a half-capacity tunnel can be built now for £560,000.

It has to be decided whether to build the full-capacity tunnel now, or a half-capacity tunnel now and a second half-capacity tunnel 20 years later.

If interest rates are not included, the comparison appears as follows:

Full capacity now	Half capacity now
£690,000 now	£560,000 now
	£560,000 20 years later

Since the £560,000 initial investment now is common to both alternatives, it has to be decided whether to spend £130,000 extra now on the full-capacity tunnel, or £560,000 in 20 years' time on a further half-capacity tunnel.

If interest is taken at 8 per cent, the PW of £560,000 in 20 years' time is

$$\frac{560{,}000}{1.08^{20}}$$

or

$$560{,}000 \times {}^{8\%}\text{SPPWF}^{20} = £120{,}142$$
$$(0.21454)$$

Total PW of half capacity including renewal after 20 years is £560,000 + £120,142 = £680,142.

This indicates that the half-capacity plan would be better. However, maintenance and running costs must be included in order to make an accurate evaluation.

It is estimated that lining repair costs of the two tunnels every 10 years are as follows:

Full-capacity tunnel	£23,000
Half-capacity tunnel	£20,000

It has also been estimated that, due to friction, pumping costs will be greater by £2,000 each year if one half-capacity tunnel is used, and they will be £4,200/year more when two half-capacity tunnels are in use.

Assuming that the aqueduct scheme will have a life of 80 years, the PW will now be as follows:

Full capacity now

Investment	£690,000	now	£690,000

Lining repairs:

$$23{,}000\left[\frac{1}{1.08^{10}}+\frac{1}{1.08^{20}}+\frac{1}{1.08^{30}}+\frac{1}{1.08^{40}}+\frac{1}{1.08^{50}}+\frac{1}{1.08^{60}}+\frac{1}{1.08^{70}}\right]$$

or, using compound interest factors,

23,000 × [summation of present worth factors = 0.858] = 19,752
+ 690,000

Total PW £709,752

Half capacity now
First tunnel
Investment £560,000 now £560,000
Lining repairs 20,000 × 0.8588 £17,176
Extra pumping costs:

$$P = \frac{2{,}000 \times (1.08^{80} - 1)}{0.08 \times 1.08^{80}}$$

or

$P = 2{,}000\ {}^{8\%}\mathrm{USPWF}^{80}$ (from section 12.16) £24,947
12.4735

Second tunnel
Investment:

$\dfrac{560{,}000}{1.08^{20}}$ or $560{,}000 \times {}^{8\%}\mathrm{SPPWF}^{20}$ £120,142
0.2145

Lining and repairs:

$$20{,}000\left[\frac{1}{1.08^{30}} + \frac{1}{1.08^{40}} + \frac{1}{1.08^{50}} + \frac{1}{1.08^{60}} + \frac{1}{1.08^{70}}\right]$$

0.0994 0.0460 0.0213 0.0099 0.0046

or, from section 12.16, 0.1812 × 20,000 £3,624
Extra pumping costs:

$$P_{20} = 4{,}200 \times \frac{1.08^{60} - 1}{0.08 \times 1.08^{60}} \quad \text{or} \quad 4{,}200 \times {}^{8\%}\mathrm{USPWF}^{60} = £51{,}981$$

12.3765

The present worth of these pumping costs at the time of commencement of construction of the first tunnel is

£51,981 × $^{8\%}\mathrm{SPPWF}^{20}$ £11,150
0.2145

Total PW £737,017
Full-capacity costs £709,754

Half-capacity costs £737,039 =	first tunnel	£560,000
	lining and repairs	£17,176
	extra pumping costs	£24,947
	second tunnel	£120,120
	lining and repairs	£3,624
	extra pumping costs	£11,150
		£737,017

Therefore a full-capacity tunnel is preferable.

Case study 5

Proposed road improvement scheme. An existing 50 km stretch of single-lane road will carry 175 vehicles per day at the start of proposed construction operations for its improvement. The number of vehicles are expected to increase by 12 per cent per year compounded. Two alternative schemes are under consideration.

Scheme 1 (road improvement). Improve the existing road including a two-lane rolled asphalt pavement, minor realignments and drainage improvements. Construction costs are estimated to be £8 million.

Scheme 2 (reconstruction). Reconstruct the road including a two-lane rolled asphalt pavement, drainage improvements and major realignments, shortening the length of road by 6 km. Construction costs are estimated to be £12 million.

The cost of maintenance of the existing road per km per year is £70 + £6.5N where N is the traffic flow of vehicles per day. Construction duration for both schemes will be 2 years and the reduction of operating costs will start at the end of construction work. Average operating costs per vehicle are estimated to be £0.13/km per day on the existing road and £0.09/km per day on the rolled asphalt road. Maintenance costs of the rolled asphalt road will be £550/km per year. Assume that maintenance costs during the 2-year construction period will be similar to that of the existing road.

Using PW calculations and an interest rate of 10 per cent determine whether construction work should proceed, and if so, which scheme will be the most economical. Assume that the flow of benefits and costs will extend over a period of 40 years from the start of proposed construction operations.

Solution

For the existing road and both schemes determine the present value of:

1. Operating cost of vehicles.
2. Road maintenance costs.
3. Road construction costs.

1. *Operating costs of vehicles*
   This is determined using equation [9] in Appendix 1 (section 12.13).

$$P = C\left[\frac{\left(\frac{1+t}{1+i}\right)^{n+1} - \frac{1+t}{1+i}}{\frac{1+t}{1-i} - 1}\right]$$

Accumulated present worth factor (APWF): i = 10 per cent = 0.1; t = 12 per cent = 0.12.

2. *Road maintenance costs*
   Fixed cost: sum of present worth of each year's maintenance is

$$a\left[\frac{(1+i)^n - 1}{i(1+i)^n}\right] \quad \text{or} \quad a \times {}^{i}\text{USPWF}^{n}$$

Variable cost (cost and traffic flow): accumulated PW of maintenance costs will be

$$P = \text{cost per vehicle per km per year}\left[\frac{\left(\frac{1+t}{1+i}\right)^{n+1} - \frac{1+t}{1+i}}{\frac{1+t}{1+i} - 1}\right]$$

3. *Road construction cost*
   Fixed sum given in present value figures. Consider the total costs of retaining the existing road for 40 years. Operating cost:

   $0.13 \times 175 \times 365 \times 50 \times 60^{(\text{APW})} = £24{,}911{,}250$

   Road maintenance:
   Fixed price; 40 years at £70/km per year
   $70 \times 50 \times {}^{10\%}\text{USPWF}^{40} = £34{,}227$
   (9.779)
   Variable cost; maintenance and traffic:

   $P = 6.5 \times 175 \times 50 \times 60 = £3{,}412{,}500$
   Total maintenance costs = £3,446,727

Scheme 1

Operating cost of vehicles:
Two years on existing road:

$P = 175 \times 365 \times 50 \times 0.13 \times {}^{12\%}(\text{APW})^2 = £852{,}994$
2.05448

Thirty-eight years on rolled asphalt road (3 to 40); necessary to use 40-year cost less 2 years' cost:

$$P = 175 \times 365 \times 50 \times 0.09 \times \left(\frac{1.01818^{41} - 1.01818}{0.01818} - 2.0548\right)$$

$= £16,402,969$

Total operating costs = £17,255,963.

Road maintenance costs

Fixed price – 2 years on existing road:

$$P = 70 \times 50 \times \underset{(1.73554)}{{}^{10\%}\text{USPWF}^2} = £6,074$$

Variable cost on existing road

$$P = 6.5 \times 175 \times 50 \times \frac{1.01818^3 - 1.01818}{0.01818} = £116,871$$

Maintenance cost = £122,945.

Thirty-eight years on rolled asphalt road:

$$P = 550 \times 50 \times \left[\underset{(9.779)}{{}^{10\%}\text{USPWF}^{40}} - \underset{(1.736)}{{}^{10\%}\text{USPWF}^{2}}\right] = £221,183$$

Therefore total road maintenance = £344,128.

Scheme 2

Operating costs of vehicles

Two years on existing road – 50 km. Thirty-eight years on rolled asphalt – 44 km, same as for 2 years on existing road (scheme 1):

$P = £852,156$

Thirty-eight years on rolled asphalt (as for scheme 1 except that on 44/50 km of section of scheme 1). Therefore from scheme 1

$$£16,402,969 \times \frac{44}{50} = £14,434,612$$

Total vehicle operating costs = £15,287,606.

Road maintenance costs

Two years on existing road (as for scheme 1) = £122,945. Thirty-eight years on rolled asphalt (as for scheme 1 except that 44/50 km section of scheme 1):

$$P = 221{,}183 \times \frac{44}{50} = £194{,}641$$

Total road maintenance cost = £315,468.

Summary

Existing road	Operating cost	£24,911,250
	Road maintenance	£3,446,727
	Construction cost	—
	Total	£28,357,977
Scheme 1	Operating cost	£17,255,963
	Road maintenance	£344,128
	Construction cost	£8,000,000
	Total	£25,600,091
Scheme 2	Operating cost	£15,287,606
	Road maintenance	£317,568
	Construction cost	£12,000,000
	Total	£27,605,174

Therefore construction should proceed on scheme 1.

Case study 6

Airfield facilities in a developing country are under review. It is felt that present installations will not be capable of handling the increasing tourist traffic in the future. The level of aircraft movements is expected to grow at a rate of 8 per cent per year from the present level of 100 movements/day. Annual maintenance costs for the existing facilities are £200,000 + £100N, where N is the number of movements per day. Saturation level for the existing facilities is 252 movements per day.

Two schemes have been proposed. Scheme A: recondition the present facilities when saturation is reached, at a cost of £2 million (maintenance costs will be at the same rate, before and after reconditioning). Scheme B: shut down the existing airfield now and sell the land for industrial development for £3 million. Construct a new airport, with 600 movements/day capacity, at a cost of £5 million. Maintenance costs for the new facilities will run at £75,000/year.

Use PW calculations to determine which scheme should be implemented if a 20-year life is expected from both schemes, and the interest rate prevailing is 10 per cent.

Scheme A

Recondition when saturation is reached. Cost £2 million. Saturation is reached in x years.

252 $= {}^{x}\text{USCAF}^{8\%} = 100$
Therefore
USCAF = 2.52, i.e. $x = 12$ years
$\therefore$ PW of recondition costs $= {}^{12}\text{SPPWF}^{10\%} \times 2 \times 10^6$
(0.31863)
= £637,260

Maintenance costs (same as before and after). Costs £200,000 + £100N. Fixed price cost per km:

$$\text{PW (fixed)} = a\,\frac{(1+i^n)-1}{i\,(1+i)^n} \quad i = 10 \text{ per cent} \quad a = 200{,}000. \qquad \text{Life} = 20 \text{ years} = n.$$

Therefore

$$\text{PW (fixed costs)} = 200{,}000 \times \frac{1.1^{20}-1}{0.1(1.1)^{20}} = £1{,}702{,}700$$

Use accumulated PW factor (variable price equation):

$r = 100, \quad v = 100, \quad t = 8$ per cent, $\quad n = 20, \quad i = 10$ per cent

$$\text{Present worth of variable costs (PMD)} = r \times v \left[\frac{\left(\frac{1+t}{1+i}\right)^{n+1} - \left(\frac{1+t}{1+i}\right)}{\left(\frac{1+t}{1+i}\right) - 1}\right]$$

$$= 100 \times 100 \times \left[\frac{\left(\frac{1.08}{1.1}\right)^{21} - \left(\frac{1.08}{1.1}\right)}{\left(\frac{1.08}{1.1}\right) - 1}\right]$$

$$= £165{,}876$$

Total PW = £2,504,592.

Scheme B
Shut down and sell land for £3 million. New airport cost £5 million. (NPW = £2 million). Maintenance costs:

$$£75{,}000 \times \frac{1.1^{20}-1}{0.1(1.1)^{20}} = £638{,}513$$

Total PW = £2,638,513. Therefore, scheme A is best.

12.13 Appendix 1 – mathematics of compounding and discounting

Symbols

P = the original principal, a present sum of money that is lent or borrowed;
i = the interest rate per unit of time;
n = number of units of time or the interest periods;
I = simple interest, the total value payable for the use of the money at simple interest;
S = the sum of money at the end of n units of time compounded at interest i, comprising the principal plus the interest payable;
R = the uniform series end-of-period payment or receipt which extends for n periods.

Simple interest is rarely used in engineering economy studies except where very short time periods are involved.

Interest, $I = P \cdot i \cdot n$. ... [1]

Compound interest

Compounding interest is the process of paying interest on both the principal and interest of previous time periods. When interest is paid yearly the interest is said to be compounded annually; when paid every 6 months it is said to be compounded semi-annually, etc.

Period	*Principal at start of period*	*Interest earned during period*	*Amount at end of period*
1	P	Pi	$P+Pi = P(1+i)$
2	$P(1+i)$	$P(1+i)i$	$P+Pi+P(1+i)i = P(1-i)^2$
3	$P(1+i)^2$	$P(1+i)^2i$	$P(1+i)^2+P(1+i)^2i = P(1+i)^3$
4	$P(1+i)^3$	$P(1+i)^3i$	$P(1+i)^3+P(1+i)^3i = P(1+i)^4$
n	$P(1+i)^{n-1}$	$P(1+i)^{n-1}i$	$P(1+i)^{n-1}+P(1+i)^{n-1}i = P(1+i)^n$

General formula

$S = P(1+i)^n$ ... [2]

The factor $(1+i)^n$ is called the 'compound amount factor' (CAF). N.B. Here, n refers to the number of periods for which the interest i applies. 6 per cent compounded semi-annually for 4 years is equivalent to $S = P(1+0.03)^8$.

Example
The sum of £560 is invested today at 7 per cent interest; what amount will have accumulated after 8 years? £560 $(1+0.07)^8$ = £962.

Present worth

In order to determine the value, at the present time, of a sum of money that will be received or spent at some future date it is possible to rearrange formula [2].

$$P = \frac{S}{(1+i)^n} \qquad \ldots [3]$$

Here, $1/(1+i)^n$ is referred to as the 'present worth factor' (PWF).

Example
A man has £49,693 in the bank now due to a deposit made 10 years ago. What was the deposit? Interest rate 12 per cent.

$£49,693\,(1+i)^{-n} = £16,000$

Annuities

A series of equal payments, which are made at the end of equal periods, is known as an 'annuity'. Examples are:

1. Debt payment by means of a series of uniform periodic payments (e.g. hire-purchase, etc.).
2. Accumulation of a lump sum from a series of uniform periodic payments (e.g. a sinking fund for an endowment mortgage).
3. Receipt of uniform periodic payments in lieu of a single payment (e.g. retirement annuity).
4. Allowance for the future decrease in value of property.
5. Determination of a series of uniform periodic payments that would be equivalent to several unequal payments made at unequal intervals of time (e.g. investment appraisal).

Considering a payment R made at the end of n period at interest i to give an accumulated final sum of S (Fig. 12.7).

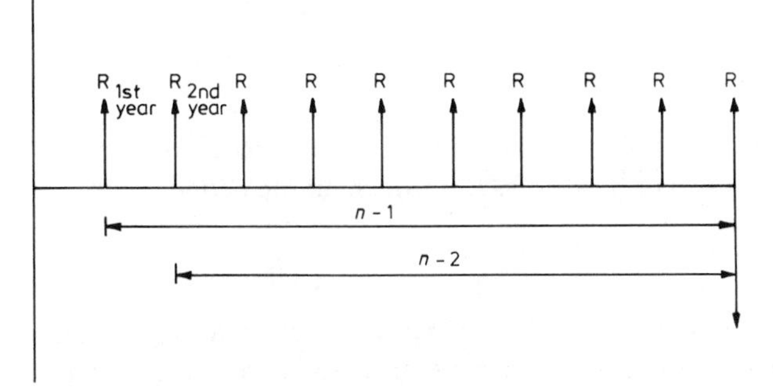

Fig. 12.7 Net present value

Payment made each year; the first payment earns interest for $(n-1)$ years. The second payment earns interest for $(n-2)$ years, and so on.

Value of first year's payment after $n-1$ years $= R(1+i)^{n-1}$
Value of second year's payment after $n-2$ years $= R(1+i)^{n-2}$
Final payment (earns no interest) $= R$

Therefore

$$S = R(1+i)^{n-1} + R(1+i)^{n-2} \ldots + R$$

$$= R[1+(1+i) + (1+i)^2 \ldots (1+i)^{n-1}] \qquad \ldots \text{(A)}$$

Multiply both sides of the equation by $(1+i)$

$$S(1+i) = R[(1+i) + (1+i)^2 \ldots (1+i)^n] \qquad \ldots \text{(B)}$$

Subtract (A) from (B):

$$iS = R[(1+i)^n - 1]$$

$$S = R\frac{(1+i)^n - 1}{i} \qquad \ldots [4]$$

The factor:

$$\frac{(1+i)^n - 1}{i}$$

is known as the 'uniform series compound amount factor' (USCAF).

Example
Given an interest rate of 11 per cent per year, what sum would have accumulated after 6 years if £1,000 were invested at the end of each year for 6 years?

$$£1{,}000\left(\frac{(1+0.11)^6 - 1}{0.11}\right) = 1{,}000 \times 7.912 = £7{,}912$$

Similarly,

$$R = S\left[\frac{i}{(1+i)^n - 1}\right] \qquad \ldots [5]$$

is known as the 'sinking fund deposit factor' where a fund, S, is provided at the end of a given period by means of a series of payments, R, throughout the period.

Example
A uniform annual investment is to be made in order to provide sufficient capital at the end of 7 years for replacement of a hydraulic excavator. If the interest rate is 16 per cent, determine the annual investment needed to provide for £35,000.

$$£35{,}000 \frac{0.16}{(1+0.16)^7 - 1} = 35{,}000 \times 0.08761$$

$$= £3{,}066.4$$

In considering the principal or capital, P, the following equations can be derived.

$$R = S\left[\frac{i}{(1+i)^n - 1}\right] \quad \text{and} \quad S = P(1+i)^n$$

$$R = P(1+i)^n\left[\frac{i}{(1+i)^n - 1}\right]$$

$$R = P\frac{i(1+i)^n}{(1+i)^n - 1} \quad \text{(USCRF)} \qquad \ldots [6]$$

Example: equivalent annual cost
With an interest rate of 14 per cent what uniform end-of-period payments must be made for 25 years to repay an initial debt of £20,000. (Could be mortgage repayments.)

$$£20{,}000 \frac{0.14(1+0.14)^{25}}{(1+0.14)^{25} - 1} = £20{,}000 \times 0.14549 = £2{,}910/\text{year}$$

or

$$R = P\left[\frac{i}{(1+i)^n - 1} + i\right]$$

$$= P\ [\text{sinking fund factor} + \text{interest}]$$

Long period returns (perpetuity)

The formula for capital recovery can be written in the form

$$R = P\left[\frac{i}{(1+i)^n - 1} + i\right]$$

As the number of periods (n) increases the denominator becomes large in relation to i, and the expression tends towards P_2. Over a long period, $R = P_2$, thus at 12 per cent $R = P \cdot 0.12$ and at 15 per cent $R = P \cdot 0.15$.

$$P = R\frac{(1+i)^n - 1}{i(1+i)^n} \quad \text{(USPWF)} \qquad \ldots [7]$$

Example
Your rich uncle has offered you £10,000 today or £2,500 per year for 5 years. If the interest rate is 12 per cent which would you choose?

$$2500 \times \left[\frac{(1+0.12)^5-1}{0.12(1+0.12)^5}\right] = £9011 \text{ or using section 12.16}$$

$$2500 \times \begin{pmatrix}\text{USPWF}\\ 3.6047\end{pmatrix} = 9011$$

Accept £10,000 now!

Uniform gradient formula

Some costs, particularly maintenance costs, tend to rise each year as the asset ages. For example, from past records the estimated maintenance cost for a machine might be £2,000 for the first year of service, rising each year thereafter by £500. If the machine has a 4-year life the cash flow can be shown diagrammatically as in Fig. 12.8.

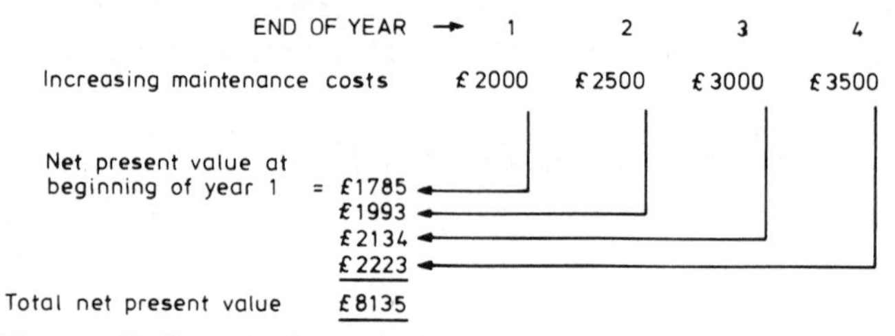

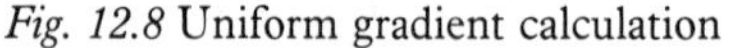

Fig. 12.8 Uniform gradient calculation

For appraisal purposes it may be necessary to determine the equivalent annual cost or NPV of these costs.

Net present value at 12 per cent interest.

$$\text{NPV} = (2{,}000 \times {}^{12\%}\underset{(0.89285)}{\text{SPPWF}^1}) + (2{,}500 \times {}^{12\%}\underset{(0.79719)}{\text{SPPWF}^2}) + (3{,}000 \times {}^{12\%}\underset{(0.71178)}{\text{SPPWF}^3})$$

$$+ (3{,}500 \times {}^{12\%}\underset{(0.63551)}{\text{SPPWF}^4})$$

$$1{,}785 \quad + \quad 1{,}993 \quad + \quad 2{,}134 \quad + \quad 2{,}223$$

$$\text{NPV} = £8{,}137.6$$

Equivalent annual cost (EAC) = $£8137.6 \times {}^{12\%}\text{USCRF}^4 = £2{,}678$.

This is a uniform gradient problem. It is easily solved over such a low number of years. When problems extend over many years this process becomes very cumbersome. It is advisable to use a simple formula to calculate the equivalent annual cost of a uniform gradient.

Let g = annual gradient. Taking the compound amount at the end of the nth period of each single payment we obtain:

$$S = (n-1)g + (n-2)g(1+i) + (n-3)g(1+i)^2 + \ \ldots\ 2g(1+i)^{n-3} + g(1+i)^{n-2}$$

Factoring out g,

$$S = g[(n-1) + (n-2)(1+i) + (n-3)(1+i)^2 + \ \ldots\ 2(1+i)^{n-3} + (1+i)^{n-2}] \qquad \ldots \text{(A)}$$

Multiplying both sides of the equation by $(1+i)$,

$$(1+i)S = g[(n-1)(1+i) + (n-2)(1+i)^2 + (n-3)(1+i)^3 + \ \ldots\ + 2(1+i)^{n-2} + (1+i)^{n-1}] \qquad \ldots \text{(B)}$$

Subtracting equation (A) from equation (B),

$$\begin{aligned}(1+i)S - S = {} & g[(n-1)(1+i) + (n-2)(1+i)^2 \\ & + (n-3)(1+i)^3 + \ \ldots\ + 2(1+i)^{n-2} + (1+i)^{n-1}] \\ & - g[(n-1(+ (n-2)(1+i) + (n-3)(1+i)^2 \\ & + \ \ldots\ + 2(1+i)^{n-3} + (1+i)^{n-2}] \\ iS = {} & g[(1+i) + (1+i)^2 + (1+i)^3 + \ \ldots\ + (1+i)^{n-2} + (1+i)^{n-1} - (n-1)]\end{aligned}$$

Rearranging the last term,

$$iS = g[1 + (1+i) + (1+i)^2 + (1+i)^3 + \ \ldots\ + (1+i)^{n-2} + (1+i)^{n-1}] - ng$$

The term in brackets is USCAF, which we have previously shown is equal to

$$\left[\frac{(1+i)^n - 1}{i}\right]$$

$$iS = g\left[\frac{(1+i)^n - 1}{i}\right] - ng$$

$$S = \frac{g}{i}\left[\frac{(1+i)^n - 1}{i}\right] - \frac{ng}{i}$$

To find the uniform series equivalent, we multiply S by the sinking fund payment factor.

$$R = \left[\frac{i}{(1+i)^n - 1}\right]\left[\frac{g}{\mathrm{i}}\left[\frac{(1+i)^n - 1}{\mathrm{i}}\right] - \frac{ng}{\mathrm{i}}\right]$$

$$R = \frac{g}{i} - \frac{ng}{i}\left[\frac{i}{(1+i)^n - 1}\right] = g\left[\frac{1}{i} - \frac{n}{(1+i)^n - 1}\right] \qquad \ldots [8]$$

The term

$$\left[\frac{1}{i} - \frac{n}{(1+i)^n - 1}\right]$$

the uniform gradient conversion factor (represented by the letters UGF), is used to convert the gradient series to a uniform series.

Using this formula the previous example can easily be solved, i.e.

$$£500\left[\frac{1}{0.12} - \frac{4}{(1+0.12)^4 - 1}\right] + £2{,}500 = £2{,}679$$

Accumulated PW of compounded operating or maintenance costs.

Let C = cost to operate present number of vehicles for 1 year on 1 km of road.
Let t = annual percentage traffic increase.

At end of:	*Cost*	*Present worth*
Year 1	$C(1+t)$	$C\frac{(1+t)}{(1+i)}$
Year 2	$C(1+t)^2$	$C\left[\frac{(1+t)}{(1+i)}\right]^2$
Year n	$C(1+t)^n$	$C\left[\frac{(1+t)}{(1+i)}\right]^n$

Then total PW of vehicle-operating costs are

$$P_0 = C\left[\frac{(1+t)}{(1+i)} + \left(\frac{1+t}{1+i}\right)^2 + \ldots + \left(\frac{1+t}{1+i}\right)^n\right]$$

Let

$$\frac{1+t}{1+i} = r$$

$$P_0 = C[r + r^2 + r^3 + \ldots r^n]$$

This is a geometric progression with first term = r.

Common ratio = r. Sum to n terms of GP = $\frac{a(1-r^n)}{1-r}$

$$P_0 = C\frac{r(1-r^n)}{1-r}$$

$$P_0(1-r) = C\,r\,(1-r^n)$$
$$P_0(r-1) = C\,r\,(r^n-1)$$

$$P_0 = C\frac{(r^{n+1}-r)}{r-1}$$

$$P_0 = C\left[\frac{\left(\frac{1+t}{1+i}\right)^{n+1} - \frac{1+t}{1+i}}{\frac{1+t}{1+i} - 1}\right] \qquad \ldots [9]$$

12.14 Summary

Factor name	*Abbreviation*	*Formula*
1. Simple interest	SI	*Pin*
2. Single-payment compound amount factor	SPCAF	$(1+i)^n$
3. Single-payment present worth factor	SPPWF	$\frac{1}{(1+i)^n}$
4. Uniform series compound amount factor	USCAF	$\frac{(1+i)^n-1}{i}$
5. Uniform series sinking fund deposit factor	USSFDF	$\frac{i}{(1+i)^n-1}$
6. Uniform series capital recovery factor	USCRF	$\frac{i(1+i)^n}{(1+i)^n-1}$
7. Uniform series present worth factor	USPWF	$\frac{(1+i)^n-1}{i(i+i)^n}$
8. Uniform gradient factor	UGF	$\left[\frac{1}{i} - \frac{n}{(1+i)^n-1}\right]$
9. Accumulated present worth of operating or maintenance costs factor	APWF	$\left[\frac{\left(\frac{1+t}{1+i}\right)^{n+1} - \frac{1+t}{1+i}}{\frac{1+t}{1+i} - 1}\right]$

12.15 Problems

1. The SPPWF for i = 16 per cent, n = 8 is 0.30502. For the same i and n, calculate: (a) SPCAF; (b) USCAF; (c) USPWF; (d) USCRF; (e) USSFDF. Check solution by use of section 12.16.

   [*Ans*: (1) 3.2784; (b) 14.240; (c) 4.3435; (d) 0.23022; (e) 0.07022]

2. Your rich uncle wishes to provide you with an income of £400 per year for 45 years. If you can invest his money at 12 per cent interest what sum must he give you to provide this income?

   [*Ans*: £33,132]

3. An excavator is purchased for the sum of £14,000 by a plant hire company. As a result of the first year's operation of the equipment a profit of £4,000 is made. In the second and third years of the excavator's operation, profits of £6,500 and £5,500 respectively are made. After 3 years' operation the excavator is scrapped and has no residual value. What is the return on the initial investment?

   [*Ans*: 6.65 per cent]

4. A joinery manufacturer is about to purchase a new planning machine. The company has used two types of machine for this process in the past. *Type 1* cost £14,000, has an estimated life of 5 years, and no salvage value. *Type 2* cost £20,500, has an estimated life of 7 years, and a £2,500 salvage value.

   Based upon past experiences with these two types of machines, the following operating and maintenance costs are anticipated:

Year of life	*Type 1*	*Type 2*
1	1,500	1,500
2	2,300	1,750
3	3,500	2,300
4		3,500
5		4,200

   Using an interest rate of 11 per cent, compare the desirability of the two types of machines using:
   (a) Equivalent annual costs method.
   (b) Present worth method.
   [*Ans*: EAC method: machine 1 £5,351, machine 2 £6,058. NPW method: machine 1 £19,777, machine 2 £28,268. Select machine 1.]

5. A length of motorway pavement costs £10,000/year to maintain. What expenditure for a new pavement is justified if no maintenance will be required for the first 3 years, £2,000 per year for the next 8 years and £7,500 a year thereafter. Interest rate is 15 per cent.

   [*Ans*: expenditure justified = £51,147]

6. There are two feasible methods of producing window-framing material. One will require the purchase of machine A for £25,000 and the expenditure of £8,000 per year in operating costs. The machine has an estimated life of 24 years with a salvage value of £19,000. Machine B has an estimated life of 12 years and £8,000 salvage value. Operating costs for machine B are estimated to be £6,400 per year.

   Using the present worth method determine which machine should be recommended as the most economic. Assume that the services of the machine are required for 24 years and cost of capital is 15 per cent.

   [*Ans*: machine A: NPW = £76,121; machine B: NPW = £61,953]

7. A joinery manufacturer requires additional storage space for his products. Three building types are under construction. Estimated data for each type are given below.

	Structural frame and cladding	*Initial cost (£)*	*Life of building (years)*	*Maintenance costs*	*Salvage value (£)*
Type 1	Timber-framed and timber cladding	430,000	35	£2,000/year from year 1 to 10; £4,000 from year 11 to 35	20,000
Type 2	Steel-framed and proprietary pre-formed cladding	490,000	40	£1,500/year from year 1 to 40	80,000
Type 3	Pre-cast concrete with brick infill	500,000	70	£3,000/year from year 8 to year 10	10,000

   Which is the most economical proposition if the prevailing interest rate is 12 per cent?

   [*Ans*: EAC: type 1 = £55,168; type 2 = £60,834; type 3 = £61,378]

8. The cladding panels to an office block are expected to have maintenance and repair expenses of £2,000 in the first year. It is anticipated that these expenses will increase £500/year each year thereafter during the cladding's expected life of 40 years. What is the present worth of these expenses for the 40-year life of the cladding using interest at 10 per cent?

   [*Ans*: £6,458]

9. Your company is considering investing capital in one of four alternative projects and requires a rate of return of at least 11 per cent. The estimated data of each project are given as follows:

	Initial investment	*Annual income*
Project 1	£15,000	£4,000
Project 2	£35,000	£8,000
Project 3	£55,000	£15,000
Project 4	£90,000	£20,000

Life of each scheme = 5 years.

Using incremental analysis, determine the optimum investment for your company's capital.

[*Ans*: project 3]

10. Two contractors submit programmes with their tenders for a site clearance scheme (Fig. 12.9). The contractors' prices differ by negligible amounts, being itemised as follows:

1.	Remove scrub	£1,500
2.	Strip topsoil	£3,900
3.	Bulk excavation	£150,000
4.	Lay drains	£9,000
5.	Replace topsoil	£3,600
6.	Erect fences	£1,050

Because of the simple nature of the work, payment is made at the end of the month in which it is carried out and it may be assumed that the work done in each month bears a linear relationship with the total for that item. The short-term interest rate for the client's money is 12 per cent.

Which tender should he accept and what will be the saving over the other?

[*Ans*: accept contractor B's tender, £2,206 cheaper in real terms]

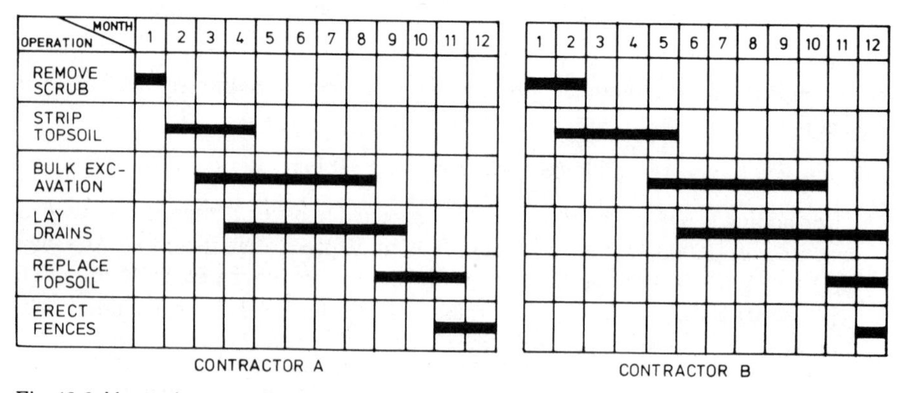

Fig. 12.9 Alternative programme for site clearance scheme

11. A local authority is considering two possible routes for an electricity mains cable. *Route1* involves laying the cable around the perimeter of a lake, 20 km in length.

Initial cost is estimated to be £5,000/km, annual maintenance cost of £500/km and a salvage value at the end of 20 years of £1,500/km.

Route 2 involves cutting across the lake by laying the cable on the lake bed 8 km long. Initial cost will be £2,000/km, annual maintenance cost of £100/km and salvage value at the end of 20 years will be £1,500/km.

The annual power loss will be £400/km for both routes. Compare the costs of the two routes if the interest rate is 12 per cent.

[*Ans*: route 1: NPV = £231,340; route 2: NPV = £188,634]

12. The owner of two quarries agrees to sell his stone on the following conditions. The buyer is to remove the stone from certain parts of both quarries according to a fixed schedule of volume, price and time. The schedule for each quarry is as follows:

Quarry A	*Quantity per year* (m^3)	*Time period*	*Price per* m^3
Phase 1	15,000	Year 1 to 10	35p
Phase 2	40,000	Year 10 to 15	70p

Quarry B	*Quantity per year* (m^3)	*Time period*	*Price per* m^3
Phase 1	25,000	Year 1 to 18	50p
Phase 2	65,000	Year 1 to 5	80p
	45,000	Year 5 to 12	80p

On the basis of equal year-end payments during each period by the buyer, determine the PW of the quarry to the owner on the basis of 12 per cent interest.

[*Ans*: total PW of quarry £433,448]

13. A local authority intends to build an office block and shopping centre on the same site. You have been asked to evaluate two alternative schemes for construction.
Scheme 1. Construction of the office block now and the construction of the shopping centre at the end of 5 years. The cost of the office block and shopping centres are £1,700,000 and £750,000 respectively. Estimated maintenance costs are £12,000/year for the first 5 years and £20,000 thereafter.
Scheme 2. This involves the construction of one large structure to facilitate both purposes. Estimated cost is £3,300,000. Maintenance costs are estimated at £35,000/year throughout the period. If the building will be usable for 40 years and interest is charged at 14 per cent, compute the PW of each plan.

[*Ans*: scheme 1: £3,564,633; scheme 2: £3,548,500]

14. A new fleet of transport vehicles is to replace an existing service. The vehicles are to be purchased in two stages. The first stage will cost £200,000 now and will have a salvage value of £60,000 after 5 years; the second stage will cost £300,000 5 years from now and will have a salvage value of £80,000 10 years from now. The initial demand for these vehicles is expected to be 1 million tonne km/year and will increase at an estimated rate of 4.5 per cent per year, compounded. Operating costs will be £0.08/tonne km. Using DCF calculations with a 15 per cent per year discount rate, calculate the charge which should be made per tonne km for the revenue to equal the cost over a 10-year period from now. Assume that the discount rate allows for inflation, and that the operating costs include taxes.

    [*Ans*: charge per tonne km = £0.13]

15. In attempting to arrive at some quantitative basis for recommending a type of construction, a contractor has estimated fixed and variable costs (for a building containing between 200 and 600 m^2) as follows:

	Concrete and brick	*Steel and brick*	*Timber and brick*
Initial cost per m^2	£280	£290	£350
Annual maintenance	£6,600	£6,000	£5,000
Annual heating and light, etc.	£4,400	£3,600	£3,000
Estimated life (years)	30	30	30
Estimated salvage value as a percentage of initial cost	Nil	5	5
Interest rate 12 per cent			

    Which form of construction should be chosen for a building of: (a) 200 m^2; (b) 400 m^2; (c) 600 m^2.

    [*Ans*: (a) timber and brick; (b) steel and brick; (c) concrete and brick]

16. It is proposed that a road should be reconstructed at a total cost of £30 million shortening its length from 200 to 180 km and reducing vehicle operating costs from £40 to £36/km per year per vehicle. The cost of maintenance is independent of traffic flow and is £1,000/km per year for the existing road and £9,000/km per year for the improved road. It is estimated that there will be 400 vehicles/day on the road at the beginning of the 2-year construction period, and that this will increase at the rate of 14.5 per cent per year (compounded). The cost of maintenance and vehicle operating costs during the construction period are the same as those for the existing road.

    Assume that the flow of costs and benefits will extend over a period of 20 years from the start of construction. Using DCF calculations and a 9 per cent net rate of interest, find whether the proposed reconstruction will be economical.

    [*Ans*: reconstruction will be more economical]

12.16 Appendix 2 – Compounding and discounting tables

4% COMPOUND INTEREST FACTORS

Period	*Single payment compound amount factor*	*Single payment present worth factor*	*Uniform series compound amount factor*	*Uniform series sinking fund deposit factor*	*Uniform series present worth factor*	*Uniform series capital recovery factor*
	SPCAF	*SPPWF*	*USCAF*	*USSFDF*	*USPWF*	*USCRF*
	Future value of £1	*Present value of £1*	*Future value of uniform series of £1*	*Uniform series whose future value is £1*	*Present value of uniform series of £1*	*Uniform series with present value of £1*
n	$(1+i)^n$	$\frac{1}{(1+i)^n}$	$\frac{(1+i)^n-1}{i}$	$\frac{i}{(1+i)^n-1}$	$\frac{(1+i)^n-1}{i(1+i)^n}$	$\frac{i(1+i)^n}{(1+i)^n-1}$
1	1.0400	0.96153	1.000	1.00000	0.9615	1.04000
2	1.0815	0.92455	2.039	0.49019	1.8860	0.53019
3	1.1248	0.88899	3.121	0.32034	2.7750	0.36034
4	1.1698	0.85480	4.246	0.23549	3.6298	0.27549
5	1.2166	0.82192	5.411	0.18462	4.4518	0.22462
6	1.2653	0.79031	6.632	0.15076	5.2421	0.19076
7	1.3159	0.75991	7.898	0.12660	6.0020	0.16660
8	1.3685	0.73069	9.214	0.10852	6.7372	0.14852
9	1.4233	0.70258	10.582	0.09449	7.4353	0.13449
10	1.4802	0.67556	12.006	0.08329	8.1108	0.12329
11	1.5394	0.64958	13.486	0.07414	8.7604	0.11414
12	1.6010	0.62459	15.025	0.06655	9.3850	0.10655
13	1.6650	0.60057	16.626	0.06014	9.9856	0.10014
14	1.7316	0.57747	18.291	0.05466	10.5631	0.09466
15	1.8009	0.55526	20.023	0.04994	11.1183	0.08994
16	1.8729	0.53390	21.824	0.04582	11.6522	0.08582
17	1.9479	0.51337	23.697	0.04219	12.1656	0.08219
18	2.0258	0.49362	25.645	0.03899	12.6592	0.07899
19	2.1068	0.47464	27.671	0.03613	13.1339	0.07613
20	2.1911	0.45638	29.778	0.03358	13.5903	0.07358
21	2.2787	0.43883	31.969	0.03128	14.0291	0.07128
22	2.3699	0.42195	34.247	0.02919	14.4511	0.06919
23	2.4647	0.40572	36.617	0.02730	14.8568	0.06730
24	2.5633	0.39012	39.082	0.02558	15.2469	0.06558
25	2.6658	0.37511	41.645	0.02401	15.6220	0.06401
26	2.7724	0.36068	44.311	0.02256	15.9827	0.06256
27	2.8833	0.34681	47.084	0.02123	16.3295	0.06123
28	2.9987	0.33347	49.967	0.02001	16.6630	0.06001
29	3.1186	0.32065	52.966	0.01887	16.9837	0.05887
30	3.2433	0.30831	56.084	0.01783	17.2920	0.05783
35	3.9460	0.25341	73.652	0.01357	18.6646	0.05357
40	4.8010	0.20828	95.025	0.01052	19.7927	0.05052
45	5.8411	0.17119	121.029	0.00826	20.7200	0.04826
50	7.1066	0.14071	152.667	0.00655	21.4821	0.04655
55	8.6463	0.11565	191.159	0.00523	22.1086	0.04532
60	10.5196	0.09506	237.990	0.00420	22.6234	0.04420
65	12.7987	0.07813	294.968	0.00339	23.0466	0.04339
70	15.5716	0.06421	364.290	0.00274	23.3945	0.04274
75	18.9452	0.05278	448.631	0.00222	23.6804	0.04222
80	23.0497	0.04338	551.244	0.00181	23.9153	0.04181
85	28.0436	0.03565	676.090	0.00147	24.1085	0.04147
90	34.1193	0.02930	827.983	0.00120	24.2627	0.04120
95	41.5113	0.02408	1012.784	0.00098	24.3977	0.04098
100	50.5049	0.01980	1237.623	0.00080	24.5049	0.04080

5% COMPOUND INTEREST FACTORS

Period	*Single payment compound amount factor*	*Single payment present worth factor*	*Uniform series compound amount factor*	*Uniform series sinking fund deposit factor*	*Uniform series present worth factor*	*Uniform series capital recovery factor*
	SPCAF	*SPPWF*	*USCAF*	*USSFDF*	*USPWF*	*USCRF*
	Future value of £1	*Present value of £1*	*Future value of uniform series of £1*	*Uniform series whose future value is £1*	*Present value of uniform series of £1*	*Uniform series with present value of £1*
n	$(1+i)^n$	$\frac{1}{(1+i)^n}$	$\frac{(1+i)^n-1}{i}$	$\frac{i}{(1+i)^n-1}$	$\frac{(1+i)^n-1}{i(1+i)^n}$	$\frac{i(1+i)^n}{(1+i)^n-1}$
1	1.0500	0.95238	1.000	1.00000	0.9523	1.05000
2	1.1024	0.90702	2.049	0.48780	1.8594	0.53780
3	1.1576	0.86383	3.152	0.31720	2.7232	0.36720
4	1.2155	0.82270	4.310	0.23201	3.5459	0.28201
5	1.2762	0.78352	5.525	0.18097	4.3294	0.23097
6	1.3400	0.74621	6.801	0.14701	5.0756	0.19701
7	1.4071	0.71068	8.142	0.12281	5.7863	0.17281
8	1.4774	0.67683	9.549	0.10472	6.4632	0.15472
9	1.5513	0.64460	11.026	0.09069	7.1078	0.14069
10	1.6288	0.61391	12.577	0.07950	7.7217	0.12950
11	1.7103	0.58467	14.206	0.07038	8.3064	0.12038
12	1.7958	0.55683	15.917	0.06282	8.8632	0.11282
13	1.8856	0.53032	17.712	0.05645	9.3935	0.10645
14	1.9799	0.50506	19.598	0.05102	9.8986	0.10102
15	2.0789	0.48101	21.578	0.04634	10.3796	0.09634
16	2.1828	0.45811	23.657	0.04226	10.8377	0.09226
17	2.2920	0.43629	25.840	0.03869	11.2740	0.08869
18	2.4066	0.41552	28.132	0.03554	11.6895	0.08554
19	2.5269	0.39573	30.539	0.03274	12.0853	0.08274
20	2.6532	0.37688	33.065	0.03024	12.4622	0.08024
21	2.7859	0.35894	35.719	0.02977	12.8211	0.07799
22	2.9252	0.34184	38.505	0.02597	13.1630	0.07597
23	3.0715	0.32557	41.430	0.02413	13.4885	0.07413
24	3.2250	0.31006	44.501	0.02247	13.7986	0.07247
25	3.3863	0.29530	47.727	0.02095	14.0939	0.07095
26	3.5556	0.28124	51.113	0.01956	14.3751	0.06956
27	3.7334	0.26784	54.669	0.01829	14.6430	0.06829
28	3.9201	0.25509	58.402	0.01712	14.8981	0.06712
29	4.1161	0.24294	62.322	0.01604	15.1410	0.06604
30	4.3219	0.23137	66.438	0.01505	15.3724	0.06505
35	5.5160	0.18129	90.320	0.01107	16.3741	0.06107
40	7.0399	0.14204	120.799	0.00827	17.1590	0.05827
45	8.9850	0.11129	159.700	0.00626	17.7740	0.05626
50	11.4673	0.08720	209.347	0.00477	18.2559	0.05477
55	14.6356	0.06832	272.712	0.00366	18.6334	0.05366
60	18.6791	0.05353	353.583	0.00282	18.9292	0.05282
65	23.8398	0.04194	456.797	0.00218	19.1610	0.05218
70	30.4264	0.03286	588.528	0.00169	19.3246	0.05169
75	38.8326	0.02575	756.653	0.00132	19.4849	0.05131
80	49.5614	0.02017	971.228	0.00102	19.5964	0.05102
85	63.2543	0.01580	1245.086	0.00080	19.6838	0.05080
90	80.7303	0.01238	1594.607	0.00062	19.7522	0.05062
95	103.0346	0.00970	2040.693	0.00049	19.8058	0.05049
100	131.5012	0.00760	2610.025	0.00038	19.8479	0.05038

6% COMPOUND INTEREST FACTORS

Period	*Single payment compound amount factor*	*Single payment present worth factor*	*Uniform series compound amount factor*	*Uniform series sinking fund deposit factor*	*Uniform series present worth factor*	*Uniform series capital recovery factor*
	SPCAF	*SPPWF*	*USCAF*	*USSFDF*	*USPWF*	*USCRF*
	Future value of £1	*Present value of £1*	*Future value of uniform series of £1*	*Uniform series whose future value is £1*	*Present value of uniform series of £1*	*Uniform series with present value of £1*
n	$(1+i)^n$	$\frac{1}{(1+i)^n}$	$\frac{(1+i)^n-1}{i}$	$\frac{i}{(1+i)^n-1}$	$\frac{(1+i)^n-1}{i(1+i)^n}$	$\frac{i(1+i)^n}{(1+i)^n-1}$
1	1.0600	0.94339	1.000	1.00000	0.9433	1.06000
2	1.1235	0.88999	2.059	0.48543	1.8333	0.54543
3	1.1910	0.83961	3.183	0.31410	2.6730	0.37410
4	1.2624	0.79209	4.374	0.22859	3.4651	0.28859
5	1.3382	0.74725	5.637	0.17739	4.2123	0.23739
6	1.4185	0.70496	6.975	0.14336	4.9173	0.20336
7	1.5036	0.66505	8.393	0.11913	5.5823	0.17913
8	1.5938	0.62741	9.897	0.10103	6.2097	0.16103
9	1.6894	0.59189	11.491	0.08702	6.8016	0.14702
10	1.7908	0.55839	13.180	0.07586	7.3600	0.13586
11	1.8982	0.52678	14.971	0.06679	7.8868	0.12679
12	2.0121	0.49696	16.869	0.05927	8.3838	0.11927
13	2.1329	0.46883	18.882	0.05296	8.8526	0.11296
14	2.2609	0.44230	21.015	0.04748	9.2949	0.10758
15	2.3965	0.41726	23.275	0.04296	9.7122	0.10296
16	2.5403	0.39364	25.672	0.03895	10.1058	0.09895
17	2.6927	0.37136	28.212	0.03544	10.4772	0.09544
18	2.8543	0.35034	30.905	0.03235	10.8276	0.09235
19	3.0255	0.33051	33.759	0.02962	11.1581	0.08962
20	3.2071	0.31180	36.785	0.02718	11.4699	0.08718
21	3.3995	0.29415	39.992	0.02500	11.7640	0.08500
22	3.6035	0.27750	43.392	0.02304	12.0415	0.08304
23	3.8197	0.26179	46.995	0.02127	12.3033	0.08127
24	4.0489	0.24697	50.815	0.01967	12.5503	0.07967
25	4.2918	0.23299	54.864	0.01822	12.7833	0.07822
26	4.5493	0.21981	59.156	0.01690	13.0031	0.07690
27	4.8223	0.20736	63.705	0.01569	13.2105	0.07569
28	5.1116	0.19563	68.528	0.01459	13.4061	0.07459
29	5.4183	0.18455	73.639	0.01357	13.5907	0.07357
30	5.7434	0.17411	79.058	0.01264	13.7648	0.07264
35	7.6860	0.13010	111.434	0.00897	14.4982	0.06897
40	10.2857	0.09722	154.761	0.00646	15.0462	0.06646
45	13.7646	0.07265	212.743	0.00470	15.4558	0.06470
50	18.4201	0.05428	290.335	0.00344	15.7618	0.06344
55	24.6503	0.04056	394.172	0.00253	15.9905	0.06253
60	32.9876	0.03031	533.128	0.00187	16.1614	0.06187
65	44.1449	0.02265	719.082	0.00139	16.2891	0.06139
70	59.0759	0.01692	967.932	0.00103	16.3845	0.06103
75	79.0569	0.01264	1300.948	0.00076	16.4558	0.06076
80	105.7959	0.00945	1746.599	0.00057	16.5091	0.06057
85	141.5788	0.00706	2342.981	0.00042	16.5489	0.06042
90	189.4645	0.00527	3141.075	0.00031	16.5786	0.06031
95	253.5462	0.00394	4209.103	0.00023	16.6009	0.06023
100	339.3020	0.00294	5638.367	0.00017	16.6175	0.06017

7% COMPOUND INTEREST FACTORS

Period	*Single payment compound amount factor*	*Single payment present worth factor*	*Uniform series compound amount factor*	*Uniform series sinking fund deposit factor*	*Uniform series present worth factor*	*Uniform series capital recovery factor*
	SPCAF	*SPPWF*	*USCAF*	*USSFDF*	*USPWF*	*USCRF*
	Future value of £1	*Present value of £1*	*Future value of uniform series of £1*	*Uniform series whose future value is £1*	*Present value of uniform series of £1*	*Uniform series with present value of £1*
n	$(1+i)^n$	$\frac{1}{(1+i)^n}$	$\frac{(1+i)^n-1}{i}$	$\frac{i}{(1+i)^n-1}$	$\frac{(1+i)^n-1}{i(1+i)^n}$	$\frac{i(1+i)^n}{(1+i)^n-1}$
1	1.0700	0.93457	1.000	1.00000	0.9345	1.07000
2	1.1448	0.87343	2.069	0.48309	1.8080	0.55309
3	1.2250	0.81629	3.214	0.31105	2.6243	0.38105
4	1.3107	0.76289	4.439	0.22522	3.3872	0.29522
5	1.4025	0.71298	5.750	0.17389	4.1001	0.24389
6	1.5007	0.66634	7.153	0.13979	4.7665	0.20979
7	1.6057	0.62274	8.654	0.11555	5.3892	0.18555
8	1.7181	0.58200	10.259	0.09746	5.9712	0.16746
9	1.8384	0.54393	11.977	0.08348	6.5152	0.15348
10	1.9671	0.50834	13.816	0.07237	7.0235	0.14237
11	2.1048	0.47509	15.783	0.06335	7.4986	0.13335
12	2.2521	0.44401	17.888	0.05590	7.9426	0.12590
13	2.4098	0.41496	20.140	0.04965	8.3576	0.11965
14	2.5785	0.38781	22.550	0.04434	8.7454	0.11434
15	2.7590	0.36244	25.129	0.03979	9.1079	0.10979
16	2.9521	0.33873	27.888	0.03585	9.4466	0.10585
17	3.1588	0.31657	30.840	0.03242	9.7632	0.10242
18	3.3799	0.29586	33.999	0.02941	10.0590	0.09941
19	3.6165	0.27650	37.378	0.02675	10.3355	0.09675
20	3.8696	0.25841	40.995	0.02439	10.5940	0.09439
21	4.1405	0.24151	44.865	0.02228	10.8355	0.09228
22	4.4304	0.22571	49.005	0.02040	11.0612	0.09040
23	4.7405	0.21094	53.436	0.01871	11.2721	0.08871
24	5.0723	0.19714	58.176	0.01718	11.4693	0.08718
25	5.4274	0.18424	63.249	0.01581	11.6535	0.08581
26	5.8073	0.17219	68.676	0.01456	11.8257	0.08456
27	6.2138	0.16093	74.483	0.01341	11.9867	0.08342
28	6.6488	0.15040	80.697	0.01239	12.1371	0.08239
29	7.1142	0.14056	87.346	0.01144	12.2776	0.08144
30	7.6122	0.13136	94.460	0.01058	12.4090	0.08058
35	10.6765	0.09366	138.236	0.00723	12.9476	0.07723
40	14.9744	0.06678	199.635	0.00500	13.3317	0.07500
45	21.0024	0.04761	285.749	0.00349	13.6055	0.07349
50	29.4570	0.03394	406.528	0.00245	13.8007	0.07245
55	41.3149	0.02420	575.928	0.00173	13.9399	0.07173
60	57.9464	0.07125	813.520	0.00122	14.0391	0.07122
65	81.2728	0.01230	1146.755	0.00087	14.1099	0.07087
70	113.9893	0.00877	1614.134	0.00061	14.1603	0.07061
75	159.8760	0.00625	2269.657	0.00044	14.1963	0.07044
80	224.2343	0.00445	3189.062	0.00031	14.2220	0.07031
85	314.5002	0.00317	4478.575	0.00022	14.2402	0.07022
90	441.1029	0.00226	6287.185	0.00015	14.2533	0.07015
95	618.6696	0.00161	8823.852	0.00011	14.2626	0.07011
100	867.7162	0.00115	12381.661	0.00008	14.2692	0.07008

8% COMPOUND INTEREST FACTORS

Period	*Single payment compound amount factor*	*Single payment present worth factor*	*Uniform series compound amount factor*	*Uniform series sinking fund deposit factor*	*Uniform series present worth factor*	*Uniform series capital recovery factor*
	SPCAF	*SPPWF*	*USCAF*	*USSFDF*	*USPWF*	*USCRF*
	Future value of £1	*Present value of £1*	*Future value of uniform series of £1*	*Uniform series whose future value is £1*	*Present value of uniform series of £1*	*Uniform series with present value of £1*
n	$(1+i)^n$	$\frac{1}{(1+i)^n}$	$\frac{(1+i)^n-1}{i}$	$\frac{i}{(1+i)^n-1}$	$\frac{(1+i)^n-1}{i(1+i)^n}$	$\frac{i(1+i)^n}{(1+i)^n-1}$
1	1.0800	0.92592	1.000	1.00000	0.9259	1.08000
2	1.1663	0.85733	2.079	0.48076	1.7832	0.56076
3	1.2597	0.79383	3.246	0.30803	2.5770	0.38803
4	1.3604	0.73502	4.506	0.22192	3.3121	0.30192
5	1.4693	0.68058	5.866	0.17045	3.9927	0.25045
6	1.5868	0.63016	7.335	0.13631	4.6228	0.21631
7	1.7138	0.58349	8.922	0.11207	5.2063	0.19207
8	1.8509	0.54026	10.636	0.09401	5.7466	0.17401
9	1.9990	0.50024	12.487	0.08007	6.2468	0.16007
10	2.1589	0.46319	14.486	0.06902	6.7100	0.14902
11	2.3316	0.42888	16.645	0.06007	7.1389	0.14007
12	2.5181	0.39711	18.977	0.05269	7.5360	0.13269
13	2.7196	0.36769	21.495	0.04652	7.9037	0.12652
14	2.9371	0.34046	24.214	0.04129	8.2442	0.12129
15	3.1721	0.31524	27.152	0.03682	8.5594	0.11682
16	3.4259	0.29189	30.324	0.03297	8.8513	0.11297
17	3.7000	0.27026	33.750	0.02962	9.1216	0.10962
18	3.9960	0.25024	37.450	0.02670	9.3718	0.10670
19	4.3157	0.23171	41.446	0.02412	9.6035	0.10412
20	4.6609	0.21454	45.761	0.02185	9.8181	0.10185
21	5.0338	0.19865	50.422	0.01983	10.0168	0.09983
22	5.4365	0.18394	55.456	0.01803	10.2007	0.09803
23	5.8714	0.17031	60.893	0.01642	10.3710	0.09642
24	6.3411	0.15769	66.764	0.01497	10.5287	0.09497
25	6.8484	0.14601	73.105	0.01367	10.6747	0.09367
26	7.3963	0.13520	79.954	0.01250	10.8099	0.09250
27	7.9880	0.12518	87.350	0.01144	10.9351	0.09144
28	8.6271	0.11591	95.338	0.01048	11.0510	0.09048
29	9.3172	0.10732	103.965	0.00961	11.1584	0.08961
30	10.0626	0.09937	113.283	0.00882	11.2577	0.08882
35	14.7853	0.06763	172.316	0.00580	11.6545	0.08580
40	21.7245	0.04603	259.056	0.00386	11.9246	0.08386
45	31.9204	0.03132	386.505	0.00258	12.1084	0.08258
50	46.9016	0.02132	573.770	0.00174	12.2334	0.08174
55	68.9138	0.01451	848.923	0.00117	12.3186	0.08117
60	101.2570	0.00987	1253.213	0.00079	12.3765	0.08079
65	148.7798	0.00672	1847.247	0.00054	12.4159	0.08054
70	218.6063	0.00457	2720.079	0.00036	12.4428	0.08036
75	321.2045	0.00311	4002.556	0.00024	12.4610	0.08024
80	471.9547	0.00211	5886.934	0.00016	12.4735	0.08016
85	693.4564	0.00144	8655.705	0.00011	12.4819	0.08011
90	1018.9149	0.00098	12723.936	0.00007	12.4877	0.08007
95	1497.1203	0.00066	18701.503	0.00005	12.4916	0.08005
100	2199.7612	0.00045	27484.515	0.00003	12.4943	0.08003

9% COMPOUND INTEREST FACTORS

Period	*Single payment compound amount factor*	*Single payment present worth factor*	*Uniform series compound amount factor*	*Uniform series sinking fund deposit factor*	*Uniform series present worth factor*	*Uniform series capital recovery factor*
	SPCAF	*SPPWF*	*USCAF*	*USSFDF*	*USPWF*	*USCRF*
	Future value of £1	*Present value of £1*	*Future value of uniform series of £1*	*Uniform series whose future value is £1*	*Present value of uniform series of £1*	*Uniform series with present value of £1*
n	$(1+i)^n$	$\frac{1}{(1+i)^n}$	$\frac{(1+i)^n-1}{i}$	$\frac{i}{(1+i)^n-1}$	$\frac{(1+i)^n-1}{i(1+i)^n}$	$\frac{i(1+i)^n}{(1+i)^n-1}$
1	1.0900	0.91743	1.000	1.00000	0.9174	0.09000
2	1.1880	0.84168	2.089	0.47846	1.7591	0.56846
3	1.2950	0.77218	3.278	0.30505	2.5312	0.39505
4	1.4115	0.70842	4.573	0.21866	3.2397	0.30866
5	1.5386	0.64993	5.984	0.16709	3.8896	0.25709
6	1.6771	0.59626	7.523	0.13291	4.4859	0.22291
7	1.8280	0.54703	9.200	0.10869	5.0329	0.19869
8	1.9925	0.50186	11.028	0.09067	5.5348	0.18067
9	2.1718	0.46042	13.021	0.07679	5.9952	0.16679
10	2.3673	0.42241	15.192	0.06582	6.4176	0.15582
11	2.5804	0.38753	17.560	0.05694	6.8051	0.14694
12	2.8126	0.35553	20.140	0.04965	7.1607	0.13965
13	3.0658	0.32617	22.953	0.04356	7.4869	0.13356
14	3.3417	0.29924	26.019	0.03843	7.7861	0.12843
15	3.6424	0.27453	29.360	0.03405	8.0606	0.12405
16	3.9703	0.25186	33.003	0.03029	8.3125	0.12029
17	4.3276	0.23107	36.973	0.02704	8.5436	0.11704
18	4.7171	0.21199	41.301	0.02421	8.7556	0.11421
19	5.1416	0.19448	46.018	0.02173	8.9501	0.11173
20	5.6044	0.17843	51.160	0.01954	9.1285	0.10954
21	6.1088	0.16369	56.764	0.01761	9.2922	0.10761
22	6.6585	0.15018	62.873	0.01590	9.4424	0.10590
23	7.2578	0.13778	69.531	0.01438	9.5802	0.10438
24	7.9110	0.12640	76.789	0.01302	9.7066	0.10302
25	8.6230	0.11596	84.700	0.01180	9.8225	0.10180
26	9.3991	0.10639	93.323	0.01071	9.9289	0.10071
27	10.2450	0.09760	102.723	0.00973	10.0625	0.09973
28	11.1671	0.08954	112.968	0.00885	10.1161	0.09885
29	12.1721	0.08215	124.135	0.00805	10.1982	0.09805
30	13.2676	0.07537	136.307	0.00733	10.2736	0.09733
35	20.4139	0.04898	215.710	0.00463	10.5668	0.09463
40	31.4094	0.03183	337.882	0.00295	10.7573	0.09295
45	48.3272	0.02069	525.858	0.00190	10.8811	0.09190
50	74.3575	0.01344	815.083	0.00122	10.9616	0.09122
55	114.4082	0.00874	1260.091	0.00079	11.0139	0.09079
60	176.0312	0.00568	1944.791	0.00051	11.0479	0.09051
65	270.8459	0.00369	2998.288	0.00033	11.0700	0.09033
70	416.7300	0.00239	4619.223	0.00021	11.0844	0.09021
75	641.1908	0.00155	7113.232	0.00014	11.0937	0.09014
80	986.5515	0.00101	10950.572	0.00009	11.0998	0.09009
85	1517.9319	0.00065	16854.798	0.00005	11.1037	0.09005
90	2335.5264	0.00042	25939.182	0.00003	11.1063	0.09003
95	3593.4969	0.00027	39916.632	0.00002	11.1080	0.09002
100	5529.0406	0.00018	61422.673	0.00001	11.1091	0.09001

10% COMPOUND INTEREST FACTORS

Period	*Single payment compound amount factor*	*Single payment present worth factor*	*Uniform series compound amount factor*	*Uniform series sinking fund deposit factor*	*Uniform series present worth factor*	*Uniform series capital recovery factor*
	SPCAF	*SPPWF*	*USCAF*	*USSFDF*	*USPWF*	*USCRF*
	Future value of £1	*Present value of £1*	*Future value of uniform series of £1*	*Uniform series whose future value is £1*	*Present value of uniform series of £1*	*Uniform series with present value of £1*
n	$(1+i)^n$	$\frac{1}{(1+i)^n}$	$\frac{(1+i)^n-1}{i}$	$\frac{i}{(1+i)^n-1}$	$\frac{(1+i)^n-1}{i(1+i)^n}$	$\frac{i(1+i)^n}{(1+i)^n-1}$
1	1.1000	0.90909	1.000	1.00000	0.9090	1.10000
2	1.2099	0.82644	2.099	0.47619	1.7355	0.57619
3	1.3309	0.75131	3.309	0.30211	2.4868	0.40211
4	1.4640	0.68301	4.640	0.21547	3.1698	0.31547
5	1.6105	0.62092	6.105	0.16379	3.7907	0.26379
6	1.7715	0.56447	7.715	0.12960	4.3552	0.22960
7	1.9487	0.51315	9.487	0.10540	4.8684	0.20540
8	2.1435	0.46650	11.435	0.08744	5.3349	0.18744
9	2.3579	0.42409	13.579	0.07364	5.7590	0.17364
10	2.5937	0.38554	15.937	0.06274	6.1445	0.16274
11	2.8531	0.35049	18.531	0.05396	6.4950	0.15396
12	3.1384	0.31863	21.384	0.04676	6.8136	0.14676
13	3.4522	0.28966	24.522	0.04077	7.1033	0.14077
14	3.7974	0.26333	27.974	0.03574	7.3666	0.13574
15	4.1772	0.23939	31.772	0.03147	7.6060	0.13147
16	4.5949	0.21762	35.949	0.02781	7.8237	0.12781
17	5.0544	0.19784	40.544	0.02466	8.0215	0.12466
18	5.5599	0.17985	45.599	0.02193	8.2014	0.12193
19	6.1159	0.16350	51.159	0.01954	8.3649	0.11954
20	6.7274	0.14864	57.274	0.01745	8.5135	0.11745
21	7.4002	0.13513	64.002	0.01562	8.6486	0.11562
22	8.1402	0.12284	71.402	0.01400	8.7715	0.11400
23	8.9543	0.11167	79.543	0.01257	8.8832	0.11257
24	9.8497	0.10152	88.497	0.01129	8.9847	0.11129
25	10.8347	0.09229	98.347	0.01016	9.0770	0.11016
26	11.9181	0.08390	109.181	0.00915	9.1609	0.10915
27	13.1099	0.07627	121.099	0.00825	9.2372	0.10825
28	14.4209	0.06934	134.209	0.00745	9.3065	0.10745
29	15.8630	0.06303	148.630	0.00672	9.3696	0.10672
30	17.4494	0.05730	164.494	0.00607	9.4269	0.10607
35	28.1024	0.03558	271.024	0.00368	9.6441	0.10368
40	45.2592	0.02209	442.592	0.00225	9.7790	0.10225
45	72.8904	0.01371	718.904	0.00139	9.8628	0.10139
50	117.3908	0.00851	1163.908	0.00085	9.9148	0.10085
55	189.0591	0.00528	1880.591	0.00053	9.9471	0.10053
60	304.4816	0.00328	3034.816	0.00032	9.9671	0.10032
65	490.3706	0.00203	4893.706	0.00020	9.9796	0.10020
70	789.7468	0.00126	7887.468	0.00012	9.9873	0.10012
75	1271.8952	0.00078	12708.952	0.00007	9.9921	0.10007
80	2048.4000	0.00048	20474.000	0.00004	9.9951	0.10004
85	3298.9687	0.00030	32979.687	0.00003	9.9969	0.10003
90	5313.0221	0.00018	53120.221	0.00001	9.9981	0.10001
95	8556.6753	0.00011	85556.753	0.00001	9.9988	0.10001
100	13780.6110	0.00007	137796.110	0.00000	9.9992	0.10000

11% COMPOUND INTEREST FACTORS

Period	*Single payment compound amount factor*	*Single payment present worth factor*	*Uniform series compound amount factor*	*Uniform series sinking fund deposit factor*	*Uniform series present worth factor*	*Uniform series capital recovery factor*
	SPCAF	*SPPWF*	*USCAF*	*USSFDF*	*USPWF*	*USCRF*
	Future value of £1	*Present value of £1*	*Future value of uniform series of £1*	*Uniform series whose future value is £1*	*Present value of uniform series of £1*	*Uniform series with present value of £1*
n	$(1+i)^n$	$\frac{1}{(1+i)^n}$	$\frac{(1+i)^n-1}{i}$	$\frac{i}{(1+i)^n-1}$	$\frac{(1+i)^n-1}{i(1+i)^n}$	$\frac{i(1+i)^n}{(1+i)^n-1}$
1	1.1100	0.90090	1.000	1.00000	0.9009	1.11000
2	1.2320	0.81162	2.109	0.47393	1.7125	0.58393
3	1.3676	0.73119	3.342	0.29921	2.4437	0.40921
4	1.5180	0.65873	4.709	0.21232	3.1024	0.32232
5	1.6850	0.59345	6.227	0.16057	3.6958	0.27057
6	1.8704	0.53464	7.912	0.12637	4.2305	0.23637
7	2.0761	0.48165	9.783	0.10221	4.7121	0.21221
8	2.3045	0.43392	11.859	0.08432	5.1461	0.19432
9	2.5580	0.39092	14.163	0.07060	5.5370	0.18060
10	2.8394	0.35218	16.722	0.05980	5.8892	0.16980
11	3.1517	0.31728	19.561	0.05112	6.2065	0.16112
12	3.4984	0.28584	22.713	0.04402	6.4923	0.15402
13	3.8832	0.25751	26.211	0.03815	6.7498	0.14815
14	4.3104	0.23199	30.094	0.03322	6.8918	0.14322
15	4.7845	0.20900	34.405	0.02906	7.1908	0.13906
16	5.3108	0.18829	39.189	0.02551	7.3791	0.13551
17	5.8950	0.16963	44.500	0.02247	7.5487	0.12147
18	6.5435	0.15282	50.395	0.01984	7.7016	0.12984
19	7.2633	0.13767	56.939	0.01756	7.8392	0.12756
20	8.0623	0.12403	64.202	0.01557	7.9633	0.12557
21	8.9491	0.11174	72.265	0.01383	8.0750	0.12383
22	9.9335	0.10066	81.214	0.01231	8.1757	0.12231
23	11.0262	0.09069	91.147	0.01097	8.2664	0.12097
24	12.2391	0.08170	102.174	0.00978	8.3481	0.11978
25	13.5854	0.07260	114.413	0.00874	8.4217	0.11874
26	15.0798	0.06631	127.998	0.00781	8.4880	0.11781
27	16.7386	0.05974	143.078	0.00698	8.5478	0.11698
28	18.5798	0.05382	159.817	0.00625	8.6016	0.11625
29	20.6236	0.04848	178.397	0.00560	8.6501	0.11560
30	22.8911	0.04368	199.020	0.00502	8.6937	0.11502
35	38.5748	0.02592	341.589	0.00292	8.8552	0.11292
40	65.0008	0.01538	581.825	0.00171	8.9510	0.11171
45	109.5302	0.00912	986.638	0.00101	9.0079	0.11101
50	184.5647	0.00541	1668.770	0.00059	9.0416	0.11059
55	311.0023	0.00321	2818.203	0.00035	0.0616	0.11035
60	524.0570	0.00190	4755.064	0.00021	9.0735	0.11021
65	883.0665	0.00113	8018.787	0.00012	9.0806	0.11012
70	1488.0185	0.00067	13518.350	0.00007	9.0847	0.11007
75	2507.3976	0.00039	22785.432	0.00004	9.0872	0.11004
80	4225.1109	0.00023	38401.008	0.00002	9.0887	0.11002
85	7119.5571	0.00014	64714.155	0.00001	9.0896	0.11001
90	11996.8680	0.00008	109053.340	0.00000	9.0901	0.11000
95	20215.4180	0.00004	183767.430	0.00000	9.0904	0.11000
100	34064.1570	0.00002	309665.060	0.00000	0.0906	0.11000

12% COMPOUND INTEREST FACTORS

Period	*Single payment compound amount factor*	*Single payment present worth factor*	*Uniform series compound amount factor*	*Uniform series sinking fund deposit factor*	*Uniform series present worth factor*	*Uniform series capital recovery factor*
	SPCAF	*SPPWF*	*USCAF*	*USSFDF*	*USPWF*	*USCRF*
	Future value of £1	*Present value of £1*	*Future value of uniform series of £1*	*Uniform series whose future value is £1*	*Present value of uniform series of £1*	*Uniform series with present value of £1*
n	$(1+i)^n$	$\frac{1}{(1+i)^n}$	$\frac{(1+i)^n-1}{i}$	$\frac{i}{(1+i)^n-1}$	$\frac{(1+i)^n-1}{i(1+i)^n}$	$\frac{i(1+i)^n}{(1+i)^n-1}$
1	1.1200	0.89285	1.000	1.00000	0.8928	1.12000
2	1.2543	0.79719	2.119	0.47169	1.6900	0.59169
3	1.4049	0.71178	3.374	0.29634	2.4018	0.41634
4	1.5735	0.63551	4.779	0.20923	3.0373	0.32923
5	1.7623	0.56742	6.352	0.15740	3.6047	0.27740
6	1.9738	0.50663	8.115	0.12322	4.1114	0.24322
7	2.2106	0.45234	10.089	0.09911	4.5637	0.21911
8	2.4759	0.40388	12.299	0.08130	4.9676	0.20130
9	2.7730	0.36061	14.775	0.06767	5.3282	0.18767
10	3.1058	0.32197	17.548	0.05698	5.6502	0.17698
11	3.4785	0.28747	20.654	0.04841	5.9376	0.16841
12	3.8959	0.25667	24.133	0.04143	6.1943	0.16143
13	4.3634	0.22917	28.029	0.03567	6.4235	0.15567
14	4.8871	0.20461	32.392	0.03087	6.6281	0.15087
15	5.4735	0.18269	37.279	0.02682	6.8108	0.14682
16	6.1303	0.16312	42.753	0.02339	6.9739	0.14339
17	6.8660	0.14564	48.883	0.02045	7.1196	0.14045
18	7.6899	0.13003	55.749	0.01793	7.2496	0.13793
19	8.6127	0.11610	63.439	0.01576	7.3657	0.13576
20	9.6462	0.10366	72.052	0.01387	7.4694	0.13387
21	10.8038	0.09255	81.698	0.01224	7.5620	0.13224
22	12.1003	0.08264	92.502	0.01081	7.6446	0.13081
23	13.5523	0.07378	104.602	0.00955	7.7184	0.12955
24	15.1786	0.06588	118.155	0.00846	7.7843	0.12846
25	17.0000	0.05882	133.333	0.00749	7.8431	0.12749
26	19.0400	0.05252	150.333	0.00665	7.8956	0.12665
27	21.3248	0.04689	169.373	0.00590	7.9425	0.12590
28	23.8838	0.04186	190.698	0.00524	7.9844	0.12524
29	26.7499	0.03738	214.582	0.00466	8.0218	0.12466
30	29.9599	0.03337	241.332	0.00414	8.0551	0.12414
35	52.7996	0.01893	431.663	0.00231	8.1755	0.12231
40	93.0509	0.01074	767.091	0.00130	8.2437	0.12130
45	163.9875	0.00609	1358.229	0.00073	8.2825	0.12073
50	289.0021	0.00346	2400.017	0.00041	8.3044	0.12041
55	509.3204	0.00196	4236.003	0.00023	8.3169	0.12023
60	897.5966	0.00111	7471.638	0.00013	8.3240	0.12013
65	1581.8719	0.00063	13173.932	0.00007	8.3280	0.12007
70	2787.7987	0.00035	23223.322	0.00004	8.3303	0.12004
75	4913.0538	0.00020	40933.781	0.00002	8.3316	0.12002
80	8658.4794	0.00011	72145.661	0.00001	8.3323	0.12001
85	15259.1980	0.00006	127151.650	0.00000	8.3327	0.12000
90	26891.9150	0.00003	224090.950	0.00000	8.3330	0.12000
95	47392.7240	0.00002	394931.030	0.00000	8.3331	0.12000
100	83522.2210	0.00001	696010.170	0.00000	8.3332	0.12000

14% COMPOUND INTEREST FACTORS

Period	*Single payment compound amount factor*	*Single payment present worth factor*	*Uniform series compound amount factor*	*Uniform series sinking fund deposit factor*	*Uniform series present worth factor*	*Uniform series capital recovery factor*
	SPCAF	*SPPWF*	*USCAF*	*USSFDF*	*USPWF*	*USCRF*
	Future value of £1	*Present value of £1*	*Future value of uniform series of £1*	*Uniform series whose future value is £1*	*Present value of uniform series of £1*	*Uniform series with present value of £1*
n	$(1+i)^n$	$\frac{1}{(1+i)^n}$	$\frac{(1+i)^n-1}{i}$	$\frac{i}{(1+i)^n-1}$	$\frac{(1+i)^n-1}{i(1+i)^n}$	$\frac{i(1+i)^n}{(1+i)^n-1}$
1	1.1400	0.87719	1.000	1.00000	0.8771	1.14000
2	1.2995	0.76946	2.139	0.46728	1.6466	0.60728
3	1.4815	0.67497	3.439	0.29073	2.3216	0.43073
4	1.6889	0.59208	4.921	0.20320	2.9137	0.34320
5	1.9254	0.51936	6.610	0.15128	3.4330	0.29128
6	2.1949	0.45558	8.535	0.11715	3.8886	0.25715
7	2.5022	0.39963	10.730	0.09319	4.2883	0.23319
8	2.8525	0.35055	13.232	0.07557	4.6388	0.21557
9	3.2519	0.30750	16.085	0.06216	4.9463	0.20216
10	3.7072	0.26974	19.337	0.05171	5.2161	0.19171
11	4.2262	0.23661	23.044	0.04339	5.4527	0.18339
12	4.8179	0.20755	27.270	0.03666	5.6602	0.17666
13	5.4924	0.18206	32.088	0.03116	5.8423	0.17116
14	6.2613	0.15971	37.581	0.02660	6.0020	0.16660
15	7.1379	0.14009	43.842	0.02280	6.1421	0.16280
16	8.1372	0.12289	50.980	0.01961	6.2650	0.15961
17	9.2764	0.10779	59.117	0.01691	6.3728	0.15691
18	10.5751	0.09456	68.394	0.01462	6.4674	0.15462
19	12.0556	0.08294	78.969	0.01266	6.5503	0.15266
20	13.7434	0.07276	91.024	0.01098	6.6231	0.15098
21	15.6675	0.06382	104.768	0.00954	6.6869	0.14954
22	17.8610	0.05598	120.435	0.00830	6.7429	0.14830
23	20.3615	0.04911	138.297	0.00723	6.7920	0.14723
24	23.2122	0.04308	158.658	0.00630	6.8351	0.14630
25	26.4619	0.03779	181.870	0.00549	6.8729	0.14549
26	30.1665	0.03314	208.332	0.00480	6.9060	0.14480
27	34.3899	0.02907	238.499	0.00419	6.9351	0.14419
28	39.2044	0.02550	272.889	0.00366	6.9606	0.14366
29	44.6931	0.02237	312.093	0.00320	6.9830	0.14320
30	50.9501	0.01962	356.786	0.00280	7.0026	0.14280
35	98.1001	0.01019	693.572	0.00144	7.0700	0.14144
40	188.8834	0.00529	1342.024	0.00074	7.1050	0.14074
45	363.6790	0.00274	2590.564	0.00038	7.1232	0.14038
50	700.2329	0.00142	4994.520	0.00020	7.1326	0.14020

16% COMPOUND INTEREST FACTORS

Period	*Single payment compound amount factor*	*Single payment present worth factor*	*Uniform series compound amount factor*	*Uniform series sinking fund deposit factor*	*Uniform series present worth factor*	*Uniform series capital recovery factor*
	SPCAF	*SPPWF*	*USCAF*	*USSFDF*	*USPWF*	*USCRF*
	Future value of £1	*Present value of £1*	*Future value of uniform series of £1*	*Uniform series whose future value is £1*	*Present value of uniform series of £1*	*Uniform series with present value of £1*
n	$(1+i)^n$	$\frac{1}{(1+i)^n}$	$\frac{(1+i)^n-1}{i}$	$\frac{i}{(1+i)^n-1}$	$\frac{(1+i)^n-1}{i(1+i)^n}$	$\frac{i(1+i)^n}{(1+i)^n-1}$
1	1.1600	0.86206	1.000	1.00000	0.8620	1.16000
2	1;3455	0.74316	2.159	0.46296	1.6052	0.62296
3	1.5608	0.64065	3.505	0.28525	2.2458	0.44525
4	1.8106	0.55229	5.066	0.19737	2.7981	0.35737
5	2.1003	0.47611	6.877	0.14540	3.2742	0.30540
6	2.4363	0.41044	8.977	0.11138	3.6847	0.27138
7	2.8262	0.35382	11.413	0.08761	4.0385	0.24761
8	3.2784	0.30502	14.240	0.07022	4.3435	0.23022
9	3.8029	0.26295	17.518	0.05708	4.6065	0.21708
10	4.4114	0.22668	21.321	0.04690	4.8332	0.20690
11	5.1172	0.19541	25.732	0.03886	5.0286	0.19886
12	5.9360	0.16846	30.850	0.03241	5.1971	0.19241
13	6.8857	0.14522	36.786	0.02718	5.3423	0.18718
14	7.9875	0.12519	43.671	0.02289	5.4675	0.18289
15	9.2655	0.10792	51.659	0.01935	5.5754	0.17935
16	10.7480	0.09304	60.925	0.01641	5.6684	0.17641
17	12.4676	0.08020	71.673	0.01395	5.7487	0.17395
18	14.4625	0.06914	84.140	0.01188	5.8178	0.17188
19	16.7765	0.05960	98.603	0.01014	5.8774	0.17014
20	19.4607	0.05138	115.379	0.00866	5.9288	0.16866
21	22.5744	0.04429	134.840	0.00741	5.9731	0.16741
22	26.1863	0.03818	157.414	0.00635	6.0113	0.16635
23	30.3762	0.03292	183.601	0.00544	6.0442	0.16544
24	35.2364	0.02837	213.977	0.00467	6.0726	0.16467
25	40.8742	0.02446	249.213	0.00401	6.0970	0.16401
26	47.4141	0.02109	290.088	0.00344	6.1181	0.16344
27	55.0003	0.01818	337.502	0.00296	6.1363	0.16296
28	63.8004	0.01567	392.502	0.00254	6.1520	0.16254
29	74.0085	0.01351	456.303	0.00219	6.1655	0.16219
30	85.8498	0.01164	530.311	0.00188	6.1771	0.16188
35	180.3140	0.00554	1120.712	0.00089	6.2153	0.16089
40	378.7210	0.00264	2360.756	0.00042	6.2334	0.16042
45	795.4436	0.00125	4965.272	0.00020	6.2421	0.16020
50	1670.7033	0.00059	10435.645	0.00009	6.2462	0.16009

18% COMPOUND INTEREST FACTORS

Period	*Single payment compound amount factor*	*Single payment present worth factor*	*Uniform series compound amount factor*	*Uniform series sinking fund deposit factor*	*Uniform series present worth factor*	*Uniform series capital recovery factor*
	SPCAF	*SPPWF*	*USCAF*	*USSFDF*	*USPWF*	*USCRF*
	Future value of £1	*Present value of £1*	*Future value of uniform series of £1*	*Uniform series whose future value is £1*	*Present value of uniform series of £1*	*Uniform series with present value of £1*
n	$(1+i)^n$	$\frac{1}{(1+i)^n}$	$\frac{(1+i)^n-1}{i}$	$\frac{i}{(1+i)^n-1}$	$\frac{(1+i)^n-1}{i(1+i)^n}$	$\frac{i(1+i)^n}{(1+i)^n-1}$
1	1.180	0.84745	1.000	1.00000	0.8474	1.18000
2	1.3923	0.71818	2.179	0.45871	1.5656	0.63871
3	1.6430	0.60863	3.572	0.27992	2.1742	0.45992
4	1.9387	0.51578	5.215	0.19173	2.6900	0.37173
5	2.2877	0.43710	7.154	0.13977	3.1271	0.31977
6	2.6995	0.37043	9.441	0.10591	3.4976	0.28591
7	3.1854	0.31392	12.141	0.08236	3.8115	0.26236
8	3.7588	0.26603	15.326	0.06524	4.0775	0.24524
9	4.4354	0.22545	19.085	0.05239	4.3030	0.23239
10	5.2338	0.19106	23.521	0.04251	4.4940	0.22251
11	6.1759	0.16191	28.755	0.03477	4.6560	0.21477
12	7.2875	0.13721	34.931	0.02862	4.7932	0.20862
13	8.5993	0.11628	42.218	0.02368	4.9095	0.20368
14	10.1472	0.09854	50.818	0.01967	5.0080	0.19967
15	11.9737	0.08351	60.965	0.01640	5.0915	0.19640
16	14.1290	0.07077	72.938	0.01371	5.1623	0.19371
17	16.6722	0.05997	87.068	0.01148	5.2223	0.19148
18	19.6732	0.05083	103.740	0.00963	5.2731	0.18963
19	23.2144	0.04307	123.413	0.00810	5.3162	0.18810
20	27.3930	0.03650	146.627	0.00681	5.3527	0.18681
21	32.3237	0.03093	174.020	0.00574	5.3836	0.18574
22	38.1420	0.02621	206.344	0.00484	5.4099	0.18484
23	45.0076	0.02221	244.486	0.00409	5.4321	0.18409
24	53.1089	0.01882	289.494	0.00345	5.4509	0.18345
25	62.6686	0.01595	342.603	0.00291	5.4669	0.18291
26	73.9489	0.01352	405.271	0.00246	5.4804	0.18246
27	87.2597	0.01146	479.220	0.00208	5.4918	0.18208
28	102.9665	0.00971	566.480	0.00176	5.5016	0.18176
29	121.5005	0.00823	669.447	0.00149	5.5098	0.18149
30	143.3706	0.00697	790.947	0.00126	5.5168	0.18126
35	327.9971	0.00304	1816.650	0.00055	5.5386	0.18055
40	750.3780	0.00133	4163.211	0.00024	5.5481	0.18024
45	1716.6831	0.00058	9531.572	0.00010	5.5523	0.18010
50	3927.3551	0.00025	21813.083	0.00004	5.5541	0.18004

20% COMPOUND INTEREST FACTORS

Period	*Single payment compound amount factor*	*Single payment present worth factor*	*Uniform series compound amount factor*	*Uniform series sinking fund deposit factor*	*Uniform series present worth factor*	*Uniform series capital recovery factor*
	SPCAF	*SPPWF*	*USCAF*	*USSFDF*	*USPWF*	*USCRF*
	Future value of £1	*Present value of £1*	*Future value of uniform series of £1*	*Uniform series whose future value is £1*	*Present value of uniform series of £1*	*Uniform series with present value of £1*
n	$(1+i)^n$	$\frac{1}{(1+i)^n}$	$\frac{(1+i)^n-1}{i}$	$\frac{i}{(1+i)^n-1}$	$\frac{(1+i)^n-1}{i(1+i)^n}$	$\frac{i(1+i)^n}{(1+i)^n-1}$
1	1.200	0.8333	1.000	1.00000	0.833	1.20000
2	1.440	0.6944	2.200	0.45455	1.528	0.65455
3	1.728	0.5787	3.640	0.27473	2.106	0.47473
4	2.074	0.4823	5.368	0.18629	2.589	0.38629
5	2.488	0.4019	7.442	0.13438	2.991	0.33438
6	2.986	0.3349	9.930	0.10071	3.326	0.30071
7	3.583	0.2791	12.916	0.07742	3.605	0.27742
8	4.300	0.2326	16.499	0.06061	3.837	0.26061
9	5.160	0.1938	20.799	0.04808	4.031	0.24808
10	6.192	0.1615	25.959	0.03852	4.192	0.23852
11	7.430	0.1346	32.150	0.03110	4.327	0.23110
12	8.916	0.1122	39.581	0.02526	4.439	0.22526
13	10.699	0.0935	48.497	0.02062	4.533	0.22062
14	12.839	0.0770	59.196	0.01689	4.611	0.21680
15	15.407	0.0649	72.035	0.01388	4.675	0.21388
16	18.488	0.0541	87.442	0.01144	4.730	0.21144
17	22.186	0.0451	105.931	0.00944	4.775	0.20944
18	26.623	0.0376	128.117	0.00781	4.812	0.20781
19	31.948	0.0313	154.740	0.00646	4.843	0.20646
20	38.338	0.0261	186.688	0.00536	4.870	0.20536
21	46.005	0.0217	225.026	0.00444	4.891	0.20444
22	55.206	0.0181	271.031	0.00369	4.909	0.20369
23	66.247	0.0151	326.237	0.00307	4.925	0.20307
24	79.497	0.0126	392.484	0.00255	4.937	0.20255
25	95.396	0.0105	471.981	0.00212	4.948	0.20212
26	114.475	0.0087	567.377	0.00176	4.956	0.20176
27	137.371	0.0073	681.853	0.00147	4.964	0.20147
28	164.845	0.0061	819.223	0.00122	4.970	0.20122
29	197.814	0.0051	984.068	0.00102	4.975	0.20102
30	237.376	0.0042	1181.882	0.00085	4.979	0.20085
35	590.668	0.0017	2948.341	0.00034	4.992	0.20034
40	1469.772	0.0007	7343.858	0.00014	4.997	0.20014
45	3657.262	0.0003	18281.310	0.00005	4.999	0.20005
50	9100.438	0.0001	45497.191	0.00002	4.999	0.20002

25% COMPOUND INTEREST FACTORS

Period n	*Single payment compound amount factor* *SPCAF* *Future value of £1* $(1 + i)^n$	*Single payment present worth factor* *SPPWF* *Present value of £1* $\frac{1}{(1 + i)^n}$	*Uniform series compound amount factor* *USCAF* *Future value of uniform series of £1* $\frac{(1 + i)^n - 1}{i}$	*Uniform series sinking fund deposit factor* *USSFDF* *Uniform series whose future value is £1* $\frac{i}{(1 + i)^n - 1}$	*Uniform series present worth factor* *USPWF* *Present value of uniform series of £1* $\frac{(1 + i)^n - 1}{i(1 + i)^n}$	*Uniform series capital recovery factor* *USCRF* *Uniform series with present value of £1* $\frac{i(1 + i)^n}{(1 + i)^n - 1}$
1	1.250	0.8000	1.000	1.00000	0.800	1.25000
2	1.562	0.6400	2.250	0.44444	1.440	0.69444
3	1.953	0.5120	3.812	0.26230	1.952	0.51230
4	2.441	0.4096	5.766	0.17344	2.362	0.42344
5	3.052	0.3277	8.207	0.12185	2.689	0.37185
6	3.815	0.2621	11.259	0.08882	2.951	0.33882
7	4.768	0.2097	15.073	0.06634	3.161	0.31634
8	5.960	0.1678	19.842	0.05040	3.329	0.30040
9	7.451	0.1342	25.802	0.03876	3.463	0.28876
10	9.313	0.1074	33.253	0.03007	3.571	0.28007
11	11.642	0.0859	42.566	0.02349	3.656	0.27349
12	14.552	0.0687	54.208	0.01845	3.725	0.26845
13	18.190	0.0550	86.949	0.01150	3.824	0.26150
14	22.737	0.0440	86.949	0.01150	3.824	0.26150
15	28.422	0.0352	109.687	0.00912	3.859	0.25912
16	35.527	0.0281	138.109	0.00724	3.887	0.25724
17	44.409	0.0225	173.636	0.00576	3.910	0.25576
18	55.511	0.0180	218.045	0.25459	3.928	0.25459
19	69.389	0.0144	273.556	0.00366	3.942	0.25366
20	86.736	0.0115	342.945	0.00202	3.945	0.25292
21	108.420	0.0092	429.681	0.00233	3.963	0.25233
22	135.525	0.0074	538.101	0.00186	3.970	0.25186
23	169.407	0.0059	673.626	0.00148	3.976	0.25148
24	211.758	0.0047	843.033	0.00119	3.981	0.25119
25	264.698	0.0038	1054.791	0.00095	3.985	0.25095
26	330.872	0.0030	1319.489	0.00076	3.988	0.25076
27	413.590	0.0024	1650.361	0.00061	3.990	0.25061
28	516.988	0.0019	2063.952	0.00048	3.992	0.25048
29	646.236	0.0015	2580.939	0.00039	3.994	0.25039
30	807.794	0.0012	3227.174	0.00031	3.995	0.25031
35	2645.190	0.0004	9856.761	0.00010	3.998	0.25010
40	7523.164	0.0001	30088.655	0.00003	3.999	0.25003

30% COMPOUND INTEREST FACTORS

Period	Single payment compound amount factor	Single payment present worth factor	Uniform series compound amount factor	Uniform series sinking fund deposit factor	Uniform series present worth factor	Uniform series capital recovery factor
	SPCAF	SPPWF	USCAF	USSFDF	USPWF	USCRF
	Future value of £1	Present value of £1	Future value of uniform series of £1	Uniform series whose future value is £1	Present value of uniform series of £1	Uniform series with present value of £1
n	$(1+i)^n$	$\frac{1}{(1+i)^n}$	$\frac{(1+i)^n-1}{i}$	$\frac{i}{(1+i)^n-1}$	$\frac{(1+i)^n-1}{i(1+i)^n}$	$\frac{i(1+i)^n}{(1+i)^n-1}$
1	1.300	0.7692	1.000	1.00000	0.769	1.30000
2	1.690	0.5917	2.300	0.43478	1.361	0.73478
3	2.197	0.4552	3.990	0.25063	1.816	0.55063
4	2.856	0.3501	6.187	0.16163	2.166	0.46163
5	3.713	0.2693	9.043	0.11058	2.436	0.41058
6	4.827	0.2072	12.756	0.07839	2.643	0.37839
7	6.275	0.1594	17.583	0.05687	2.802	0.35687
8	8.157	0.1226	23.858	0.04192	2.952	0.34192
9	10.604	0.0943	32.015	0.03124	3.019	0.33124
10	13.780	0.0725	42.619	0.02346	3.092	0.32346
11	17.922	0.0558	56.405	0.01773	3.147	0.31773
12	23.298	0.0429	74.327	0.01345	3.190	0.31345
13	30.288	0.0330	97.625	0.01024	3.223	0.31024
14	39.374	0.0254	127.913	0.00782	3.249	0.30782
15	51.186	0.0195	167.286	0.00598	3.268	0.30598
16	66.542	0.0150	218.427	0.00458	3.283	0.30458
17	86.504	0.0116	285.014	0.00351	3.295	0.30351
18	112.455	0.0089	371.518	0.00269	3.304	0.30269
19	146.192	0.0068	483.973	0.00207	3.311	0.30207
20	190.050	0.0053	630.165	0.00159	3.316	0.30159
21	247.065	0.0040	820.215	0.00122	3.320	0.30122
22	321.184	0.0031	1067.280	0.00094	3.323	0.30094
23	417.539	0.0024	1388.464	0.00072	3.325	0.30072
24	542.801	0.0018	1806.003	0.00055	3.327	0.30055
25	705.641	0.0014	2348.803	0.00043	3.329	0.30043
26	917.333	0.0011	3054.444	0.00033	3.330	0.30033
27	1192.533	0.0008	3971.778	0.00025	3.331	0.30025
28	1550.293	0.0006	5164.311	0.00019	3.331	0.30019
29	2015.381	0.0005	6714.604	0.00015	3.332	0.30015
30	2619.996	0.0004	8729.985	0.00011	3.332	0.30011
35	9727.860	0.0001	32422.868	0.00003	3.333	0.30003

40% COMPOUND INTEREST FACTORS

Period	*Single payment compound amount factor*	*Single payment present worth factor*	*Uniform series compound amount factor*	*Uniform series sinking fund deposit factor*	*Uniform series present worth factor*	*Uniform series capital recovery factor*
	SPCAF	*SPPWF*	*USCAF*	*USSFDF*	*USPWF*	*USCRF*
	Future value of £1	*Present value of £1*	*Future value of uniform series of £1*	*Uniform series whose future value is £1*	*Present value of uniform series of £1*	*Uniform series with present value of £1*
n	$(1+i)^n$	$\frac{1}{(1+i)^n}$	$\frac{(1+i)^n-1}{i}$	$\frac{i}{(1+i)^n-1}$	$\frac{(1+i)^n-1}{i(1+i)^n}$	$\frac{i(1+i)^n}{(1+i)^n-1}$
1	1.400	0.7143	1.000	1.00000	0.714	1.40000
2	1.960	0.5102	2.400	0.41667	1.224	0.81667
3	2.744	0.3644	4.360	0.22936	1.589	0.62936
4	3.842	0.2603	7.104	0.14077	1.849	0.54077
5	5.378	0.1859	10.946	0.09136	2.035	0.49136
6	7.530	0.1328	16.324	0.06126	2.168	0.46126
7	10.541	0.0949	23.853	0.04192	2.263	0.44192
8	14.758	0.0678	34.395	0.02907	2.331	0.42907
9	20.661	0.0484	49.153	0.02034	2.379	0.42034
10	28.925	0.0346	69.814	0.01432	2.414	0.41432
11	40.498	0.0247	98.739	0.01013	2.438	0.41013
12	56.694	0.0176	139.235	0.00718	2.458	0.40718
13	79.371	0.0126	195.929	0.00510	2.469	0.40510
14	111.120	0.0090	275.300	0.00363	2.478	0.40363
15	155.568	0.0064	386.420	0.00259	2.484	0.40259
16	217.795	0.0046	541.988	0.00185	2.489	0.40185
17	304.913	0.0033	759.784	0.00132	2.492	0.40132
18	426.879	0.0023	1064.697	0.00094	2.494	0.40094
19	597.630	0.0017	1491.576	0.00067	2.496	0.40067
20	836.683	0.0012	2089.206	0.00048	2.497	0.40048
21	1171.356	0.0009	2925.889	0.00034	2.498	0.40034
22	1639.898	0.0006	4097.245	0.00024	2.498	0.40024
23	2295.857	0.0004	5737.142	0.00017	2.499	0.40017
24	3214.200	0.0003	8032.999	0.00012	2.499	0.40017
25	4499.880	0.0002	11247.199	0.00009	2.499	0.40009
26	6299.831	0.0002	15747.079	0.00006	2.500	0.40006
27	8819.764	0.0001	22046.910	0.00005	2.500	0.40005

References and bibliography

Part I Construction planning studies

Ackoff, R. L. and **Sasieni, M. W.** (1968) *Fundamentals of Operations Research*, Wiley, New York.

American Society of Civil Engineers (1979) *Readings in Cost Engineering*, Vol. 1, New York.

Armstrong, B. (1981) *Programming Building Contracts*, Northwood Books, London.

Brech, E. F. L. (ed.) (1975) *Construction Management in Principle and Practice*, Longman, London.

Burch, T., *Planning and Organisational Problems Associated with Confined Sites*, Chartered Institute of Building. Site Management Paper No. 76.

Burman, P. J. (1972) *Precedence Networks for Project Planning and Control*, McGraw-Hill (UK), Maidenhead.

Byrne, B. (1967) 'The basic principles of site organisation', *Building Trades Journal.*

BS 6046: 1981 Use of network techniques in project management. (Part 2).

Calvert, (1981) *Introduction to Building Management*, Butterworths, London.

Chartered Institute of Building (1980) *Programmes in Construction: A guide to good practice*, CIOB, London.

Construction Industry Training Board, *Precedence Diagram Planning*, Pitman, London.

Cooke, B. (1981) *Contract Planning and Contractural Procedures*, Macmillan, London.

Erskine-Murray, P. E. (1972) *Construction Planning: Mainly a question of how*, Chartered Institute of Building.

Halpin, D. W. and **Woodhead, R. W.** (1980) *Construction Management*, Wiley, New York.

Hollins, R. J. (1971) *Production and Planning Applied to Building*, George Godwin, London.

Institution of Civil Engineers (1971) *Civil Engineering Procedures*, ICE, London.

Lang, D. W. (1977) *Critical Path Analysis* (2nd edn), (Teach Yourself Books) Hodder and Stoughton, London.

Levin, R. I. and **Kirkpatrick, C. A.** (1977) *Planning and Control with PERT/CPM*, McGraw-Hill, New York.

Loomba, P. N. (1976) *Management – A Quantitative Perspective*, Macmillan, New York.

Lumsden, P. (1968) *The Line-of-Balance Method*, Pergamon Press Ltd, Oxford.

Makower, M. S. and **Williamson, E.** (1975) *Teach Yourself Operational Research*, Hodder and Stoughton, London.

Morell, A. J. H. (ed.) (1967) *Problems of Stocks and Storage*, Monograph no. 4, Oliver and Boyd, Edinburgh.

Moskowitz, H. and **Wright, G.** (1979) *Operations Research Techniques for Management*, Prentice Hall, Englewood Cliffs, New Jersey.

National Building Agency (1970) *Programming House Building by Line of Balance*, NBA, London.

Oxley, R. and **Poskitt, J.** (1980) *Management Techniques Applied to the Construction Industry* (3rd edn.), Granada, London.

Phillips, D. T. and **Garcia-Diaz, A.** (1981) *Fundamentals of Network Analysis*, Prentice Hall, Englewood Cliffs, New Jersey.

Pigott, P. T. (1971) *Project Planning for Builders: 1. Bar Charts 2. Networks*, An Foras Forbartha, Dublin.

Pilcher, R. (1976) *Principles of Construction Management* (2nd edn.), McGraw-Hill (UK), Maidenhead.

Reynaud, C. B. (1967) *The Critical Path Network Analysis Applied to Building*, George Godwin, London.

Thompson, P. (1981) *Organisation and Economics of Construction*, McGraw-Hill (UK), Maidenhead.

Part II Construction economy studies

Alfred, A. M. and **Evans, J. B.** (1971) *Appraisal of Investment Projects by Discounted Cash Flow*, Chapman and Hall, London.

Barish, N. N. (1962) *Economic Analysis*, McGraw-Hill, New York.

Cooke, B. and **Jepson, W. B.** (1979) *Cost and Financial Control for Construction Firms*, Macmillan, London.

Institution of Civil Engineers (1969) *An Introduction to Engineering Economics*, ICE, London.

Lumby, S. (1981) *Investment Appraisal and Related Decision*, Thomas Nelson, Walton-on-Thames.

Merrett, A. J. and **Sykes, A.** (1962) *The Finance and Analysis of Capital Projects*, Longman, London.

Pigott, P. T. (1971) *Finance for Builders*, An Foras Forbartha, Dublin.

Pilcher, R. (1973) *Appraisal and Control of Project Costs*, McGraw-Hill (UK), Maidenhead.

Riggs, J. L. (1968) *Economic Decision Models for Engineers and Managers*, McGraw-Hill, USA.

Sizer, J. (1969) *An Insight into Management Accountancy*, Penguin Books, Harmondsworth.

Stone, P. A. (1980) *Building Design Evaluation Costs-in-use*, E. and F. N. Spon, London.

Wright, M. G. (1967) *Discounted Cash Flow*, McGraw-Hill (UK), Maidenhead.

Index